CONCISE
DICTIONARY OF SCIENCE

CONCISE
DICTIONARY OF SCIENCE

PHYSICS – MATHEMATICS – NUCLEONICS

ASTRONOMY – CHEMISTRY

by

FRANK GAYNOR

PHILOSOPHICAL LIBRARY

New York

15 East 40th Street, New York

Printed in the United States of America

INTRODUCTION

In the course of the past quarter century science and technology have revolutionized life on earth to an extent far greater and more significant than ever before in the entire history of man. That period has seen the advent of the Atomic Age; of the Age of Jets; of earth-circling satellites and of rockets to the moon. It saw the advent of penicillin and other life-saving antibiotics, of cortisone and its synthetic derivatives, the fractionation of blood into its multifarious life-giving substances, and the development of anti-polio vaccine. Radio-astronomy has led to the discovery of a vast, invisible universe as great, if not greater, in extent than the visible universe, the dimensions of which have also been doubled within the past decade. With giant accelerators of nuclear particles we have penetrated deeper into the vast universe of the infinitesimal nucleus of the atom, and with powerful radiations from the atom's nucleus we are penetrating ever deeper into the infinitesimal-infinite universe of the nucleus of the living cell, basic constituent of all things living.

Along with the tremendous expansion of the domain of pure and applied science has also come a profound change in the public attitude to the subject. No longer is interest in science confined only to a few living in a world apart. In fact, one of the most dramatic aspects of the Twentieth Century is the awakening of interest in science among the masses of the people, together with the realization among the many that knowledge of science has become an essential of everyday existence.

This awakening of interest among the great masses of the people, and their eagerness to gain a better understanding of the underlying principles and concepts of science, have created new demands on lexicography, the science of words. The growth of science, and the increasing emergence of new subdivisions of older branches of science, such as virology, enzymology, cytogenetics, radiochemistry, high-energy physics, have been paralleled by a corresponding extension of scientific terminology, every advance bringing with it new terms or extensions of the meaning of old terms.

This Concise Dictionary of Science comes at last at a time when the need for such a volume has become most urgent for both scientist and layman alike. It has the merit of conciseness, an essential requisite, happily not gained at the expense of accuracy. It is a highly welcome addition to the field of scientific education for young and old, professional and non-professional, scientist and layman.

William L. Laurence
The New York Times

A

A

(1) The symbol for the element *argon.* (2) An abbreviation for *Angstrom unit* (also Å).

A-bomb

See *atomic bomb.*

ab-

A prefix which when placed before the name of a practical electrical unit denotes the corresponding unit in the CGS electromagnetic system of units; e.g., *abvolt,* the absolute unit of potential.

abampere

The electromagnetic unit of current in the CGS system. 1 abampere is defined as the current which produces a magnetic intensity of 2π oersteds at the center of a circuit in a one-turn circular conductor of a radius of 1 cm. in a vacuum. 1 abampere = 10 amperes.

Abbe number

The reciprocal of the *dispersive power* of a substance. (Also called *constringence.*)

abcoulomb.

The electromagnetic unit of charge, or quantity of electricity, in the CGSsystem. I abcoulomb is defined as the quantity of electricity which passes any given point of an electric circuit in one second when the current is 1 abampere. 1 abcoulomb = 10 coulombs.

aberration

In optics, a defect of optical images attributable to the fact that no lens system is able to form an absolutely perfect image. The chief varieties of such defects are: *astigmatism, coma, chromatic aberration (lateral* or *longitudinal), curvature of field, lens distortion,* and *spherical aberration.*

abfarad

The CGS electromagnetic unit of capacitance. One abfarad is the capacitance of a condenser when a charge of one abcoulomb produces a difference of potential of one abvolt between the terminals. 1 abfarad = 10^9 farads.

abhenry
The electromagnetic unit of inductance in the CGS system. 1 abhenry represents the inductance in a circuit in which a current changing at the rate of 1 abampere per second produces an electromotive force of 1 abvolt. 1 abhenry = 10 henrys.

abmho
The electromagnetic unit of conductance in the CGS system. I abmho represents the conductance of a conductor when with an unvarying difference of potential of 1 abvolt between its terminals, the current in the conductor is 1 abampere. 1 abmho = 10^9 mhos.

abohm
The electromagnetic unit of resistance in the CGS system. 1 abohm represents the resistance of a conductor when with an unvarying current of 1 abampere flowing through it, the difference of potential between the terminals is 1 abvolt. 1 abohm = 10^{-9} ohm.

abrasion
The wearing away of the surface of an object by the cutting action of solids.

abrasive (*noun*)
A substance used to smooth, wear down and polish solids. Abrasives may be natural (e.g., diamond, corundum, emery, etc.) or manufactured (e.g., tungsten or boron carbide, chromic or ferric oxide, etc.).

abrasive (*adj.*)
Possessing the ability to cause abrasion.

abrasive hardness
The resistance of a material to surface scratching or wear by rubbing or grinding.

abscissa
The perpendicular from a given point on a graph or a map grid to the vertical center line or ordinate. A perpendicular from a given point to the *y* axis. Often referred to as the *x* axis or *x* coordinate.

absolute
Pure, unmixed, exact, nonrelative.

absolute altimeter
An electronic instrument which indicates altitude above the surface of the earth, in contradistinction to the readings of

aneroid altimeters which indicate altitude above sea level.

absolute constant

A *constant* which always has the same value whenever it is employed or referred to (e.g., 10, $\sqrt{7}$, π), in contradistinction to an *arbitrary constant.*

absolute density

In chemistry, mass per unit volume (usually grams per cubic centimeter).

absolute error

The actual inaccuracy in measurement or calculation. If a quantity equal to x is given inaccurately by an amount m and thereby becomes $(x \pm m)$ then m is called the absolute error. Or if a stick 5 ft. long is measured 5 ft. 2 inches it has an absolute error of 2 inches.

absolute humidity

In meteorology, (1) the mass of water vapor present per unit volume of space, i.e., the density of water vapor; usually expressed in grams per cubic meter or grains per cubic foot, (2) the gas pressure exerted by the water vapor per unit area, expressed in barometric units. (These two definitions are actually equivalent.)

absolute index of refraction

The index of refraction when the light enters a medium from a vacuum.

absolute inequality

An *inequality* (q.v.) that holds true for all values of the terms involved.

absolute joule

A unit of energy or work. 1 absolute joule $= 10^7$ ergs.

absolute magnitude

The magnitude of a star as it would appear if viewed from a distance of 10 parsecs (32.6 light-years), where its parallax would be 0.1 of one second.

absolute number

A number used without connection to any particular object, as 4, 7, 9, etc.; when applied to something, as "4 men," 7 books," "9 inches," etc., they become *concrete numbers.*

absolute potential

The absolute electric potential at a point is the work per unit charge which is required in order to bring a positive charge of electricity from an infinite distance to that point.

absolute system of units

A system in which a few units are selected as basic ones from which all the others are derived.

absolute temperature

Temperature according to a scale in which the lowest mark is the *absolute zero*. To calculate temperatures according to the absolute scale, add 273.1 to the centigrade value or 459.8 to the Fahrenheit value.

absolute term

In an equation, the term which does not contain the unknown quantity. Thus in $(x^2 + 2ax + 4a)$ the absolute term is 4a.

absolute unit

Any derived unit which is referred directly to the *fundamental units.*

absolute value

The absolute value of a number is the arithmetical number which remains when the sign of a positive or a negative number is omitted. Example: the absolute value of +2, or of −2 is 2, usually written as $|2|$.

absolute value of a complex number

The numerical length of the vector which represents the complex number in question; also called the *modulus* of the complex number.

absolute volt

The absolute electromagnetic unit of electromotive force, defined as the electromotive force which is induced when a conductor 1 centimeter in length is moved at a velocity of 1 centimeter per second perpendicularly across a uniform magnetic field of unit flux density (1 gauss); 10^8 absolute volts = 1 volt. (Referred to as *abvolt* for short.)

absolute zero

The temperature at which all molecular motion ceases, −273.1° C.

absolutely convergent series

A convergent series of terms the corresponding series of the absolute values of which is likewise convergent. In other words, if the Σx_n of positive real numbers $x_1 + x_2 + x_3 + \ldots\ldots x_n + \ldots\ldots$ is convergent, we say that the series Σx_n of real numbers $x_1 + x_2 + x_3 + \ldots\ldots x_n + \ldots\ldots$ is absolutely convergent. Thus the series $\frac{1}{1^2} + \frac{1}{2^2} + \frac{1}{3^2}$ is absolutely convergent.

absorbent
Having the capacity to take up another substance.

absorber
(1) A sheet of material placed between the source of an electromagnetic radiation and the place where it is detected or measured. When this sheet contains elements chosen especially to reduce the intensity of radiation of certain undesired wavelengths, the absorber is usually called a *filter.* (2) In a nuclear reactor, a substance which absorbs neutrons without reproducing them. (3) In general, a medium, substance or functional part that takes up matter or energy.

absorption
In chemistry, the penetration of a liquid or solid into or through another liquid or solid, the particles thus absorbed being molecular in size.

absorption bands
Dark bands in the spectrum of radiant energy, resulting from the absorption of particular wavelengths of the radiation by the constituents of the medium through which it passes.

absorption coefficient
The rate of the absorption of radiation by the substance traversed by it; a constant which is characteristic of the nature and energy of the radiation and of the material absorbing it.

absorption edge
(1) The sharp edge-like discontinuity in the graph of the absorption coefficient of x-rays in a given substance, in which the absorption coefficient is indicated as ordinate, and the wavelength as abscissa. (2) The wavelength corresponding to such sharp discontinuity.

absorption factor
The *coefficient of absorption.*

absorption of light
(1) In optics, the process by which a substance or object "soaks up" all the component colors of a beam of white light except certain specific colors which it reflects or transmits. (2) The result of this process. (Cf. *selective absorption.*)

absorption of radiation
A process by which electromagnetic or corpuscular radiation vanishes by being incorporated in a substance exposed to it, usually being transformed into another form of energy. A phenomenon analogous to the *absorption of light* (q.v.).

absorption spectrum
The spectrum of radiation which has passed through a selectively absorptive substance, characterized by the presence of dark lines or gaps which are due to the absorption of certain wavelengths.

absorptivity
Absorptive power, expressed in terms of the magnitude of that portion of a radiant energy which is absorbed by the substance of the body upon which the radiation normally impinges. It equals the ratio of the radiant energy absorbed to the originally incident amount.

abstract
Unrelated, detached, standing alone.

abstract mathematics
That part of mathematics which deals with pure numbers, orders, or forms which are unattached to any objects or things. Thus 3, 5, 12 are pure numbers as against 3 dogs, 5 pounds, 12 miles, which are concrete.

abvolt
Short for *absolute volt* (q.v.).

abwatt
A unit of power. 1 abwatt represents 1 abampere through a difference of potential of 1 abvolt, and equals 10^{-7} watt.

A-C
An abbreviation for *alternating current.*

Ac
(1) The symbol for the element *actinium.* (2) The chemical symbol for *acetic,* for *acetyl,* and for *acyl.*

AcA, AcB, AcC, AcC′, AcD, AcU, AcX.
The symbols for *actinium A, actinium B, actinium C, actinium C′, actinium D, actinium uranium* and *actinium X* respectively.

accelerated motion
Motion in which the velocity of the moving object or system is changing. (If the rate of the change of velocity is constant, the motion is referred to as *uniformly accelerated motion.*)

acceleration
The rate of change in velocity (either linear or angular) per unit time, customarily expressed in centimeters or feet per second per second.

acceleration of gravity
See *standard gravity.*

accelerator
In chemistry, a substance which increases the speed of a chemical reaction.

accelerator of nuclear particles
Apparatus employed to increase the velocity of sub-atomic particles. (*Betatron, bevatron, cyclotron, linear accelerator, proton synchrotron, synchro-cyclotron, synchrotron.*)

accessory elements
See *trace elements.*

accumulator
See *storage battery.*

accuracy
Degree of the conformity which the result of a calculation or operation shows with a certain standard. The term relates to the quality of a result, and is distinct from *precision* which relates to the quality or refinement of the operation by which the result is obtained.

acerate
Sharp.

acetate
A salt or an ester of an acetic acid.

acetic
Containing CH_3CO-; symbol: Ac.

acetic acid
A pungent, colorless, liquid compound (CH_3COOH) obtained by oxidation of alcohol, distillation of wood, or by processing acetylene. Vinegar is a dilute solution formed by fermentation of alcoholic liquids such as cider or wine. Also called *ethanoic acid.*

acetyl
The monovalent acyl radical CH_3CO-; symbol: Ac.

acetylation
The introduction of an acetyl group into an organic compound.

acetylene
A colorless, poisonous, highly inflammable gas, generated by the action of water on calcium carbide.

acetylene welding
Fusion welding, using acetylene and oxygen.

achromat
An *achromatic lens.*

achromatic
Without color.

achromatic lens
An optical lens which has been corrected partly for chromatic aberration, so that the focal length does not depend on the color of light being used. Also called *achromat.*

acicular
Sharp, needle-pointed, slender, needle-shaped.

acid
(1) A chemical compound containing hydrogen which can be replaced by metallic elements and which produces hydrogen ions in solution. It neutralizes *bases* to form *salts,* and is usually characterized by a sour taste and by turning blue litmus red. (2) Any substance capable of giving *protons* (H^+) to a base (*Brönsted-Lowry theory*).

acid anhydride
An oxide of a nonmetal, which forms an acid with water; the result of an intermolecular reaction between organic acids involving the loss of water.

acid-ash value
In chemistry, the excess of mineral ions.

acid hydrogen
That hydrogen of a compound which can be transferred as proton (H^+) to a base.

acid number
An expression of the amount of free acid in fats, oils, waxes and resins, expressed as the number of milligrams of potassium hydroxide (KOH) needed to neutralize one gram of the material being tested. Also called *acid value.*

acid oxide
An oxygen compound of a nonmetal. (Also called *acidic oxide.*)

acid rock
In geological terminology, an igneous rock with a preponderant silica content (e.g., granite).

acid value
See *acid number.*

acidic oxide
See *acid oxide.*

acidimetry
The ascertainment of the amount of acid present in a solution by means of *titration.*

acnode

A point on a curve in the vicinity of which there is no other point on the curve. Also called *isolated point* or *conjugate point.*

acoustic

(1) Generating or transmitting sound, or designed or able to do so. (2) Generated by sound. (3) Having the properties or characteristics of sound.

acoustic grating

A row of rods of equal size arranged at uniform intervals, performing the same function with reference to sound as an optical *diffraction grating* does with respect to light.

acoustic impedance

The acoustic impedance of a sound medium on a given surface situated in a wave front is the complex quotient of the sound pressure (i.e., the force per unit area) on the surface by the flux (i.e., volume velocity of linear velocity multiplied by the area), through the surface. It equals the *mechanical impedance* divided by the square of the area of the surface considered, and is expressed in *acoustic ohms.*

acoustic ohm

A unit of acoustic impedance, reactance or resistance; one acoustic ohm is the magnitude of the acoustic impedance, reactance or resistance when a sound pressure of one dyne per square centimeter produces a volume velocity of one cubic centimeter per second.

acoustic reactance

The imaginary component of the *acoustic impedance,* measured in *acoustic ohms.*

acoustic resistance

The real component of the *acoustic impedance,* measured in acoustic ohms.

acoustic shadow

The effect produced by an obstacle placed in the path of a sound wave, analogous in many respect to an optical shadow.

acoustical

Pertaining to or linked with sound or acoustics. (Cf. *acoustic.*)

acoustics

The science of sound and hearing, including the production and transmission of sound as well as the forces and influences which may affect the sound waves and vibrations and their perception.

acting force

The force exerted by one body on another, and which according to *Newton's third law of motion* causes the second body to exert an equal force *(reacting force)* on the first one.

actinic radiation

Radiation able to produce a chemical change while not necessarily able to affect the human sense of sight.

actinides

The elements with atomic numbers above 88. According to many authorities, these elements occupy one single place in the extended periodic table, in the same group into which the *rare earth elements* ("lanthanides") are classified.

actinium

A rare radioactive element, first member of the actinium group of the extended periodic table, atomic number 89, atomic weight 227. Discovered by Debierne in 1899. Symbol: Ac.

actinium A (AcA)

A radioactive isotope of polonium ($_{84}Po^{215}$).

actinium B (AcB)

A radioactive lead isotope ($_{82}Pb^{211}$).

actinium C (AcC)

A radioactive bismuth isotope ($_{83}Bi^{211}$).

actinium C' (AcC')

A radioactive polonium isotope ($_{84}Po^{211}$).

actinium C" (AcC")

A radioactive thallium isotope ($_{81}Tl^{207}$).

actinium D (AcD)

The stable lead isotope ($_{82}Pb^{207}$) which is the end product of the uranium-actinium series.

actinium emanation

See *actinon.*

actinium group

See *actinides.*

actinium K (AcK)

A radioactive francium isotope ($_{87}Fr^{223}$).

actinium series

See *uranium-actinium series.*

actinium uranium (AcU)

The natural radioactive uranium isotope $_{92}U^{235}$, parent of the *uranium-actinium series.*

actinometer

An instrument for measuring the actinic effects of light or other electromagnetic radiation. The term is applied also to the instrument for measuring the direct heating power of the rays of the sun.

actinometry

The science and technique of measuring radiant energy, particularly that of the sun, in its thermal, chemical and luminous aspects.

actinon (An)

A gaseous radioactive isotope of radon ($_{86}Rn^{219}$). Atomic number 86, atomic weight 219. (Also called *actinium emanation.*)

activation

The process or act of initiating a chemical or physical reaction. In nuclear physics, the inducement of artificial radioactivity.

activation energy

In chemistry and physics, the energy necessary for starting or maintaining a reaction.

activator

A substance whose presence is necessary for the initiation of activity, e.g. in enzyme reactions.

activity

(1) Chemically, the relative rate or degree of readiness or ease with which an element or compound or agent reacts. (2) An excited state of a gaseous substance due to ionization. (3) The relative concentration of free ions; the degree of interionic activity. (4) In nuclear physics, the magnitude or rate of disintegration of a radioactive emission in terms of its observable effect. (5) Often used, loosely, as a synonym for *radioactivity*. (See also *optical activity*.)

activity series of metals

The activity series is a table of the metals arranged in the order of their decreasing chemical activity.

Note 1. This table is practically the same as the electrochemical series of the elements. However, non-metals are not included in the table given here.

Note 2. The activity series represents the ease with which the metals give up one or more electrons to form positive ions.

THE ACTIVITY SERIES

Occurrence	Metal	Reaction	Symbol	Oxides
These metals are never found free in nature.	Potassium	These metals replace hydrogen from water and dilute acids.	K	These metals form oxides which are stable.
	Sodium		Na	
	Calcium		Ca	
	Magnesium		Mg	
	Aluminum		Al	
	Manganese		Mn	
	Zinc		Zn	
	Chromium		Cr	
Rarely found free in nature.	Iron		Fe	
	Nickel		Ni	
	Tin		Sn	
	Lead		Pb	
	Hydrogen		H	
Frequently found free in nature.	Copper	These metals are attacked by oxidizing acids.	Cu	
	Bismuth		Bi	
	Antimony		Sb	
	Mercury		Hg	These metals form oxides which decompose on heating. Because of this they are frequently called the "Noble Metals."
	Silver		Ag	
Generally found free in nature.	Platinum	These metals are not attacked by common acids.	Pt	
	Gold		Au	

actual mechanical advantage
In mechanics, a measure of the utility of a machine; it is the ratio of the output force exerted by the machine on a load to the input force which the operator exerts on the machine itself. (Abbreviation: AMA.) Cf. *ideal mechanical advantage.*

acute angle
Any angle that is smaller numerically than a right angle.

acute-angled triangle
A triangle all angles of which are acute angles.

acyclic
Not containing any rings or any ring-system.

acyl
An organic radical (RCO–), obtained when the hydroxyl group is removed from an organic acid; symbol: Ac.

acylation
Introduction of an *acyl group* into an organic compound.

addition
The method of finding the sum of two or more figures or terms. It is the uniting of two or more numbers or quantities into one number or quantity. The word used for uniting two quantities is "plus" and the sign or symbol used to mean plus is +. Thus 5 plums + 3 plums + 8 plums = 16 plums.

addition compound
In chemistry, a *compound* formed by adding an atom or atom group to a molecule.

additive property
Any physical or chemical property dependent on the quantity of the substance involved, as, e.g., on the number of atoms.

adhesion
The interaction of the molecules of the surfaces of two closely adjacent dissimilar substances which causes them to cling together. Adhesion is thus the force and the result of the attraction between *dissimilar* molecules. (Cf. *cohesion*.)

adiabatic
Thermally insulated or self-contained. The term is used in thermodynamics referring to change in volume without loss or gain of heat.

adiabatic calorimeter
A *calorimeter* which theoretically remains unaffected by its surroundings and neither gains nor loses heat.

adiabatic contraction
A decrease in volume without loss or gain of heat.

adiabatic curve
A curve or graph showing the relation between pressure and volume in substances having or assumed to have the property of adiabatic expansion and contraction.

adiabatic process
A process in which no exchange of heat occurs between the substance and its environment.

adjacent
Adjoining, neighboring, bordering on.

adjacent angles
Two angles that have a common side and a common vertex: when three straight lines are drawn from a point the angles on either side of the straight line between the other two straight lines are called adjacent angles.

adsorbate
In the phenomenon of *adsorption,* the substance which is taken up at a solid surface.

adsorption
A surface phenomenon in which one substance is taken up at the solid surface of another, resulting in a change of concentration at the interface. It is the process of the adherence of molecules to an impermeable surface, specifically the formation of a layer of a foreign substance on the surface of a solid or liquid. (See also *molecular adsorption.*)

aerial (*adj.*)
In, from, or relating to the air or the atmosphere.

aerial (*noun*)
The elevated conductors of an open *antenna* system. (Popular usage employs the term, *aerial,* as a synonym for what is properly called an antenna.)

aerodynamic
Pertaining to the laws of the motion of air or other gases.

aerodynamic drag
The action of the air resistance that slows down a body moving through air.

aerodynamics
That branch of physics which treats of the laws of the motion of air and other gases ("aeriform fluids").

aerology
That branch of meteorology which is concerned with the study of the free atmosphere throughout its vertical extent.

aerometer
An instrument for measuring the weight and density of air and gases.

aeropause
An upper region of the atmosphere in which the atmosphere ceases to function for manned or unmanned flight.

aerosol
A dispersion of particles of a solid or solids in a gas.

aerospace
An adjective, meaning of or pertaining to the earth's envelope of atmosphere and the space above it, the two considered as a single realm for activity in the launching, guidance, and control of ballistic missiles, earth satellites, dirigible space vehicles, and the like.

affinity
Chemical potential; the selective tendency or ability of atoms to unite with one rather than another element.

afterglow
The luminosity persisting in a rarefied gas after the passage through it of an electrodeless discharge.

Ag
The symbol for the element *silver*.

aggregate
The total of all points or numbers that satisfy a certain condition. A sum of particles.

aggregation
The gathering or clustering of atoms or molecules which make up a body of matter. (See *state of aggregation*.)

air resistance
The frictional resistance offered by air to the motion of bodies passing through it.

Al
The symbol for the element *aluminum*.

albedo
The ratio of light reflecting from an unpolished surface to the light that falls upon it. (The term is used in particular in reference to light reflected from celestial objects.)

albumen
A specific protein, such as white of egg; a general term for a class of proteins.

alcohol
(1) The generic designation of a class of organic substances containing a hydroxyl (OH) group, of very slight acid properties owing to the union with certain carbon hydrogen radicals. An alcohol is derived from a hydrocarbon by replacing one or more of the hydrogen atoms by the hydroxyl radical (OH).—(2) Specifically, the popular term for *ethyl alcohol* or *ethanol* (C_2H_6O), obtained by fermentation or synthetically.

alcohol lamp
A lamp used in chemical laboratories where gas is not available for a Bunsen burner. It consists of the alcohol container and a wick, and is sometimes provided with a snuffer cap for putting out the flame.

alcoholysis
The splitting of a compound through the cleavage of a C–C bond induced by the addition of alcohol.

aldehyde
The intermediate stage of oxidation of an alcohol to an acid, characterized by the CHO radical.

aldehyde
A generic term for a class of compounds derived from a hydrocarbon by oxidation, i.e., by the substitution of an oxygen atom for two hydrogen atoms; they contain the –CHO group.

algebra
That part of mathematical science in which letters, characters or symbols, instead of or in addition to numbers, are used for the calculations. The numerical value of the letters, characters or symbols may be different for different calculations.

algebraic
Relating to algebra.

algebraic curve
A curve the equation of which is a polynomial one, i.e., of the form $f(x, y) = 0$, where x and y are the coordinates. Algebraic curves of a degree greater than 2 are referred to as *higher plane curves* (which classification includes the *transcendental curves*, too).

algebraic equivalent
An algebraic expression which is equal to a series. Thus

$$\frac{1}{(1 - x)^2} = 1 + 2x + 3x^2 + 4x^3 + \ldots\ldots \text{ when x is limited.}$$

algebraic expression
Any quantity expressed by literal and arithmetical numbers combined by means of the fundamental operations defined for algebra.

algebraic plane curve
See *algebraic curve*.

algebraic sum
The sum of two numbers in algebra. To be distinguished from "arithmetical sum" in which the sum of the two numbers is always greater than either of them. In algebra, the sum of two numbers can be smaller than one of them if one is negative. Example: $5 + (-2)$ equals 3.

alidade
An instrument used to measure vertical elevations and distances.

alignment array
See *end-on array of antennas*.

alignment chart
See *nomogram.*

aliphatic
Said of those carbon compounds characterized by straight, branched, or cyclic carbon chains, and lacking the benzenoid ring structure.

aliquot
In mathematics, part of a number or quantity which will divide the number or quantity without a remainder; a submultiple. Thus 5 is an aliquot of 15.

aliquot
In chemistry, a measured proportion of the total volume of a solution.

alkali
A generic term for the oxides or hydroxides of the *alkali metals.* Alkalis are very active bases.

alkali compounds
In geological terminology, any compound containing potassium or sodium.

alkali metals
The monovalent, electropositive metallic elements lithium, sodium, potassium, rubidium, caesium and francium, in Group I of the periodic system.

alkalimetry
The ascertainment of the amount of alkali that is present in a given solution, by means of *titration.*

alkaline
Having the property of producing hydroxyl ions in an aqueous solution.

alkaline earth
An oxide of one of the *alkaline earth metals.*

alkaline earth metals
The divalent, electropositive metallic elements calcium, strontium and barium in Group II of the periodic system. Magnesium and radium (also beryllium) are included sometimes as well.

alkaloid
An organic base in which at least one nitrogen atom forms part of a cyclic system. Alkaloids have names ending in *-ine* (indicating the presence of the *amine group*).

alkane
A generic term for a group of organic compounds, members

of the saturated aliphatic hydrocarbon series, having the general formula C_nH_{2n+2} and names ending in *ane*.

alkene

A generic term for a group of organic compounds, members of the unsaturated aliphatic hydrocarbon series, which contain double bonds, have the general formula C_nH_{2n} and names ending in *ene*.

alkyl

A monovalent radical, product of the removal of a hydrogen atom from an aliphatic hydrocarbon.

alkyl halide

An organic compound consisting of an alkyl group and a halogen atom.

alkylene

An organic radical derived from an unsaturated aliphatic hydrocarbon.

alkyne

A generic term for a group of organic compounds, members of the unsaturated aliphatic hydrocarbon series, which contain double bonds, and have the general formula C_nH_{2n-2}. (Also called *alkine*.)

allomerism

Similarity or identity of crystalline form, but with a difference in chemical composition.

allomers

Substances which are *allomeric*, i.e., have the same crystalline form, but differ in chemical composition (as $KClO_4$ and $KMnO_4$).

allomorphism

Similarity or identity in chemical composition, but with a difference in crystalline form.

allomorphs

Substances which are *allomorphic*, i.e., have the same chemical composition, but differ in crystalline form.

allotriomorphic crystal

A crystal which has taken its shape from its surroundings, i.e., one the normal growth of which has been restricted, such as one found within masses of metals where mutual interference interrupts growth.

allotrope

One of two or more isomeric forms of an element (e.g., diamond, graphite and amorphous carbon are the three allotropes, or *allotropic forms,* of the element carbon.)

allotropism, allotropy

The existence, or capability of existence, of an element in two or more forms (*allotropes*) possessing a different set of predominantly physical properties, specifically different atomic lattice arrangements (e.g. carbon occurring in nature as diamond, as graphite, etc.).

alloy

A substance that has metallic properties and consists of two or more elements, or of metallic and nonmetallic elements, which are miscible with each other when molten, and do not separate into distinct layers when solid. (In the commercial usage, the term "alloy" would also include the case where some separation into distinct layers has occurred.) (Cf. *alloy steel.*)

alloy steel

A steel to which one or more alloying elements have been deliberately added with the object of conferring particular properties upon it.

alpha (α), beta (β), gamma (γ) lines

The lines in an X-ray series in the order of increasing frequency.

alpha-decay (α-decay)

Radioactive change by emission of an α particle. Since the α particle has a mass of 4 a.m.u. and a positive charge of 2 electronic charges, the emission of an α particle produces a daughter element which has an atomic weight less by 4 than that of the parent element and an atomic number less by 2 than the atomic number of the parent.

alpha emitter

A radioactive element, whether natural or artificial, which changes into another element by *α-decay*

alpha iron

An allotropic form of iron, the atoms of which are arranged in a *body centered cubic space lattice.*

alpha particle

(α, He^{++}, ${}_2He^4$). An elementary particle, discovered by Rutherford in 1904, identified as the nucleus of a helium atom which has lost its 2 planetary electrons, and is therefore positively charged. Each alpha particle consists of two protons and two neutrons, has a mass of 6.644×10^{-24} grams, or 4.00279 a.m.u., e/m = 4820 e.m.u. per gram, and has a positive electric charge which is numerically the double of the electronic charge, i.e. about 9.6050×10^{-10} e.s.u. Alpha particles are ejected by certain disintegrating radioactive substances (*alpha emitters*).

alpha radiator

See *alpha emitter.*

alpha radioactivity

See *alpha decay, alpha emitter.*

alpha ray

A stream of *alpha particles;* it is deflected by an electric or magnetic field, and is capable of affecting a photographic plate and of producing dense ionization.

alternate angles

If two straight lines are cut by another straight line, the angles formed by the two straight lines with the straight line that cuts them, but lying on opposite sides of this line, are called alternate angles.

alternate exterior angles

Alternate angles neither of which lies between the two lines cut by a transversal.

alternate interior angles

Angles lying on opposite sides of a transversal and between the two lines cut by it.

alternate segments of a circle

The two segments of a circle on either side of their common chord.

alternating current (A-C)

A periodic current, i.e., one in which the charge flow reverses periodically, and the average value of which over a period is zero.

Alternating Gradient Synchrotron (A.G.S.)

A new machine which is now being built at Brookhaven. It will have a circumference of about ½ mile. The protons will be accelerated by means of a 110-ft.-long, 50 million-electron-volt accelerator. The protons will be accelerated around the ½-mile-long vacuum chamber. The top energy will be approximately 25 billion electron volts. Scheduled for completion and operation for the year 1960.

alternating series

A series in which the signs of the terms alternate.

altitude of a cone
The perpendicular distance from the base to the vertex.
altitude of a cylinder
The perpendicular distance between the bases.
altitude of a frustum of a cone or pyramid
The perpendicular distance between the bases.
altitude of a polygon
The perpendicular distance between any side chosen as base and a line drawn parallel to that side through the corner or vertex most distant from that side.
altitude of a prism
The perpendicular distance between the bases.
altitude of a pyramid
The perpendicular distance from the base to the vertex.
altitude of a rhombus
The perpendicular distance from any side to the opposite vertex.
altitude of a star
The angle formed at the point of observation by the straight line from the star to that point with the horizontal line passing through that point. It is also called the angle of elevation.
altitude of a trapezium
The perpendicular distance between the two parallel sides.
altitude of a triangle
The perpendicular distance from a vertex to the opposite side which is regarded as the *base*.
altitude of a zone
The perpendicular distance between the bases.
alum
A generic term for a group of double salts, viz., double sulfates of the monovalent metal and a trivalent metal with a definite amount of *water of hydration*.
aluminothermic process
A process for welding that makes use of the extreme reactivity of aluminum in reducing oxides of certain metals. (Also called *Goldschmidt process*.)
aluminum
A silvery white trivalent metallic element, in Group III of the periodic table, discovered by Wöhler in 1828, of high tensile strength, good corrosion resistance and electrical conductivity. Atomic number 13, atomic weight 26.98, density 2.70 at

20°C., melting-point 658.7°C., boiling point 1800°C. Symbol: Al.

Am

(1) The symbol for the element *americium*. (2) The symbol for the radical *ammonium*, or for the radical *amyl*.

amalgam

An alloy which contains mercury.

amalgamation

In a strict sense an alloy of mercury with some other metal. Sometimes, although improperly used, in the sense of an alloy in general.

ambient temperature

The temperature of the surrounding medium, such as gas or liquid, which comes into contact with the heated parts of apparatus.

ambiguous

Having more than one meaning. With reference to a mathematical problem, having more than one solution.

americium

An artificially produced transuranic element, not found to occur in nature. Atomic number 95, mass number of the most stable known isotope 243, valence 3, 4, 5 and 6. First produced by G. T. Seaborg and his collaborators (R. A. James and L. O. Morgan) by cyclotronic bombardment of ${}_{92}U^{238}$ with high-energy (40 Mev) alpha particles. Symbol: Am.

amicron

Any particle whose size is less than 5 $m\mu$, which is not visible under the ultramicroscope.

amide

A generic term for organic compounds derived from ammonia by replacement of one, two or all three hydrogen atoms by radicals.

amido

The monovalent $-NH_2$ radical occurring in an acid group.

amine

A generic term for a group of organic compounds which may be regarded as derivatives of the hydrocarbons in which a hydrogen atom has been replaced by the monovalent basic amine group ($-NH_2$), or as derivatives of ammonia in which one or more hydrogen atoms have been replaced by alkyl groups. They are named by suffixing *amine* to the names of the alkyl radicals involved (e.g., dimethylamine).

amino
The monovalent basic $-NH_2$ group.

amino acids
Compounds with an amino and a carboxyl group.

ammeter
An instrument for measuring electric current, with a scale calibrated in amperes. (Ammeters with scales calibrated in milliamperes or microamperes are called *milliammeter* and *microammeter,* respectively.)

ammonia
A colorless, sharp-odored, highly water-soluble gas, NH_3.

ammonium
A monovalent radical, $-NH_4$; it has basic properties and resembles an alkali metal radical; symbol: Am.

ammonolysis
The removal of one atom of hydrogen from an ammonia molecule.

amorphous
Shapeless; said of a body lacking regular crystalline structure.

amount
In simple interest, the sum of the money lent (or principal) plus the interest on the money lent.

amp
The abbreviation for *ampere.*

amperage
The quantity of electric current expressed in amperes.

ampere
The practical unit of the flow of electric current. 1 ampere = the current that will flow with a potential of 1 volt through a conductor of a resistance of 1 ohm (i.e. 1 ampere = 1 coulomb per second). 1 *absolute ampere* = 0.01 abampere.

ampere-hour
A unit of quantity of electricity; I ampere-hour is equal to a current of 1 ampere flowing for 1 hour.

ampere-hour meter
An electricity meter that registers the quantity of electricity in ampere-hours. Ampere-hour meters are sometimes adapted to register in kilowatt hours at the rated voltage of the circuit.

ampere rule
The direction of a magnetic field about a wire conducting a current is clockwise when observed by looking in the direction of the current.

Ampere's law

The magnetic intensity at any point near a current-carrying conductor can be computed on the assumption that each infinitesimal length of the conductor produces at the point an infinitesimal magnetic density. The resulting magnetic intensity at the point is the vector sum of the contributions of all the elements of the conductor.

ampere-turn

A measure of the *magnetomotive force;* one ampere-turn is one ampere of electricity flowing through one turn of a coil. Thus, the magnetomotive force in a system is the number of amperes of a current flowing in a circuit multiplied by the number of turns in the circuit (i.e., by the number of times the circuit is linked with a given path). (See *ampere-turn per inch.*)

ampere-turn per inch

A unit of magnetizing force 2.02 times greater than the *oersted.* The magnetic field in a solenoid consisting of n turns per inch of length and carrying i amperes has a strength of $n \times i$ ampere-turns. 1 ampere-turn = .4947 oersted = 1.257 *gilberts.*

amperometric value

A value determined by measurement of the electric current flowing or generated.

amphoteric

A name applied to compounds having both basic and acidic groups, or capable of ionizing either as a base or as an acid.

amphoteric ion

See *zwitterion.*

amphoteric oxide

An oxygen compound of a heavy metal.

amplitude

In physics, the maximum value of displacement in a periodic or wave motion.

amplitude of a complex number

The angle formed by the vector representing the complex number with the positive horizontal axis. Also called the *argument.*

a.m.u.

The abbreviation for *atomic mass unit.*

amyl

The monovalent $-C_5H_{11}$ radical (symbol: Am).

amylaceous

Starchy, starch-like.

anabolism
That part of the process of *metabolism* in which simple substances are combined to build complex ones, with an absorption of energy.

anaerobic
Free of air or uncombined oxygen.

analgesic
Having pain-relieving properties.

analysis
That branch of mathematics which utilizes chiefly algebra and calculus and their methods: the working out of problems, generally by algebraic sums or equations, using principles or reasoning already accepted as being correct; the dividing up of a compound expression into its simple parts or elements.

analytic geometry
Geometry subject to an analytical treatment by means of the convention of coordinates.

analytic trigonometry
That branch of trigonometry which treats of the relations and properties of trigonometrical functions.

analytical chemistry
That branch of chemistry which treats of the detection, identification and determination of substances and their components.

analyzer
In a *polariscope* or *polarimeter,* the second Nicol prism which can be said to analyze the light arriving to it from the first Nicol (the *polarizer*) and the intervening crystal section.

anaphoresis
The movement of fine particles suspended in a liquid toward the anode, produced by the application of an electric potential.

anastigmat
An *anastigmatic lens.*

anastigmatic
Free from or corrected for *astigmatism.*

anastigmatic lens
A compound lens corrected for *astigmatism.* Also called *anastigmat.*

anastigmatism
The process of making a lens combination so that both astigmatism and the curvature of the field are largely eliminated over a considerable area in the image plane.

anelastic strain

A strain which is not an elastic one, since it does not obey *Hooke's law,* but neither is plastic, for it leaves no *permanent set.*

anemometer

An apparatus for measuring the velocity of air, or any other fluid in general.

aneroid

Containing no liquid.

aneroid barometer

An instrument for measuring atmospheric pressure. It consists basically of a flat box which is partially exhausted of air, so that a partial vacuum is created in it; an external spring is used to keep it from collapsing, although there are models which require no such spring. The changes in atmospheric pressure cause the box to expand or to contract, and these expansions and contractions are indicated, through mechanical devices, on a graduated dial. The aneroid barometer is a useful instrument for measuring altitude or elevation above sea level as well.

aneroidograph

An aneroid barometer equipped with a mechanism for the continuous automatic recording of atmospheric pressure.

anesthetic

A substance used to deaden or relieve pain, by producing unconsciousness (*general* anesthetic) or by merely making a specific area of the body insensitive temporarily (*local* anesthetic)

angle

When two straight lines are drawn from a point or cut one another they are said to form or contain an *angle* (or angles). The point is called the *vertex* (v) of the angle and the straight lines are called the *arms* (a) of the angle. The size of the angle does not depend upon the length of the arms. The smaller angle which lies between the straight lines is called the *angle* and the outer angle is called the *reflex angle.*

angle in a semicircle

The angle formed at any point on the circumference of a circle by its arms which are joined to the two ends of the diameter of the circle. This angle is always a right angle.

angle of a lune
The angle at which the two half-great circles bounding a *lune* meet.

angle of convergence
The angle formed by the lines of sight of both eyes when focusing on any line, surface or object.

angle of depression
The angle formed by a horizontal straight line from the point of observation and a line passing through this point and an object below the horizontal straight line or eye level.

angle of deviation
The angle through which a refracting surface bends a ray of light, i.e., the angle between the paths of the *incident ray* and of the *refracted ray*.

angle of diffraction
The angle between the path of the incident radiation and of the diffracted beam.

angle of dip
See *magnetic dip*.

angle of elevation
The angle formed by a horizontal straight line from the point of observation and a line passing through this point and an object above the horizontal or eye level.

angle of incidence
The angle which a ray impinging on a reflecting surface forms with the perpendicular to the surface at the point of incidence.

angle of intersection
The angle between the tangents of two curves which intersect each other, at either point of intersection.

angle of polarization
See *polarizing angle*.

angle of reflection
The angle between the direction in which a radiation is reflected and the perpendicular to the reflecting surface.

angle of refraction
The angle between the direction in which a radiation is refracted and the perpendicular to the refracting surface.

angle ratio
The proportion of the sides of a right angle triangle in rela-

tion to one of the acute angles. These proportions are called the trigonometrical ratios.

angles in a segment of a circle

The angle subtended at any point on the circumference or arc of a circle having a chord of the circle as its base and the vertex on the circumference of the circle. All such angles, which are in the same segment of the circle, are equal.

angle-sum theorem

The statement that the sum of the three angles of a plane triangle is 180°.

Angstrom unit (A, Å, A.U., angstrom)

Unit of measurement of the wavelength of light and other radiations, so named in honor of the Swedish physicist Jonas Ångstrom (1814-1874). 1 Ångstrom unit equals one-tenth of a millimicron (.1 $m\mu$) . i.e., one hundred millionth of a centimeter.

angular

Relating to, consisting of, or measured by an angle or angles.

angular acceleration

The rate of change of the *angular velocity* of the motion of a body or system.

angular displacement

A vector quantity indicating change of angular position.

angular distance

The ratio of the length of an arc of a circle to its radius, expressed in radians.

angular length

With reference to waves, the wavelength expressed in radians or an equivalent angular measure. It equals 2π radians or 360 degrees multiplied by the length in wavelength units.

angular magnitude

The size of angles, expressed in *degrees, minutes, seconds* or *radians* (qq.v.).

angular momentum

The product of the moment of inertia of a rotating body or system, as measured about any axis of rotation, by the angular velocity about the same axis. It equals the product of the mass and the area swept over by the radius vector in unit time. Its value is one-half of the value of the *moment of momentum.*

angular motion

The motion of a body about a fixed axis, measured by changes in its angular direction from a fixed line.

angular ratio
See *angle ratio.*

angular speed
For a rotating body, the average angular speed is the angle through which a line passing through the axis of rotation turns in a unit time.

angular velocity
The angular velocity of a periodic quantity is the frequency multiplied by 2π. If a periodic quantity is expressed as above, the angular velocity is ω. If the periodic quantity can be considered as resulting from the uniform rotation of a vector, the angular velocity is the number of radians per second passed over by the rotating vector.

anharmonic ratio
See *cross-ratio.*

anhydride
(1) An oxide of a nonmetallic element or an organic radical, capable of forming an acid when combined with water or of being formed from an acid by the abstraction of water. (2) A compound formed from other compounds by the abstraction of water.

anhydrous
Absolutely without water; said of a compound that has lost or given up all its water.

anion
A negative ion in an electrolytic solution which is attracted to the positive pole or anode. An ion that is negatively charged.

anisotropic
Having different physical properties in different directions.

anisotropic crystal
A crystal which has unequal axes and exhibits different properties when tested along axes extending in different directions.

anisotropy
The variance of physical properties with the direction.

anisotropy energy
The excess energy required to produce magnetization to saturation in a crystal in the *hard direction.*

annealing
The process of heating steel and holding it at a suitable temperature followed by cooling at a suitable rate, with the ob-

ject of improving softness, machinability and cold-working properties, or of removing stresses and obtaining a desired structure.

annuity

A series of payments to be made at given intervals over a period of years. Also the sum of money payable annually or periodically.

anode

A positive electrode. An electrode through which current enters any conductor of the non-metallic class. Specifically, an electrolytic anode is an electrode at which negative ions are discharged, or positive ions are formed, or at which other oxidizing reactions occur.

anode effect

The sudden drop in current in an electrolytic process caused by the formation of a gaseous film over the surface of the anode.

anode ray

Stream of positively charged particles, originating from the anode of a vacuum tube.

anode slime

The residue which adheres to, or drops from, the anode in an electrolytic refining process. Also called *anode mud.*

anodic

(1) Relating to or involving the *anode.* (2) The term applied to that metal or alloy in an electrochemical cell which has the lower potential.

anodyne

A pain-relieving drug.

anolyte

The portion of an electrolyte in an electrolytic cell adjacent to an anode. If a diaphragm is present, it is the portion of an electrolyte on the anode side of the diaphragm.

anomalous

Exceptional, varying from the usual, the standard or the mean.

antecedent

The first of the two terms of a ratio. Thus in the ratio of A to B, written A : B, the term A is the antecedent. They are also the first and third terms of a proportion. Thus in *a* : b :: *c* : d, a and c are the antecedents; b and d are the *consequents.*

antenna
A conductor or a system of conductors for radiating or receiving radio waves.

antenna array
A system of antennas coupled together for directional effects. They may be arranged in the direction of transmission *(alignment array* or *end-on array)* or perpendicular to it *(broadside array)*.

anthracite
A hard, black, lustrous coal which consists mainly of fixed carbon (85-90%) and very little volatile matter. It yields a high degree of heat with very little flame, and is the most widely used coal for residential heating. (Also called *hard coal.*)

antibiotic
(1) Having the property of destroying organic life. (2) As a *noun,* a generic term for substances which inhibit or prevent the growth of a bacteria.

anticathode
The anode of an x-ray tube, situated opposite the *cathode,* which serves as the target of the stream of electrons emitted by the cathode, and as the emitter of the x-rays.

anticlockwise
An occasionally used synonym for *counterclockwise.*

anticosecant
See *arc-cosecant.*

anticosine
See *arc-cosine.*

anticotangent
See *arc-cotangent.*

antilogarithm
The number corresponding to a logarithm. Thus if the logarithm of 24 is 1.38021 then the number corresponding to the logarithm 1.38021 in the logarithm tables is 24; or 24 is the antilogarithm of 1.38021.

antimatter
"Inverted" matter; antiproton in the center and proton going around thc center.

antimony
A metallic element in Group V of the periodic table. Atomic number 51, atomic weight 121.76, specific gravity 6.62 at 20°C., melting-point 630°C., boiling-point 1380°C., valence

3 or 5. Discovered by Valentine in 1450. Antimony is a white, brittle metal, rarely found in the free state. Symbol: Sb.

antineutrino

A sub-nuclear particle, which is in relation to the neutrino as the positron is to the electron, having near zero mass, no electric charge, and a spin opposite in direction (for a given direction of the magnetic moment) to that of the spin of the neutrino.

antineutron

A particle of mass equal to that of the neutron, without electric charge, and with a magnetic moment opposite to that of the neutron.

antinodal region

In physics, the area where two waves of the same wavelength and amplitude and traveling through a medium at identical speeds in directly opposite directions cancel out the effects of each other; these areas are separated from each other by half a wavelength. (Also called *loop*.)

antinode

Any of the points, lines or surfaces of a stationary-wave system which have a maximum amplitude.

antioxidant

(1) Having the property of retarding or preventing oxidation of another substance. (2) A noun, used as a generic term for substances having such property.

antiparallel lines

A designation applied at times to lines which are parallel to each other but have opposite senses.

antiparticle

In nuclear physics, a generic term for a sub-atomic particle which is identical with another sub-atomic particle in every characteristic and property, except that its electric charge and spin or magnetic moment, while being equal in magnitude to those of the latter, are opposite in sign, respectively direction. (E.g., the *antiproton* is the antiparticle of the *proton*.)

antiproton

A sub-atomic particle, identical with a proton in every characteristic and property, except that its charge is negative (hence it is referred to also as a *negative proton*).

antisecant

See *arc-secant*.

antiseptic

(1) Pertaining to, producing or characterized by the absence of or freedom from infection-producing or putrescent bacteria. (2) As a *noun,* a generic term for substances which prevent infection by destroying putrescent bacteria.

antisine

See *arc-sine.*

anti-Stokes line

The extra lines appearing in the spectrum of light scattered from molecules of transparent substances. (See *Raman effect.*)

antitangent

See *arc-tangent.*

antitrigonometric function

A synonym for *inverse trigonometric function* (q.v.).

aperiodic

Devoid of periodicity or rhythm; occurring at irregular intervals of time.

aperture

An opening, hole or orifice. In an optical system, the lens opening through which light may pass; numerically, it is equal to the diameter of the largest entering beam of light which can travel completely through the system.

apex

Summit, highest point.

apochromatic lens

A lens consisting usually of three components, each made of a different kind of glass, corrected to a high degree for chromatic aberration with respect to two or more selected colors or wavelengths of visible light.

apothem of a regular polygon

The radius of the inscribed circle; the perpendicular distance from the center to a side.

Appleton layer

A layer of the *ionosphere* above the *Heaviside layer;* it is ionized and reflects shorter wavelengths and higher frequencies than the radio waves of 3000 kilocycles. Also called *F layer.*

applied mathematics

The study and use of mathematical principles as tools in other fields or subjects, especially in physics, chemistry, engineering, and shopwork. Also, the application of the principles of mathematics to any other subject or to practical life. For example, proving that a triangular system of braces is rigid by means of the fact that two triangles are equal if the corresponding sides are equal is "applied mathematics," while proving the triangles equal is "pure" or "abstract" mathematics.

aqua fortis

Concentrated nitric acid.

aqua regia

Nitro-hydrochloric acid, a mixture of one part nitric acid (HNO_3) with three parts hydrochloric acid (HCl); used as a solvent for noble metals, also as an etchant for stainless steel.

aqueous solution

A solution in which water is the solvent.

aqueous vapor

The vapor state of water, more commonly called *water vapor.* Odorless, colorless.

Arabic numerals

The numerals 1, 2, 3, 4, 5, 6, 7, 8, 9, so called because they are modified forms of the numerals used by the Arab countries and were introduced to Europe by the Arabs; they are assumed to have originated in India.

arbitrary

Not governed by any fixed rule or standard; chosen according to one's own will.

arbitrary constant

A *constant* which retains the same fixed value in a given example or discussion, but is not regarded as fixed in general. (See *absolute constant.*)

arbitrary unit

Any unit of measurement which is not related to any *fundamental unit.*

arc

In mathematics in general, any segment of a curve. In topology, the path between two points.

arc

In physics, a luminous glow formed by the flow of electric current through ionized air, gas, or vapor between separated electrodes or contacts, characterized by a high current density and a low potential gradient.

arc lamp

An apparatus in which an electric *arc* is used to produce a very bright light.

arc length

The length of the arc of a curve, expressed in linear units of measurement.

arc-cosecant

The angle the cosecant of which is the given number. Thus, the arc-cosecant of x, written as *arccsc* x or $csc^{-1}\ x$, equals the angle the cosecant of which is x. Also called *inverse cosecant* and *anticosecant.*

arc-cosine

The angle the cosine of which is the given number. Thus, the arc-cosine of x, written *arccos* x or $cos^{-1}\ x$, equals the angle the cosine of which is x. Also called *inverse cosine* and *anticosine.*

arc-cotangent

The angle the cotangent of which is the given number. Thus, the arc-cotangent of x, written as *arccot* x or $cot^{-1}\ x$, equals the angle the cotangent of which is x. Also called *inverse cotangent* and *anticotangent.*

arc-secant

The angle the secant of which is the given number. Thus, the arc-secant of x, written as *arcsec* x or $sec^{-1}\ x$, equals the angle the secant of which is x. Also called *inverse secant* and *antisecant.*

arc-sine

The angle the sine of which is the given number. Thus, the arc-sine of x, written *arcsin* x or $sin^{-1}\ x$, equals the angle the sine of which is x. Also called *inverse sine* and *antisine.*

arc-tangent

The angle the tangent of which is the given number. Thus, the arc-tangent of x, written *arctan* x or $tan^{-1}\ x$, equals the angle the tangent of which is x. Also called *inverse tangent* and *antitangent.*

Archimedes' principle

A body that is submerged partly or wholly in a fluid is buoyed up by a force equal to the weight of the fluid which the body or its submerged portion displaces.

area

(1) Any plane surface bounded by straight or curved lines. (2) The number of units of surface enclosed by a geometric figure.

areal

Relating to areas.

Argand diagram

The coordinate system used as frame of reference for the graphic representation of *complex numbers*. The two axes of the system are called the *real axis* (or *axis of reals*) and the *imaginary axis* (or *axis of imaginaries*), and the plane itself on which the complex numbers are plotted is called *Argand plane, complex plane,* or *z plane.*

argillaceous

Clayey; having the appearance or consistency of clay.

argon

A zero-valent element, one of the rare gases, in Group O of the periodic table, discovered by Rayleigh and Ramsay in 1894. Atomic number 18, atomic weight 39.944. Argon is a colorless, odorless, monatomic gas, and constitutes 0.95% of the atmosphere of the earth. Its melting-point is −189.2°C., its boiling-point −185.7°C., its density 1.7837 g. per liter at normal temperature and pressure. Symbol: A.

argument of a complex number

The angle of the complex number, also referred to as its *amplitude.*

argument of a function

The independent variable of the function, on the value of which that of another function or quantity depends. (Thus in the equation $x = 2y$, the value of y depends on the value assigned to x, which in this case is the argument.)

arithmetic

The study of the integers 1, 2, 3, 4, 5, . . . under the operations of addition, subtraction, multiplication, division, raising to powers, and extracting roots, and the use of the results of these studies in everyday life.

arithmetic mean

The arithmetic mean of any number of quantities is the sum of those quantities divided by their number. It is the sum of any number of consecutive terms of an arithmetic progression divided by the number of terms taken.

arithmetic progression

(A. P.) A sequence each term of which is equal to the sum of the preceding term and a constant; written: $a, a + d, a + 2d, \ldots, a + (n-1)d$, where a is called the first term, d the common difference or simply the difference, and $a + (n - 1)d$ the last or nth term. The positive integers 1, 2, 3, . . . form an arithmetic progression. The sum of the terms of an A.P. is sometimes called an A.P.

arithmetic series

The indicated sum of the terms of an *arithmetic progression.* (The reciprocals of these terms form a *harmonic series.*)

arithmetic triangle

1	1	1	1	1	1	1	1	1	1	....
1	2	3	4	5	6	7	8	9	....	
1	3	6	10	15	21	28	36	....		
1	4	10	20	35	56	84	....			
1	5	15	35	70	126	....				
1	6	21	56	126	....					
1	7	28	84	....						
1	8	36	....							
1	9	....								
1	....									

In this triangle of numbers, after writing down the original one (unity) all the other numbers are obtained by adding together the numbers above and to the left of the desired number. Read horizontally or in the vertical columns, the numbers are figurate numbers.

This triangle can also be written in the shape of a pyramid, thus:

1
1 1
1 2 1
1 3 3 1
1 4 6 4 1
1 5 10 10 5 1
1 ....

In this case the desired number is found by adding together the two numbers immediately above and to the right and left of the desired number.

These numbers are useful in certain series and when expanding binomial expressions.

arms of an angle

When two straight lines are drawn from a point or when two straight lines cut one another. These straight lines are called the arms of the angles they contain.

aromatic acid

Any organic compound in which one or more hydrogen atoms

of an aromatic nucleus have been replaced by the carboxyl group (the monovalent acidic —COOH group).

aromatic compound

An organic compound containing a closed ring of atoms.

array

An ordered arrangement of objects.

array of aerials

A system of *aerials* coupled together in order to obtain directional effects; they may be arranged in the direction of the radiation (*end-on* or *alignment array*) or in a line perpendicular to that direction (*broadside array*).

arsenic

An element in Group V of the periodic table, discovered by Schröder in 1649. Atomic number 33, atomic weight 74.91, valence 3 and 5, sublimes at 500°C. under pressure, boiling-point 615°C., specific gravity 5.73 at 15°C. Symbol As.

artificial gravity

The effects of centrifugal force proposed to be utilized to simulate those of gravity in manned rockets and space stations beyond the effective gravitational field of the earth or another celestial body.

artificial radioactivity

The radioactivity of artificial *radioisotopes,* the majority of which are β-emitters. Discovered by F. Joliot and I. Joliot-Curie (France) in 1934.

aryl

An organic radical, product of the removal of a hydrogen atom from an aromatic hydrocarbon.

asbestos

The commercial name of several fibrous silicate minerals, e.g., serpentine, tremolite, etc. Asbestos is extremely fire resistant and for that reason, as well as for its low thermal conductivity, is used as a heat insulating and fireproofing material.

aspherical lens

A lens in which the surfaces depart from the true spherical shape.

associative law

The terms of an expression connected by plus (+) or minus (−) signs can be grouped together in any way. Thus $a + b - c + d - e + f = (a + b) - c + (d - e) + f$, or $a + (b - c) + d - (e + f)$.

associative law for multiplication

The factors of a product can be grouped together in any way we please. Thus $3a \times 4bc \times d = 3 \times a \times 4 \times b \times c \times d = 4 \times d \times 3 \times b \times a \times c = (ba) \times 3\ (dc) \times 4$.

astatine

A very rare radioactive element in Group VII (halogens) of the periodic table, atomic number 85, first announced by Allison and Murphy under the name *alabamine* in 1929; also referred to formerly as *eka iodine,* it was renamed *astatine* by Segré, McKenzie and Corson who produced the isotope At^{211} in 1940, by a bombardment of bismuth with high-energy (32 mev) α particles. Symbol: At.

astigmatism

In optics, a defect of an optical image produced by an aberration in the lens as a result of which for objects off the axis the rays passing through different meridians of the lens come to focus in different planes.

Aston dark space, Aston's dark space

The narrow dark region of the discharge in a vacuum tube which can occasionally be seen between cathode glow and cathode. (Also called *primary dark space.*)

astrionics

A newly coined term (Dr. W. A. Heflin) for the art or science of adapting electronics to space flight.

astrography

That branch of astronomy which deals with the positions, composition and other characteristics of the stars.

astroid

A *hypocycloid* of four cusps.

astrometry

The branch of astronomy that deals with the measurements of the celestial bodies, especially those to determine their sizes, position and movements.

astronautics

(1) The art, skill, or activity of operating space vehicles. (2) In a broader sense, the art or science of designing, building, and operating space vehicles.

astronomic latitude

The astronomic latitude of a place on earth is the angle between the plane of the equator and the direction of gravity at the place.

astronomical triangle
A spherical triangle used for the solution of astronomical problems. Its vertices are the *zenith,* the *elevated pole,* and the *celestial body* observed; its three sides are the *co-latitude* (the complementary angle of the observer), the *polar distance* of the celestial body (the complementary angle of its declination), and the *zenith distance* of the celestial body (the complementary angle of its altitude).

astronomy
The study of the celestial bodies; their composition, motion, history, distances, etc.

astrophysics
That branch of *astronomy* which deals with the material constitution, chemical composition, temperature, etc. of the stars and other celestial bodies through the application of the laws and principles of physics; it may thus be regarded as a combination of astronomy and physics.

asymptotes
A line or lines which approach nearer to a curve than any other line, but never meet the curve, although infinitely extended. They may be curves or straight lines. Rectilinear asymptotes are the tangents to a curve at an infinite distance.

asymptotes of a hyperbola
The tangents to the hyperbola which intersect in the point of intersection of the transverse and conjugate axes.

At
The symbol for the element *astatine.*

—ate
In chemistry, a suffix denoting: (1) a salt, or (2) a compound in which a nonmetal occurs in its positive higher valence (in contradistinction to the suffix *-ite*), as in *sulfate* which is a salt of sulfur*ic* acid.

athermancy
The property of absorbing heat radiations.

atmosphere
The body of air which surrounds the earth, defined at its outer limits by the actual presence of air particles but in such few numbers that collisions between them are so rare as to make the force of gravity the only means of keeping them associated with air particles at lower altitude. (See *effective atmosphere.*)

atmosphere (atm)

The unit of pressure, based on the pressure exerted by the air on the surface of the earth, i.e. 1 atm = 1.0333 kg. per cm^2 (14.7 lb. per square inch), or about 1,013,246 dynes per cm^2.

atmospheric drag

The retarding effect of air resistance that slows down a body moving through the atmosphere.

atmospheric pressure

The force per unit area exerted by the atmosphere in any part of the atmospheric envelope.

atmospheric refraction

Refraction of light from a distant point by the atmosphere, caused by its passing obliquely through varying air densities.

atmospheric windows

The atmospheric envelope of the earth absorbs practically every frequency of the electromagnetic spectrum, except for those of visible light and shorter radio waves. Those two bands of electromagnetic radiation, the only ones to which the atmosphere of the earth is transparent, may therefore be thought of as two "windows" in an otherwise opaque atmosphere, namely: the *"optical window,"* through which we can view the universe by naked eye and through optical telescopes, making use of the wavelength and frequencies of visible light, and the *"radio window,"* through which the radio telescopes can scan the universe and furnish us with new data, "seeing" by the short radio waves.

atmospherics

Natural electromagnetic disturbances of radio reception, produced by atmospheric conditions. (In the United States, the term *static* has come into general use as a synonym for "atmospherics.")

atom

The smallest building block of chemical elements, which cannot be divided or broken up by chemical means. (The word "atom" is derived from the Greek, meaning "that which cannot be cut up".)

atom model

A schematic picture of an atom or atoms.

atomic

Relating or pertaining to an atom or to atoms in general. (Cf. *nuclear.*)

atomic battery

A device for storing and delivering the electrical energy of a radioactive product of fission.

atomic bomb

A military weapon which utilizes the induced nuclear fission chain reaction of radioactive substances as its source of energy. The United States of America was the first nation successfully to develop atom bombs, although at the end of World War II Germany was not far from reaching the same goal. The basic principle is the use of two isolated lumps of a fissionable substance (in the bomb dropped on Hiroshima uranium-235, in the later bombs plutonium was used), each in itself of subcritical mass, but their combined masses exceeding the critical mass; these two lumps are brought together, into close contact, by a mechanical arrangement so as to form one supercritical mass, or detonated by removal of the neutron-absorber.

atomic diameter

The diameter of an individual atom. Atomic diameters of elements vary within the periods of the periodic table from 0.6×10^{-8} to 5.4×10^{-8} cm.; the atoms of the rare gases have the smallest, the atoms of the alkali metals the largest diameter in each period.

atomic heat

The thermal capacity of 1 gram-atom of an element, or the quantity of heat required to raise the temperature of 1 gram-atom of the element by 1°C.

atomic mass

The mass of the atom, customarily expressed in *atomic mass units.*

atomic mass unit (a.m.u.)

The unit of the atomic mass in nuclear physics is 1/16 of the mass of one atom of the oxygen isotope ${}_8O^{16} = 1.6603 \times 10^{-24}$ gram. In terms of energy, 1 a.m.u. $= 931.8$ mev $= 1.49 \times 10^{-3}$ ergs. Cf. *atomic weight.*

atomic number

The number assigned to an element in the periodic table, hydrogen having the lowest atomic number (1) among the known elements, according to the system introduced by H. G. J. Moseley in 1914 and extended during the past decade to include the transuranic elements, too. The atomic number of every element is equal to the total number of positive nuclear charges, i.e. to the number of protons contained by the nucleus of its atoms, and also to the total number of the negatively charged extranuclear electrons present in every atom of the element. All atoms of the same element have the same atomic number (i.e. the same number of protons in their nuclei, and the same number of extranuclear electrons), but not necessarily the same atomic weight (i.e. not necessarily the same number of neutrons in their nuclei). Those atoms of the same element which have different atomic weights are called *isotopes* of the element in question. (Symbol: Z.)

atomic pile

This term was formerly generally used, and is still quite frequent in popular usage, for a nuclear chain *reactor*.

atomic structure

According to the currently accepted view, the atom consists of a central part, called *nucleus,* and a number of *electrons* (called orbital or planetary electrons) circling about the latter, like planets about a sun. The nucleus is of a high specific weight; it contains most of the mass of the entire atom (its mass is considered equal to the atomic mass), and is composed of positively charged particles, called *protons* (the number of which always equals the atomic number) and particles of 0 charge, called *neutrons* (the number of which equals the difference between the *atomic weight* and the *atomic number*). While the nucleus carries an integral number of positive charges (an integral number of protons) each of 1.6×10^{-19} coulomb, each electron carries one negative charge of 1.6×10^{-19} coulomb, and the number of orbital electrons is equal to the number of protons in the nucleus so that the atom as a whole has a net charge of 0. The electrons are arranged in successive *shells* around the nucleus; the maximum number of electrons in each shell is determined by natural laws, and the extranuclear electronic structure of the atom is characteristic of the element.

atomic volume

The volume in milliliters occupied by 1 gram-atom of the element in the solid state, i.e. the weight of 1 atom in grams divided by its specific gravity. The sizes of atoms are proportional to their atomic volumes.

atomic weight (Symbol: a.w.)

The relative weight of an atom of an element, expressed in

multiples of 1/16 of the atomic weight of the oxygen atom, which is taken arbitrarily as 16. In the chemical scale of atomic weights, the average weight of oxygen atoms as they occur in nature is taken as 16.0000, whereas the physical scale of *atomic mass units* uses the atomic weight of the most abundant *oxygen isotope* ($_8O^{16}$) as 16.0000. Any atomic weight on the physical scale can be converted to the corresponding weight on the chemical scale by dividing it by 1.000272. Atomic weights represent the statistical averages for the different isotopes of the elements.

atomization

The dispersion of a liquid into particles to form a fine spray.

atomsmasher

A popular term for *accelerators of nuclear particles* (q.v.).

atomsmashing

A popular term for *nuclear bombardment* or *nuclear fission.*

attenuation

With reference to a wave phenomenon, the decrease in displacement with distance in the direction of propagation.

attraction

The gravitational force with which one particle draws another toward itself. It is proportional to the product of their masses divided by the square of the distance between them. Usually called simply "gravitation."

A.U., Å.U.

Abbreviations for *Ångstrom unit.*

Au

The symbol for the element *gold.*

audibility

A measure of the strength of a sound or signal as compared with the strength which will produce a sound just strong enough to be audible. The customary unit for expressing audibility is the *decibel.*

aurora borealis

Streams of light appearing in the sky, usually in the polar regions. Caused by seepage of electrified matter through the earth's magnetic screen, symmetrically deflected toward the north and south poles.

audible

Within the range of hearing of the average human ear, i.e., vibrating at a rate between 20 and 20,000 cycles per second.

austenite
The allotropic form of iron (gamma iron) which has a *face centered cubic lattice,* the parameter of which increases with increasing carbon content.

autocatalysis
A chemical reaction which is accelerated by one of its own products.

autoclave
A vessel, usually constructed of thick steel plates, for carrying out chemical reactions under high pressure and at high temperatures.

automatic
Equipped with a self-acting mechanism or device that performs a required act at a predetermined time or under predetermined conditions.

automation
A word coined by the Ford Motor Company to designate automatic feeding, unloading and handling of work to, from and between production machines.

auxiliary circle of an ellipse
The circle drawn with the major axis of the ellipse as a diameter.

auxiliary elevation view
Any *elevation view* other than the front, right profile, left profile and rear views.

auxochrome
A group which when introduced into an organic compound containing a chromophor, converts the compound into a dye.

average
The sum of a number of quantities divided by the number. A mean proportion. Thus five numbers added together, say: $4 + 6 + 3 + 5 + 2 = 20$ which divided by five $= 20/5 =$ an average of 4.

average life
In nuclear physics, the average of the individual lifetimes of all the atoms of a radioactive substance, equal to the time required by the radioactivity of the substance to diminish to 0.3697 of its original value; it is the reciprocal of the *decay constant* and equals the *half-life* multiplied by 1.443. Also called *average lifetime* or *decay modulus.*

average velocity
A ratio of distance to time.

Avogadro's hypothesis

Equal volumes of all gases contain the same number of molecules at the same temperature and pressure. (Also referred to as *Avogadro's principle.*)

Avogadro's number

The number of the atoms contained in 1 gram-atom, or of the molecules contained in 1 mole: 6.0234×10^{23}.

avogram

One gram of a substance divided by *Avogadro's number.*

axial stereoisomerism

See *cis-trans isomerism.*

axial stress

The stress set up in a body in the direction of the axis.

axiom

A self-evident and generally accepted principle. An assumption or postulate. The distinction between postulate and axiom is not very sharp. Axiom refers more to the accepted truth of a theorem than does a postulate. One may postulate something that could be proved but this would not necessarily be an axiom. Examples of axioms are: things equal to the same thing are equal to each other; if equals are added to equals the results are equal; if equals are subtracted from equals the remainders are equal, etc.

axis

Any real or imaginary straight line that divides a plane figure into two equal or symmetrical parts. Also, a straight line about which a body revolves. (See *optical axis, principal axis.*)

axis of a circle of a sphere

That diameter of the sphere which passes through the center of the sphere, and is perpendicular to the plane of the circle. The extremities of the axis are the poles of the circle. The poles of a small circle are not equally distant from the plane of a circle. The nearer one is usually called the pole.

axis of a circular cone

The line which joins the vertex and the center of the base.

axis of a circular cylinder

The line which joins the centers of the bases.

axis of a crystal

Any of three or four noncoplanar, intersecting lines which are fixed with respect to the crystal. (See also *axis of symmetry of a crystal.*)

axis of a parabola
The lines drawn through the focus perpendicular to the directrix.

axis of precession
The axis of the cone described by the axis of a symmetrical rotating body with its *centroid* as *vertex.* (See *precession.*)

axis of revolution
In a *solid of revolution,* the fixed line about which the generating plane surface was revolved to produce the solid in question.

axis of rotation
(1) The line about which all particles or points of a rotating body or system circle with the same angular velocity. (2) See *axis of symmetry of a crystal.*

axis of symmetry
The center line on either side of which the two halves of a figure can be made to coincide (are equal) if folded along that line.

axis of symmetry of a crystal
An imaginary line about which a crystal can be rotated so that corresponding faces come into the same position. According to the number of times this occurs during one complete revolution, the axis is said to be of twofold, threefold, fourfold or sixfold symmetry.

axis of the celestial sphere
The earth's axis extended to the *celestial sphere.*

axis of the earth
That diameter of the earth about which it rotates.

azimuth
In a system of coordinates by which a point is located by its distance from another fixed point, the angle that the line from this point to the given point makes with a fixed line called the polar axis. This angle is called the polar angle or azimuth of the point.

azimuth
In surveying and in navigation, the most convenient and commonly used way of expressing direction. Azimuths are angles reckoned clockwise from zero degrees at magnetic, true, or grid north through 360 degrees. Thus, there are three different azimuths for a given line—magnetic azimuth, true azimuth, and grid azimuth.

B

B

The symbol for the element *boron.*

Ba

The symbol for the element *barium.*

back electromotive force

Electromotive force which opposes the normal flow of current in an electric circuit.

background

A generic term for those effects which are inevitably always present in physical detecting, measuring and recording apparatus, and above which the phenomenon studied must manifest itself in order to be detected and measured. E.g., in spectrographic analysis, the background is produced by the light source from continuous radiation, incandescent particles, unresolved band spectra and from scattered light in the spectrograph. (See also *background count.*)

background count

In radiation counters, the counts produced by an agency other than the radiation which it is intended to detect or measure, as, e.g., by other external radiation, or by radioactivity of the instrument itself.

Balfour-Stewart currents

Electric currents above the surface of the earth, named for the scientist who was first to mention them.

ballistics

The science of the motion of projectiles.

Balmer series

A series of lines in the hydrogen spectrum, made up of proportionately progressing frequencies; the equation (*Balmer's equation* or *Balmer's formula*) expressing the progression of the frequencies is:

$$\frac{1}{\lambda} = \bar{\nu} = R_H \left\{ \frac{1}{2^2} - \frac{1}{n^2} \right\},$$

where λ is the wavelength, ν the wave number of the spectral line, R_H is Rydberg's constant for hydrogen (109,677.76 cm^{-1}), and n is an integer higher than 2.

band of frequencies
The complete ranges of frequencies between two specified limits.

band spectrum
A spectrum, usually molecular, composed of broad bands (which nowadays can be resolved into spectral lines of atoms).

bar
The absolute unit of pressure in the C-G-S system. 1 bar = 1 megadyne per $cm.^2$; 1 *megabar* (or *megabarye*) = 10^6 bars = 10^6 megadynes per $cm.^2$; equal to 0.987 atm. (the pressure of 750.07 mm. of mercury at 60°C.). Some authors use "bar" or "barye" for 10^6 dynes per $cm.^2$, and "microbar" or "microbarye" for 1 dyne per $cm.^2$. Also called *barye.*

bar graph
A type of graphic representation in which varying relationships are shown by means of dark bars of varying length.

barium
A divalent metallic element, one of the alkaline earth metals, in Group II of the periodic table, discovered by Scheele in 1774. Atomic number 56, atomic weight 137.36, specific gravity 3.7, melting-point 850°C., boiling-point 1140°C., abundance in the earth's crust about 0.047%. Symbol: Ba.

Barkhausen effect
A discontinuous variation in magnetization in ferromagnetic substances as the magnetizing force is continuously increased or decreased.

barn
A unit for measuring capture *cross sections* (q.v.) of elements. 1 barn = 10^{-24} cm^2 per nucleus. (Symbol: δ.)

barometer
An instrument for measuring atmospheric pressure. The two basic types are the *mercurial* barometer (a glass tube, one end sealed and the other open, containing a liquid, usually mercury, the displacement of which indicates the measure of atmospheric pressure) and the *aneroid* barometer (a vacuum box the expansion and contraction of which under the effect of atmospheric pressure is registered on a dial).

baromil
The unit used in graduating a barometer in the CGS system. If the barometer is read at 45° latitude at sea level and its temperature is at the freezing point of water, one baromil corresponds to a pressure of one millibar; under other condi-

tions, corrections must be applied to reduce a baromil to a *millibar.*

barye

See *bar.*

base

In mathematics, any number which is to be raised to a power. Example: in x^2, x is the base.

base

In chemistry: (1) A substance containing the OH radical which dissociates to form OH^- ions when dissolved in water. (2) A substance with basic properties, i.e., which reacts with and neutralizes acids and forms a salt in water, which combines with hydrogen atoms ("accepts protons"), and turns litmus blue.

base

In geology: A substance capable of combining with silica in a rock, e.g., potash.

base line

One or more straight lines joined together, chosen by surveyors, from which all measurements and angles are taken to points or objects on either side of the line.

base metal

(1) A metal which becomes oxidized when heated in air, as copper, lead, zinc or tin, as distinct from a *precious metal,* such as gold or platinum. (2) A metal at the lower end of the *electro-chemical series,* as distinct from a *noble metal.* (3) With reference to an alloy, the metal preponderant in its composition. (4) In metal-working, the metal which is to be welded or cut.

base of a cone

The section of the conical surface created by the cutting plane.

base of a cylinder

Either of its two plane surfaces.

base of a frustum of a cone or of a pyramid

Either of the two parallel planes cutting the pyramid or cone to form the frustrum.

base of a logarithm

The number which when raised to the power indicated by the logarithm results in the given number. Conversely, the *logarithm* is the power to which that base is raised in order to obtain a desired number. Any number can be used as a base in logarithms; but the usual base is 10 for Brigg's or common logarithms; and e = 2.718282 for Napier's or hyperbolic system logarithms. Thus 10 to the power of 1.47712 equals 30, or $10^{1.47712} = 30$ in Brigg's or common logarithms. 10 is the base; 1.47712 is the logarithm; and 30 is the desired number.

base of a number system

The number of units in any digit place which denotes the next highest digit, i.e., the number the square of which denotes the digit next highest to the digit denoting units, and the successively higher powers of which denote the successively higher digits. The base of our commonly used number system is 10, and $100 = 10^2$, $2000 = 2x10^3$, etc. In general, any number in any number system equals $a + a_1\ n + a_2\ n^2 + a_3 + n^3 + \ldots$ where $a, a_1, a_2, a_3 \ldots$ are positive integers and n is the base of the system. The base of the system of numeration is also called the *radix*.

base of a plane figure

Any side of a triangle or polygon which is chosen as the side on which the figure is supposed to rest. It is the side opposite any chosen vertex which is the point furthest from that side. The height of the figure is measured by a perpendicular to the base drawn to a line passing through the vertex and parallel to the base.

base of a prism

Either of the two parallel planes.

base of a pyramid

The section of the pyramidal surface created by the cutting plane.

base of a spherical sector

The zone produced by the arc of the sector.

base of a zone

The intersection with the sphere of the planes forming the zone.

base potential

A potential substantially anodic to the standard hydrogen potential. (See *electro-chemical series.*)

basic

In chemical terminology, this adjective signifies *having the properties of a base.*

basic anhydride

The oxide of a metal, such as may be formed from a base by abstracting water from it, and which when added to water forms that base again.

basic oxide

An oxygen compound of a metal. (Cf. *amphoteric oxide.*)

basic rock

A general geological term for igneous rocks comparatively low in silica, whether or not rich in metallic bases. Occasionally also applied, loosely, to designate igneous rocks consisting chiefly of dark minerals.

basic salt

(1) In general, any salt the solution of which contains an excess of OH^- over H^+. (2) Specifically, any salt that contains the OH radical when in water solution.

battery

A combination of two or more galvanic cells electrically connected to work together to produce electrical energy. Common usage permits this designation to be applied also to a single cell used independently.

Baumé scale

A hydrometric scale for measuring the density of liquids. There are two separate scales, one of which is used with liquids heavier than water, sinks to 0 degrees in pure water, and 15 degrees in a 15% salt solution; the other, for liquids lighter than water, sinks to 0 degrees in a 10% salt solution, and to ten degrees in pure water.

Be

The symbol for the element *beryllium.*

beam

With reference to radiation, electromagnetic or corpuscular, in general, a bundle of rays. Specifically, with reference to light, a beam, shaft or column of light, consisting of a group of *pencils* of light.

bearing

The bearing of a line in surveying is the angle which the line makes with the north and south line, or its direction relative to the north-south line. The bearing of a point with reference to another point is the angle that the line through the two points makes with the north and south line.

bearing of a line

In orthographic projection, the direction which a horizontal projection of the line has with respect to the points of the compass.

bearings

Bearings are used to express directions as determined by

means of a compass. The bearing of a given line is the angle and direction which the line makes with respect to the north or south base direction line. Bearings are stated by quadrants and never exceed 90 degrees.

beat

The periodic variation of the amplitude of a wave phenomenon owing to the interference of two waves of the same nature but of different frequencies. In sound, the regular alternation of the strengthening and fading of the auditory sensation as the sound waves emitted by two acoustic sources differing slightly in frequency alternately reinforce and cancel out each other.

beating

The interaction of two or more periodic quantities of different frequencies producing a resultant that has the pulsations of amplitude called *beats.*

Becquerel effect

Production of an electromotive force when two identical electrons immersed in an electrolyte are illuminated unequally.

Beer's law

The degree of absorption of light depends on the thickness of the layer traversed and on the molecular concentration of colored substance of that layer.

bel

The fundamental division of a logarithmic scale for expressing the ratio of two amounts of power, the number of bels denoting such a ratio being the logarithm to the base 10 of this ratio. With P_1 and P_2 designating two amounts of power and N the number of bels denoting their ratio:

$$N = \log_{10} (P_1/P_2) \text{ bels.}$$

benzoyl

The monovalent acyl radical $C_6H_5CO–$, derived from benzoic acid; symbol: Bz.

berkelium

A transuranic element, atomic number 97, discovered by S. G. Thompson, A. Ghiorso and G. T. Seaborg in 1950 in Berkeley, California. Mass number of the most stable known isotope 245, valence 3 and 4. Symbol: Bk.

Bernouilli's lemniscate

See *lemniscate.*

Bernouilli's principle

In a steady flow of an incompressible liquid through a conduit of variable cross-section, the total energy of its unit volume is a constant magnitude.

Berthelot principle

Of all the chemical processes which can take place without the application of external energy to the system, the one involving the development of the highest quantity of heat is the one which is bound actually to take place. (This principle does not apply to high temperatures or to endothermic reactions.)

beryllium

A divalent metallic element in Group II of the periodic table, formerly also called *glucinium (Gl),* discovered by Vauquelin in 1797. Atomic number 4, atomic weight 9.013, specific gravity 1.85, melting-point 1300-1350°C., boiling-point about 2970°C. Abundance in the earth's crust: 1 part in 200,000; it is found in compounds only. Beryllium is very brittle, and very difficult to prepare in the pure state. Symbol: Be.

Bessemer process

A method of producing steel in which air is blown through molten pig iron contained in a refractory lined pear-shaped cylindrical vessel, open at the upper end for the escape of gases, the vessel being known as a *Bessemer converter.* The heat is produced by the oxidation of impurities silicon, manganese, and carbon, by the oxygen of the air blast.

Bessemer steel

Steel produced by the *Bessemer process.*

beta decay (β^-decay)

Radioactive change by emission of a β^- particle, i.e. an electron. Since the electron has almost no mass and has 1 negative electronic charge, the emission of an electron produces a *daughter element* which has the same atomic weight as the parent element, but an atomic number higher by 1 than that of the *parent.*

beta emitter

A radioactive element, whether natural or artificial, which changes into another element by *beta decay.*

beta iron

An obsolescent designation for nonmagnetic *alpha iron* which exists between 768° and 910°C.

beta particle (β particle) (β, β^-)

A negatively charged particle, i.e. an *electron,* ejected at high velocity from the nucleus of the atom of a beta emitter when one of its neutrons changes into a proton. (The beta particle is not an orbital electron; it is hurled out from the nucleus, not from the extranuclear electron shells.)

beta ray

A stream of *beta particles;* it can be deflected by electric and magnetic fields, and is more penetrating than an alpha ray but less active photographically and produces less ionization than the latter.

betatron

An accelerator, developed by D. W. Kerst (U.S.A.) in 1941, used in order to impart high velocities to electrons (beta particles). The electrons are emitted from the filament of a circular vacuum tube; they are caused by an electromagnet to circle under constant acceleration in a contracting orbit, and eventually to hit a target (a tungsten anti-cathode), where they produce high-energy photons or x-rays which emerge from the tube. A betatron of 5-6 Mev can produce x-rays equivalent to the gamma radiation of 10-20 g. of radium. (Also called *rheotron* and *induction (electron) accelerator.*)

bev

The abbreviation for "billion electron volts"; 1 bev = 10^9 electron volts = 1000 mev.

bevatron

The designation of a high-energy proton accelerator which imparts energies of several billions of electron volts to positively charged sub-atomic particles.

Bi

The symbol for the element *bismuth.*

bi-

A prefix indicating *two* or *double* (or *doubly*).

bias, bias voltage

Voltage applied permanently to the control grid of a vacuum tube in order to make it permanently negative.

biaxial

Having or referred to two axes. With reference to lenses, crystals, etc.: having two optic axes.

billion

In the United States and France, a thousand million (1,000,000,000); in England and all over Europe, except for France, a million millions (1,000,000,000,000).

bimetal, bi-metal

Two dissimilar metals having different specific properties,

as, e.g., different coefficients of thermal expansion, which are fused together into a sheet, strip, etc., and so arranged that the sheet or strip deflects under changes in temperature.

bimolecular

See *dimolecular.*

binary

In general, composed of two elements or components.

binary alloy

An alloy containing two principal elements.

binary compound

A combination of two different elements, or of two radicals which behave like elements, or of an element and one such radical; thus each molecule of a binary compound consists of two different kinds of atom.

binary system

In chemistry, a system containing two components.

binary system of numbers

A system of numbers using 2 as its *base.*

binding energy

The binding energy of the nucleus can be defined as the quantity of energy released when nucleons (protons and neutrons) combine to form an atomic nucleus, or conversely, as the energy required to disrupt a given nucleus. It is the energy equivalent of the *mass defect,* on the basis of 1 a.m.u. = 931 Mev. The binding energy is a negative magnitude; however, it is customary to indicate its absolute, numerical value only; the greater the numerical value of the binding energy, the greater the stability of the nucleus, since the greater is the energy required to disrupt it.

binomial

Any mathematical expression consisting of the sum or difference of two terms (e.g., $a + b$ or $a - b$).

binomial theorem

A rule that gives a general algebraic formula by which any power of a binomial quantity may be found, without the necessity of carrying out the progressive multiplications.

biochemistry

That branch of chemistry which deals with the structure and composition of animal and vegetable matter, metabolism, and other changes, reactions and processes taking place in living organisms.

biodynamics
The study of the motions of bodies and of the force acting on bodies in motion, or in the process of changing motion, as these motions or forces affect life. (Dr. W. A. Heflin.)

biophysics
Physics applied to biology.

biosphere
That part of the earth and its atmosphere in which animals and plants live. (Dr. W. A. Heflin.)

biquadratic
The square of a square, or the fourth power of a number. Thus $(a^2)^2 = a^4$.

biquadratic equation
An equation in which the unknown quantity is raised to the fourth power. Thus $ax^4 + bx^3 + cx = d$.

birectangular triangle
A *spherical triangle* containing two right angles.

birefringence
A synonym for *double refraction.*

birefringent
Producing or displaying *double refraction.*

bis-
A prefix meaning *twice.* In chemistry, it usually indicates that a molecule consists of two similar halves or compounds.

bisect
To divide or cut into two equal parts.

bisector
The straight line which cuts or divides another line, angle or figure into two equal parts.

bismuth
A metallic element in Group V of the periodic table, discovered by Valentine in 1450. Atomic number 83, atomic weight 209.00, melting-point 271°C., boiling-point 1450°C., specific gravity 9.8, valence 2, 3 or 5 (the most stable bismuth compounds are those in which the bismuth is trivalent). The last stable element in the periodic table. Symbol: Bi.

bituminous coal
The most important of the various types of coal because of its manifold uses (heating, manufacture of coke, coal gas, coal tars, etc.). It contains 70-80% carbon, the rest consists chiefly of hydrocarbons and some other volatile matter. (Also called *soft coal.*)

Bk

The symbol of the element *berkelium.*

black body

A theoretical thermodynamic surface, an ideal body, the surface of which absorbs all radiation incident on it, so that it neither reflects nor transmits any portion of it. Thus, the reflective power of this ideal body is zero, whereas its absorptive power is unity.

black-body radiation

Radiation emitted by an ideal *black body* or the inside of an enclosed cavity.

black body temperature

The temperature of a body the emissivity of which is unity, or of any body under *black-body* conditions.

black light

A colloquial term for any invisible radiation (as infrared or ultraviolet).

body

A definite portion of matter considered separately from other matter.

body-centered cubic space lattice

A *lattice* in which atoms are present at the corners of each cube or rectangular prism with one atom in the center of such cube or prism. The unit cell contains two atoms, because each corner is shared by seven other cubes.

boiling point

The temperature at which the saturation vapor pressure of a liquid is in equilibrium with the external pressure on the liquid. The boiling point therefore varies with the external pressure, and the higher the altitude, the lower is the temperature at which water boils. The normal boiling point of water, referred to standard atmospheric pressure of 760 mm. of mercury, is 100°C, i.e., 212°F.

bolometer

A metallic resistance thermometer of very high sensitivity, used for the measurement of thermal radiation.

bombardment

In physics, the process of hurling electrons, neutrons, or other sub-atomic particles, usually accelerated to high energies, at atomic nuclei, electrodes, etc., to produce desired effects or phenomena.

bond
In chemistry, a linkage between atoms. (See *coordinate bond, covalence, dative bond, valence, valence bond.*)

bond
In welding, the junction of the weld metal and the base metal.

boron
An element in Group III of the periodic table, discovered by Davy in 1808. Atomic number 5, atomic weight 10.82, specific gravity 2.45–2.54, melting-point 2300°C., boiling-point 2550°C. Symbol: B.

bound energy
That portion of the internal energy of a body or system which cannot be extracted from it as external work, i.e., the internal energy minus the *free energy.*

boundary conditions
Certain conditions which must be satisfied at a surface separating two media of different electrical or magnetic properties.

boundary layer
A layer of fluid, close to the surface of a body placed in a moving stream, in which the impact pressure is reduced as a result of the viscosity of the fluid.

Boyle's law
At a given temperature the product of the volume of a gas and the pressure (pv) is constant. Also called *Boyle and Mariotte's law.* Approximately true for moderate pressures. Otherwise stated, the volume of a fixed mass of a given gas at constant temperature is inversely proportional to the pressure.

b.p.
The abbreviation for *boiling point.*

Br
The symbol for the element *bromine.*

braces
Symbols used to group terms which are to be treated as a single expression. An example of braces: $\{ \quad \}$ (Cf. *brackets.*)

brachistochrone
An inverted *cycloid,* representing the curve of quickest descent.

brachistochronic path
The *minimum time path,* i.e., the curve of quickest descent between two points in space.

brackets
The signs ( ); { }; [ ]; ——; they mean that the expressions

within these signs must be worked out, added up or subtracted from one another before continuing to work out the sum. Thus $(4+2+3) \times \{7\times5\}=9\times35$; the horizontal line —— is also called *vinculum.*

brass

An alloy of copper and zinc, prepared in a great many varieties, the common kinds being about 60 parts copper and 40 parts zinc.

brazing

The generic term for a group of soldering or welding processes utilizing as filler metal a nonferrous metal or alloy the melting point of which is above 1000°F, but below that of the metals or alloys which are to be joined. (Often referred to as *hard soldering* when done with a torch.)

breadth

That dimension of an object which is measured at right angles to the length and to the height, i.e., from side to side. Also called *width.*

breeder reactor (breeder pile)

A nuclear chain reactor in which the transmutation produces more fissionable material than the quantity of parent material consumed by the chain reaction.

breeding

The production by fission of a radioelement at a rate higher than the consumption by the chain reaction of the parent element, e.g. the production of plutonium from uranium-235 by bombardment by fast neutrons.

bremsstrahlung

A German term, literally meaning "braking radiation," denoting the electromagnetic radiation produced by the rapid change in the velocity of an electron or another fast charged particle as it approaches an atomic nucleus and is deflected by it. (The term is applied to the process of producing such radiation as well.)

Brewster angle

That angle of incidence upon a given substance which produces a 90° shift in phase in plane-polarized light.

Brewster's law

The tangent of the *polarizing angle* of a substance equals its *index of refraction.*

Briggs' logarithms

Logarithms which have the base 10. Their modulus for conversion from natural logarithms is 0.434294. . . . Also called *common logarithms* and *decimal logarithms.*

brightness

A synonym for *luminescence.*

brilliance

(1) Degree of brightness. (2) That attribute of a color which permits it to be classed as equivalent to some member of the series of achromatic colors, or grays.

Brinell hardness

The hardness of a substance expressed in terms of the area of the indentation produced by a hardened steel ball 10 millimeters in diameter under a load of 3,000 kilograms (reduced to 500 kilograms for softer metals) applied for 30 seconds. The *Brinell hardness number* is calculated by dividing the load applied by the surface area of the indentation.

briquet, briquette

See *compact.*

British thermal unit (B.t.u.)

The heat required to raise the temperature of 1 lb. of water 1°F., when the water is at its maximum density which is at 4°C. or 39.2°F. One B.t.u. equals 778 foot-pounds of work (or energy).

brittleness

The property of a material which permits little bending or deformation without fracture. Brittleness and hardness are closely associated.

broadside array of antennas

A system of antennas coupled together for directional effects, arranged in a line perpendicular to the direction of transmission.

bromine

An element, one of the halogens, in Group VII of the periodic table, discovered by Balard in 1825. Atomic number 35, atomic weight 79.916, melting-point −7.2°C., boiling-point 58.78°C., specific gravity 3.12, valence 1, 3, 5 and 7. Symbol: Br.

Bronsted-Löwry theory

See *acid.*

bronze

An alloy of copper and tin in varying proportions.

brown coal

See *lignite.*

Brownian movement
A continuous, erratic movement of the particles in a liquid or gaseous medium, due to impacts with molecules of the latter.

B.t.u.
The abbreviation for *British thermal unit* (q.v.).

Bu
The chemical symbol for *butyl.*

buffer
Any substance the addition of which to a solution, compound or system enables the latter to resist changes in hydrogen ion concentration when an acid or alkali is added. There are also specific buffers to convey resistance to other, specific changes, and they are designated accordingly (e.g., *oxidation-reduction buffer*).

buffer capacity
That quantity of acid or alkali of a definite concentration the addition of which to a given solution will change the hydrogen ion concentration by one pH unit.

buffer reagent
A substance or a mixture the presence of which in a solution produces a resistance to changes in the concentration of a specific ion.

buffer solution
Any solution which consists of a weak acid or a weak base and one of the salts of the same acid or base, and which retains its pH unchanged by the addition of acids or bases.

buffered water
Water containing dissolved or suspended matter which resists changes in the pH value of the water.

bulk modulus
See *volumetric modulus of elasticity.*

buoyancy
The upward force exerted by a fluid on a solid immersed in it or floating on it. In water, it equals the weight of the water which the immersed or floating body displaces.

burette
A graduated cylindrical glass tube fitted with a ground glass stopcock, used for the measurement and delivery of small volumes of liquid, as, e.g., in titrations in volumetric analysis.

burning
Oxidation of a substance, where the rapidity of the reaction

produces incandescence, and thus the emission of light and heat.

burst

(1) In cosmic ray studies, an exceptionally large electric pulse visible in an ionization chamber; it indicates the emission or incidence of several or many ionizing particles simultaneously. The cause of the phenomenon may be a *cosmic ray shower* or a nuclear disintegration known as *spallation*.—(2) In radio astronomy, a sudden, very brief increase in the intensity of the radiation received.

butyl

The monovalent hydrocarbon radical C_4H_9— or $CH_3(CH_2)_3$—; symbol: Bu.

by-product

A generic term for any substance which is obtained during the production of another substance.

Bz

The chemical symbol of *benzoyl.*

C

C

(1) The symbol for the element *carbon.* (2) The symbol for *centigrade.*

c

The symbol for the velocity of light in a vacuum = 2.99796×10^{10} cm. per second.

Ca

The symbol for the element *calcium.*

cadmium

A metallic element in Group II of the periodic table, discovered by Stromeyer in 1817. Atomic number 48, atomic weight 112.41, specific gravity 8.65 at 20°C., melting-point 321°C., boiling-point 767°C., valence 1 or 2. Symbol: Cd.

caesium

See *cesium.*

cal

The abbreviation for *calorie.*

calcification

Deposition of insoluble calcium salts.

calcination

Driving off of volatile matter by applying heat without the access of air.

calcite

A natural crystal of calcium carbonate with unique properties with respect to the polarization of light.

calcium

A divalent metallic element, one of the alkaline earth metals, in Group II of the periodic table, discovered by Davy in 1808. Atomic number 20, atomic weight 40.08, specific gravity 1.55 at 20°C., melting-point 810°C., boiling-point 1170°C. Symbol: Ca.

calculate

To add, subtract, multiply or divide; to compute or to reckon.

calculus

A field of higher mathematics divided into *differential, infini-*

tesimal, and *integral calculus.* Infinitesimal calculus or ordinary calculus is so called because of its use of infinitesimal quantities. It sometimes refers only to that part of the calculus which deals with differentials and sums of infinitesimals. See differential calculus and integral calculus.

calibration

(1) The process of evaluating the scale readings of an instrument in terms of the physical quantity to be measured, or of determining the corrections to be applied to the scales of any measure (linear or circular) in order to obtain true values.—(2) The result of this process.

californium

A transuranic element, not found in nature. Atomic number 98, mass number of the most stable known isotope 248. Produced by G. T. Seaborg, S. G. Thompson, K. Street, Jr. and A. Ghiorso by cyclotronic bombardment of curium ($_{96}Cm^{242}$) with high-energy (35 Mev) alpha particles. Symbol: Cf.

calorescence

The absorption of heat radiations, their conversion into heat, and the emission of this heat by the absorbing surface.

calorie (cal)

The unit of heat in the CGS system; one calorie is the amount of heat required for raising the temperature of one gram of water through one degree C, either from 3.5°C to 4.5°C (the *standard* or *4° calorie*) or from 14.5°C to 15.5°C (the *normal* or *15° calorie*); also called *metric calorie, gram calorie* and *small calorie.* (See also *kilocalorie, pound calorie,* and *international steam table calorie.*)

calorific intensity

The temperature which can be reached under perfect operating conditions by the combustion of a specified fuel.

calorific power

The amount of heat, stated in calories, evolved in the complete combustion of one gram of any material, more particularly fuel.

calorific value

The quantity of heat, i.e., the number of calories, of a given unit mass of a substance on complete combustion. Expressed in *calories,* it is the number of grams of water the temperature of which is raised 1°C by the combustion of one gram of fuel; in *British Thermal Units,* it is the number of pounds of water the temperature of which is raised 1 °F by the complete combustion of 1 pound of fuel.

calorimeter

An instrument for measuring total heat liberated or absorbed.

calorimetry

The determinating of the total heat units in a body or involved in a specific chemical reaction. The measurement of heat exchange and thermal constants.

calory

See *calorie.*

calutron

An apparatus constructed by Lawrence, for the separation of the uranium isotopes U^{235} and U^{238}, operating on the mass spectrograph principle.

canal rays

Radiation which can be observed in the space behind the cathode in certain types of vacuum tubes; this radiation consists of streams of positive ions (positively charged atoms or molecules) which have travelled from the anode toward the cathode and passed through openings ("canals") in the latter. Discovered by Goldstein in 1886. (Also called *positive rays.*)

candela

See *candle.*

candle

A unit of luminous intensity, defined as 1/60 of the luminous intensity of one square centimeter of a black-body radiator at the temperature of the solidification of platinum, i.e., 2046 °K. (See also *international candle.*)

candle-meter

A unit of illumination, synonymous with the *meter-candle,* and equal to one *lux.*

candlepower (cp)

Candlepower is luminous intensity expressed in candles. See candle.

capacitance

That property of a system of conductors and dielectrics which permits the storage of electricity when potential differences exist between the conductors. It is expressed as the ratio of charge (quantity of electricity) to a potential difference, viz., in *farads.*

capacitor

A device, the primary purpose of which is to introduce capacitance into an electric circuit; it permits electricity to be stored temporarily. Also called *condenser.*

capacity

The capacity of a machine, apparatus or device is (1) its rated load, or (2) the maximum load of which it is capable under existing service conditions.

capillarity

The resultant of the forces of attraction acting between unlike molecules, as manifesting in surface tension or the rising of a liquid in tubes of very minute diameters. (See *capillary action.*)

capillary action

The force or action which operates when a tube of glass of very minute diameter and open at both ends is placed vertically in water, with its lower end immersed, causing the water to rise in the tube and reach a higher level on the inside of the tube than on the outside.

capture

The cohesion to or absorption by an atomic nucleus of a bombarding particle after collision, in most cases producing nuclear fission (*fission capture*) and transformation accompanied by the emission of another particle.

carat

(1) A standard of weight for precious stones. The *metric carat,* standardized in 1913, equals 200 milligrams.—(2) The standard of fineness for gold, pure gold being 24 carats.

carbide

A binary compound, consisting of carbon and a metal.

carbocyclic compound

Any organic compound the molecule of which contains a closed ring of carbon atoms.

carbohydrate

A generic term for a great many organic compounds containing carbon, hydrogen and oxygen (the proportions of the hydrogen and oxygen always 2:1, i.e., in proportions forming water).

carbon

A tetravalent element in Group IV of the periodic table. Atomic number 6, atomic weight 12.011, melting-point above 3500°C., boiling-point 4200°C., occurs in three allotropic forms, viz. diamond (specific gravity 3.52), graphite (specific gravity 2.25) and amorphous (specific gravity 0.5—1). Symbol: C.

carbon arc

The arc formed by an electric current passing from one carbon electrode to another, or from one carbon electrode to another conductor of electricity, producing very intense heat.

carbon cycle

A series of *thermonuclear reactions* releasing vast amounts of energy, that presumably occurs in the sun and other stars and is the source of the energy radiated by them. According to H. Bethe, the cycle that takes place in the stars with the capture of a hydrogen nucleus (a proton) by a carbon nucleus and consists of the following six nuclear transformations: (*1*) The carbon nucleus ($_6C^{12}$) captures the hydrogen nucleus ($_1H^1$) and changes into a nitrogen ($_7N^{13}$) nucleus; (*2*) the $_7N^{13}$ nucleus emits a positron and changes into another carbon isotope ($_6C^{13}$); (*3*) the $_6C^{13}$ captures another $_1H^1$ and becomes another nitrogen isotope ($_7N^{14}$); (*4*) the $_7N^{14}$ captures another $_1H^1$ and becomes an oxygen ($_8O^{15}$) nucleus; (*5*) the $_8O^{15}$ nucleus emits a positron and becomes still another nitrogen isotope ($_7N^{15}$); (*6*) finally, this $_7N^{15}$ captures another $_1H^1$ and changes into the same carbon isotope ($_6C^{12}$) which started the cycle, plus a helium ($_2H^4$) nucleus. Thus, in the course of the process (the duration of which is roughly 6.5 million years) four hydrogen atoms are captured, one helium atom and two positrons are liberated, and the original carbon ($_6C^{12}$) atom that started the cycle is restored in the final step, to start the next cycle, as a catalyst as it were: the net result of the six-step cycle therefore is the transformation of four hydrogen atoms into a helium atom with a loss of about 1 per cent of mass which is liberated as vast bursts of energy.

carbon dioxide (CO_2)

A heavy, colorless, incombustible gas, formed when carbon is burned in air, when limestone and magnesite are calcined, and by the action of hydrochloric acid on calcium carbonate.

carbon dioxide cycle

The continuous process consisting of the conversion of atmospheric carbon dioxide into starch and cellulose in plant tissue by *photosynthesis,* and its later return to the atmosphere as a decomposition product, i.e., through the decomposition of the dead plants or of the bodies of the dead animals which consumed the plant as food.

carbon monoxide (CO)

A colorless, poisonous, slightly water-soluble gas; a product of

incomplete combustion, formed when carbon is heated in a limited supply of air, when *carbon dioxide* is heated with carbon, or when carbon dioxide is passed over certain metals at high temperatures.

carbon steel

A steel the properties of which are determined primarily by the percentage of carbon present. Besides iron and carbon, such steels may contain a maximum of manganese up to 1.5%, silicon up to 0.5%, sulphur and phosphorus up to 0.1%, nickel up to 0.40%, chromium up to 0.30%, molybdenum up to 0.15%, copper up to 0.25%, tungsten, cobalt, aluminum up to 0.10%, and niobium, tantalum, titanium, vanadium, zirconium up to 0.05%. These alloying elements in such quantities are regarded as residual elements, but their deliberate addition in substantial amounts will relegate the steel to the *alloy steel* category.

carbonyl, carbonyl group

The divalent =CO radical.

Carborundum

A trade name for certain silicon carbide and other abrasive products including emery.

carboxyl group

The monovalent, acidic –COOH radical.

carburizing flame

See *reducing flame*.

cardinal

Principal, fundamental.

cardinal numbers

Any number which directly expresses how many digits are considered, as: one, two, three, four, etc., as distinguished from ordinal numbers: first, second, third fourth, etc.

cardioid

A one-cusped *epicycloid* which represents the locus of a point on a given circle that rolls around on an equal, but stationary circle.

Carnot cycle

The cycle of the ideal heat engine; it is a reversible cycle in which the working substance undergoes an adiabatic compression, followed by an isothermal expansion, then an adiabatic expansion, and finally an isothermal compression, bringing the values of volume, pressure and temperature back to their original magnitudes.

carrier wave

A wave, of which the frequency, amplitude or another characteristic is varied as a function of the instantaneous value of another wave, called the *modulating wave.*

Cartesian coordinate system

See *rectangular coordinate system.*

cartography

The science and technique of the construction and drawing of maps or charts, in particular of depicting and interpreting the topographic features of the surface of the earth.

cassiopeium

The name formerly given to the element lutecium. Symbol: Cp.

cast shadow

The intersection of the *umbra* of an object with any other surface or plane.

catabolism

That part of the process of *metabolism* in which complex substances are decomposed into simpler ones, with a development of energy.

catalysis

The facilitation of a chemical reaction by the addition of another substance called the catalytic agent or catalyst, which does not change ultimately but merely aids in the combining process.

catalyst

A substance which changes the rate of a chemical reaction actually in process without being consumed by the reaction and is recoverable chemically unchanged at the conclusion of it. (If the substance slows down or retards the reaction, it is referred to as a *negative catalyst.*)

catalytic

Producing, relating to or resulting from the process known as *catalysis.*

catalytic agent

See *catalyst.*

catalytic poison

A material that neutralizes the action of a *catalyst.*

cataphoresis

The movement of fine particles suspended in a liquid toward the cathode, produced by the application of an electric potential.

catenary

The curve assumed by a chain of uniform thickness and structure when suspended so as to hang freely from its two ends.

catenoid

A surface of revolution, generated by causing a *catenary* to rotate about its axis.

cathode

(1) In general, a negative electrode, or an electrode through which current leaves a conductor of the non-metallic class. (2) An *electrolytic cathode* is an electrode at which positive ions are discharged, or negative ions formed, or at which other reducing reactions occur. (3) In a *vacuum tube,* an electrode which is the primary source of an electron stream.

cathode dark space

The non-luminous portion of the glow discharge which is between the cathode glow and the negative glow in a Crookes tube or other discharge tube. Also called *Crookes dark space.*

cathode drop

The abrupt drop of electric potential at the cathode in a vacuum tube.

cathode glow

The luminosity which appears at the cathode of a gas discharge tube and envelops it when the tube is operated at a moderately low gas pressure.

cathode ray

A stream of electrons emitted from the cathode of an evacuated tube, or from the ionized region in its proximity, under the influence of an applied voltage.

cathode ray tube

A special form of vacuum tube used in various electronic applications, as in the television picture tube and as an oscilloscope tube.

cathodic

(1) Relating to or involving the *cathode.*– (2) A term applied in a general sense to denote an element below hydrogen in the *electrochemical series.*–(3) A term applied in a relative sense to signify a more positive electrode potential.

cathodoluminescence

Luminescence produced by cathode rays.

cathodophosphorescence

Phosphorescence produced by cathode rays.

catholyte
The portion of an electrolyte in an electrolytic cell adjacent to a cathode. If a diaphragm is present, it is the portion of the electrolyte on the cathode side of the diaphragm.

cation
A positive ion in an electrolytic solution which is attracted to the negative pole or cathode.

Cauchy's mean value formula
See *extended law of the mean.*

caustic
A term applied to substances (other than acids) which are corrosive to organic matter.

caustic curve
The line connecting the points of intersection of a number of reflected or refracted rays of light.

Cavalieri's theorem
The statement that if the bases of two solids lie in the same two planes, and if any plane passing between the bases parallel with them always forms equal sections, the volumes of the two solids are equal.

cavitation
The formation of cavities in fluids, when at some point in a flowing fluid the velocity reaches a magnitude that causes the pressure to approach zero.

Cd
The symbol for the element *cadmium.*

Ce
The symbol for the element *cerium.*

celestial body
A general term for all the objects that can be observed in the sky beyond the atmospheric envelope of the earth: the sun, the moon, the planets and their satellites, meteors, comets, stars, nebulae and external galaxies.

celestial equator
The great circle of the celestial sphere all points of which are 90° from the poles. It is the plane of the earth's equator projected onto the celestial sphere. Also called the *equinoctial.*

celestial mechanics
That branch of *astronomy* which deals with the motions of celestial bodies under the forces of gravitation. Also called *gravitational astronomy* and *mathematical astronomy.*

celestial meridians

Great circles of the *celestial sphere* passing through the poles. They are the earth's meridians projected on the celestial sphere. The upper branch of an observer's meridian is that half which is terminated by the poles and which passes through the zenith. When we speak of the meridian we usually mean the upper branch of the meridian.

celestial sphere

An imaginary sphere of infinite radius, centered at the center of the earth, on which the apparent motions of the celestial bodies take place.

cell

See *electric cell.*

cellulose

A carbohydrate of the general formula $(C_6H_{10}O_5)_n$, which forms the cell walls and fibrous structure of many plants, and is the chief constituent of paper and wood. It is insoluble in water and organic solvents but dissolves in ammoniacal copper solutions forming complex ions from which it may be regenerated by means of acid; it dissolves also in a mixture of alkali and carbon disulfide forming xanthogenates from which it is regenerated by the addition of acid to give various materials such as rayon and cellophane. Cellulose esters and ethers are widely used in plastics, lacquers and explosives, and as artificial silk, viscose, nitrocellulose, gun-cotton, celluloid.

Celsius scale of temperature

A synonym of *centigrade scale,* although in the scale devised by Andreas Celsius in 1742, 0° was the boiling point of water, and 100° its freezing point (the reverse of the centigrade scale now in use).

center of a curve

The point about which the curve is symmetrical.

center of a hyperbola

The point about which the hyperbola is symmetrical, i.e., the point in which the transverse axis, the conjugate axis and the asymptotes intersect each other. It is the midpoint of the line segment connecting the two foci (or the two vertices).

center of a regular polygon

A point that is equidistant from all the vertices of the polygon.

center of an ellipse

The midpoint between the two foci, which is also the midpoint between the two vertices.

center of area
A synonym for *centroid* of an area.

center of curvature
The center of the *circle of curvature.*

center of figure
See *center of volume.*

center of gravity
That single point where the entire mass of a body or a system may be assumed to be concentrated and to act. It is coincident with the *center of mass.*

center of inertia
A synonym for *center of mass.*

center of inversion
See *inversion.*

center of mass
That point in a given body, or in two or more bodies that act together in respect to another body, which represents the mean position of the matter in the body or bodies.

center of mass of an area
See *centroid of mass.*

center of mean position
See *centroid.*

center of percussion
That point of an oscillating body at which that body may be struck without producing any jar.

center of perspective
See *perspective center.*

center of similitude
The point of intersection of the lines joining the corresponding vertices of two similar polygons which have been placed in a position of having their sides parallel.

center of symmetry
A synonym for *center of inversion.* (See *inversion.*)

center of volume
The *center of mass* of a solid of uniform density regarded as occupying the volume of the solid. Also referred to as *center of figure.*

centered lens system
In optics, a combination of two or more lenses arranged to work in conjunction with each other, properly aligned so that their principal axes are coincident.

centesimal system of angular measure
A system (also called the *decimal system*), in which the right

angle is divided into one hundred equal parts, each called a *grade* (symbol: g); the grade is divided into one hundred equal parts each called a *minute* (symbol: '); and the minute is divided into one hundred equal parts each called a *second* (symbol: "). Example: 384 grades, 78 minutes, 93.465 seconds, is written 384g 78' 93" .465 (note the direction for these symbols for minutes and seconds is opposite to their direction in the *sexagesimal system*).

centi-

A prefix meaning *one one-hundredth.*

centigrade heat unit

A unit of heat; 1 C.h.u. is the 1/100 part of the amount of heat required in order to raise the temperature of 1 lb. of water from 0°C to 100°C.

centigrade scale of temperature

A temperature scale according to which the freezing-point of water is marked 0° and its boiling-point 100°. The abbreviation of centigrade is C. To convert centigrade degrees into Fahrenheit, multiply by 9, divide by 5 and add 32.

centigram (cg)

One hundredth of a gram.

centimeter (cm.)

The fundamental unit of length in the CGS system, equal to 0.01 meter, i.e. 0.394 inch.

centimeter-gram-second system

See *CGS system of units.*

centimeter of capacitance

1.13×10^{-12} farad.

centimeter of inductance

1 abhenry.

centimeter of mercury

A unit of pressure, equal to about 13,332 bars.

centipoise

A subunit for measuring viscosity, equal to one one-hundredth of a *poise.*

centrad

A small unit for measuring angles, equal to one one-hundredth of a radian.

central angle

An angle, the vertex of which is the center of the circle, and the sides of which are radii of the latter.

central angle of a regular polygon
The angle formed by two straight lines each of which connects a vertex with the center.

central conics
The ellipse and hyperbola, since they both have a center, such that all chords of the conic passing through it are bisected.

central force
A force acting on a moving body which is always directed toward a fixed point or a point the motion of which obeys certain known laws.

central quadrics
A subdivision of the *non-degenerate quadric surfaces* which includes the *ellipsoids* (real or imaginary) and the *hyperboloids* (of *one sheet* and of *two sheets*).

centric
In geometry, the fixed point on a plane from which all points on the circumference of a circle are equidistant. The middle point of anything.

centrifugal force
(1) The force which a mass, constrained to move in a circular path, exerts on the constraint in a direction along the radius. (2) In mechanics, the fictitious force which must be introduced as acting along a radius on a mass moving in a circle in order to reduce the problem to one in statics. The force operating in the opposite direction is called "centripetal force."

centrifuge
An instrument for separating liquids of different specific gravities by the use of centrifugal force.

centripetal force
The force which restrains a body in rotation from going in a straight line. It is directed toward the center of rotation. A force equal, but opposite in sign, to the centrifugal force.

centroid
In general, the point about which all the parts of a figure or of a body balance each other exactly. Also called *center of mean position* in plane figures, and *center of gravity* or *center of mass* in solids.

centroid of a plane figure
In a triangle, the point where the three *medians* intersect one another; in a circle, an ellipse, or a square, the center. Also called *center of mean position.*

centroid of a solid
The point where the entire mass of the solid could be concen-

trated without affecting the moment about any plane or line. Also called *center of mass* or *center of gravity* of the solid.

centroid of mass of an area

A point where the area could be concentrated without affecting the moment about any line in the plane.

centrosymmetry

Symmetry with respect to a point.

ceramics

The art and technique of making useful as well as decorative earthenware, clay or porcelain articles.

Cerenkov radiation

Radiation in the visible light range, produced when charged sub-atomic particles traverse a transparent medium with a velocity exceeding that of light in that medium. Also called *Cerenkov effect.*

cerium

A metallic element, one of the rare earth metals, discovered by Berzelius in 1803. Atomic number 58, atomic weight 140.13, specific gravity 6.85 at 20°C, melting-point 638°C, valence 3 or 4. Symbol: Ce.

cermet

The designation of various fused combinations of ceramics and metals by far not resistant enough to protect rockets from the tremendous temperatures that they must withstand when executing a re-entry into the earth's atmosphere.

cesium (caesium)

A monovalent metallic element, one of the alkali metals, in Group I of the periodic table, discovered by Bunsen and Kirchhoff in 1860. Atomic number 55, atomic weight 132.91, specific gravity 1.87. Symbol: Cs.

Cf

The symbol for the element *californium.*

cg

The abbreviation for *centigram.*

CGS (cgs) system of units

An absolute system for measuring physical quantities in which the fundamental units are the centimeter, gram and second. This system is primarily applicable only to mechanical units. It is extended to other fields of physical science by accepting the doctrine of the conservation of energy and by introducing a fourth unit or a property of a material. For example, in the theory of heat, the degree centigrade is taken as an additional unit.

chain reaction

In general, any self-sustaining process, whether molecular or nuclear, the products of which are instrumental in, and directly contribute to the propagation of the process. Specifically, a *fission* chain reaction, where the energy liberated or particles produced (*fission products*) by the fission of an atom cause the fission of other atomic nuclei, which in turn propagate the fission reaction in the same manner.

change of signs

If a bracket is preceded by a minus (–) sign, then all the signs inside the bracket change from (–) to (+) or (+) to (–) when the bracket is removed. A plus (+) sign in front of a bracket causes no change of sign inside the bracket when it is removed; as $-[4a - 3b + 6c]$ becomes $-4a + 3b - 6c$; but $+[2a - 5b - 7c]$ remains $2a - 5b - 7c$. If there is no sign in front of a term it is always a positive (+) term. Thus 2ab is + 2ab.

change of state

The process by which a substance passes from one state (solid, liquid or gaseous) to another, and in which marked changes in its physical properties and molecular structure occur. The changes of state are: *fusion* or *melting* (change from the solid to the liquid state), *vaporization* (change from the liquid to the gaseous state), *sublimation* (change from the solid directly to the gaseous state), *freezing* (change from the liquid to the solid state), and *condensation* (change from the gaseous to the liquid state or from the gaseous to the solid state).

characteristic

The integral part, or whole number, in front of the decimal point in logarithms. If the characteristic (also called *index*) is negative, the negative sign is placed over the characteristic; as $\bar{2}.4351$. In logarithms the characteristic of the logarithm to any number greater than unity (or one) is positive and is less by one than the number of figures on the left side of the decimal point; thus the characteristic of the logarithm of 354.25 is 2 and of 4.51 is 0. The characteristic of any number less than unity (or one) is negative and is greater by one than the number of zeros which follow the decimal point of the number; thus the characteristic of the logarithm of 0.067 is –2 and of 0.00098 is –4. (The number after the decimal point in logarithms is called the *mantissa*.)

characteristic number, characteristic value
A specific value or magnitude which complies with certain specific conditions.

characteristic radiation
Radiation emitted by an atom after the loss of an electron, on a wavelength depending solely on the element and on the energy levels involved.

characteristic x-rays
Monochromatic x-rays characteristic of a particular element and emitted by it, in addition to heterochromatic x-radiation, when it is exposed to cathode rays or x-rays of a sufficiently high energy.

Charles' law
The volume of a given mass of a gas at constant pressure is changed by 1/273 of its volume at 0°C. for each degree change in temperature. Also known as *Gay-Lussac's law*.

check
To verify the correctness of a sum or result or solution.

chemical *(adj.)*
Pertaining to, characteristic of, or resulting from chemical reactions, processes or chemistry in general.

chemical *(noun)*
A substance or compound of a definite, known molecular composition.

chemical affinity
The force that binds atoms together in molecules.

chemical change
A rearrangement of atoms or molecules into chemically different structures; the substances undergoing such change lose their original identity.

chemical compound
A substance composed of two or more elements combined in definite proportions by weight, the individual properties of the constituents having disappeared, and the latter being inseparable by physical means.

chemical constant
A constant of integration appearing in the mathematical expression of the vapor pressure of a substance.

chemical element
See *element*.

chemical equation
A symbolic representation of a chemical reaction showing the

relation between the reacting substances and the products, e.g.: $Zn + 2HCl \rightarrow ZnCl_2 + H_2$. Occasionally, this definition is extended to include the heat evolved during the reaction.

chemical equilibrium

A state of complete balance between two opposing chemical forces or reactions. In a chemical reaction, that state when the original substances are reacting at the same rate as the new substances are reacting with each other to form the original substances.

chemical equivalent

The weight of an element or radical that will combine with or displace 8 parts by weight of oxygen or one part of hydrogen, i.e., the atomic weight of an element divided by its valency.

chemical exchange reaction

A process in which an atom of a certain chemical substance exchanges places with a similar atom in a different substance.

chemical formula

A symbol or a combination of symbols. If more than one atom of a given element is present in the molecule, it is shown by a subscript. The formula shows the elements present and the number of atoms of each. E.g., the chemical formula of ordinary water, H_2O, indicates that each molecule consists of two atoms of hydrogen and one atom of oxygen.

chemical kinetics

That branch of physical chemistry which deals with the velocity of reactions and their various intermediate stages.

chemical reaction

Any chemical change or any transformation of molecules of one kind into molecules of another kind, brought about by the rearrangement of atoms when two or more elements or compounds are brought into contact under given conditions.

chemical solution

A solution in which a chemical reaction takes place between the *solute* and the *solvent,* which form a chemical compound, so that the solvent cannot be separated and recovered by distillation.

chemical symbol

A single capital letter, or a capital letter followed by a small letter, used to denote an atom of a chemical element, as H for hydrogen, Fe for iron, etc.

chemiluminescence

The emission of light during a chemical reaction.

chemisorption

Adsorption depending on the chemical attraction between a metal surface and a gas. It is the condition or process which holds the atoms of the gas to the solid surface by action of covalent forces.

chemosphere

A stratum of the atmosphere considered to begin at approximately 20 miles and to extend to 50 miles above the earth, marked for its photochemical activity. (Some meteorologists regard this stratum as an extension of the stratosphere.)

chemotherapy

The treatment or prevention of disease by chemical agents or substances.

chempure

A commercial term meaning *chemically pure,* which does not necessarily denote 100% freedom from admixtures or contamination.

chlorine

A gaseous element in Group VII of the periodic table, discovered by Scheele in 1774. Atomic number 17, atomic weight 35.457, density 3.21 grams per liter at normal temperature and pressure, melting-point –101.6°C., boiling-point –34.6°C., valence: 1, 3, 5, 7. Symbol: Cl.

chlorophyll

A green pigment present in the leaves of plants which absorbs energy from sunlight.

chord of a circle

Any straight line joining two points, which are not adjacent, on the circumference of a circle.

chord of a sphere

A straight line that connects any two points on the sphere.

chroma

Color intensity; the degree of departure from black or white. In color pigments, chroma is the quality which distinguishes an intense color from one not so intense: Bright red, vermilion, flaming scarlet, emerald green, and other light colors are examples of high chromas. When used in reference to light, chroma means the purity of one wavelength free from all others.

chromatic

Having color.

chromatic aberration

In optics, a lens defect as a result of which the lens brings light of different colors to a focus in different planes. This aberration is also referred to specifically as *longitudinal chromatic aberration* (cf. *lateral chromatic aberration*).

chromatic color
A color which exhibits hue.

chromatic sensitivity
A property of the eye which enables it to distinguish color hues; its numerically expressed magnitude equals the smallest change in wavelength of the radiation which still registers on the eye as a perceived difference in hue.

chromaticity
A term qualitatively descriptive of a color and dependent on hue as well as saturation, but without reference to brilliance.

chromatism
A synonym for *chromatic aberration.*

chromatographic analysis
Chemical analysis of mixtures of solutions through selective adsorption of the substances in solution by suitable absorbing materials.

chromatography
See *chromatographic analysis.*

chromatometer
An apparatus for measuring degrees of color.

chromium
A metallic element in Group VI of the periodic table, discovered by Vauquelin in 1797. Atomic number 24, atomic weight 52.01, specific gravity 7.13 at 20°C., melting-point 1615°C., boiling-point 2200°C., valence: 2, 3 or 6. Symbol: Cr.

chromogen
Any chemical group which by itself or when associated with some other group imparts color.

chromometer
An apparatus used in quantitative analysis by the color method.

chromophore
A group which gives color to an organic compound but does not necessarily make it a dye.

chromosphere
The layer of the sun's atmosphere above the reversing layer and below the *corona.* It is of a brilliant red color and several thousands of miles in thickness.

chromotropism
Response to color stimulation.

chronograph
An instrument for recording and measuring time intervals with extreme accuracy.

chronometer
The most exact type of portable timepiece.

chronoscope
An instrument for measuring the duration of short intervals of time.

C.h.u.
The abbreviation for *centigrade heat unit.*

cinder
See *slag.*

cipher
A synonym for *zero* or *naught* (written 0).

cir mil
The abbreviation for *circular mil.* (Plural: *cir mils.*)

circle
A *conic section* formed by a plane cutting a cone parallel to its base. It is the locus of a moving point the distance of which from the *center* is constant.

circle of a sphere
A section produced by a plane that cuts through the sphere. When the cutting plane passes through the center of the sphere, the resulting section is called *great circle of the sphere;* all other sections (i.e., those produced by cutting planes which do not pass through the center of the sphere) are called *small circles.*

circle of convergence
With reference to a *power series,* the circle within which the series is absolutely convergent, while being divergent outside of it, and being either convergent or divergent at points on the circumference.

circle of curvature
The osculating circle of any curve, i.e., the circle tangent to the curve the curvature of which is considered, on the concave side of the latter, as having the same curvature at the point where it is tangent to it.

circle segment
See *segment of a circle.*

circle sector
See *sector of a circle.*

circuit
See *electric circuit, magnetic circuit.*

circular cone
A cone the *base* of which is a circle. (If the axis is perpendicular to the base, the cone is referred to as a *right circular cone,* or as a *cone of revolution.*)

circular cylinder
A cylinder the *base* of which is a circle. (If the axis is perpendicular to an *element,* the cylinder is referred to as a *right circular cylinder* or as a *cylinder of revolution.*)

circular function
(1) A quantity relating to a circle so connected with another quantity in that circle that any alteration in the value of the one affects the value of the other. Example: the length of the circumference depends upon the length of the radius and vice versa.—(2) A synonym for *trigonometric function* (q.v.).

circular inch
A unit of area, equal to the area of a circle one inch in diameter equal to 0.7854 square inch.

circular magnetic wave
In general, any wave with circular magnetic lines of force.

circular measure
60 seconds = 1 minute.
60 minutes = 1 degree.
30 degrees = 1 sign.
90 degrees = 1 quadrant.
4 quadrants = 12 signs = 360 degrees = 1 circle.

circular mil
A unit of area, equal to the area of a circle one one-thousandth of an inch in diameter. It is the unit used for measuring the cross-sectional area of wires.

circular points at infinity
Two imaginary points at infinity, the common intersection of all circles in the plane.

circular polarization
A variety of *elliptical polarization,* producing *circularly polarized waves.*

circular system of angular measure
The system in which the unit angle is the angle subtended at the centre of a circle by an arc on the circumference equal in length to the radius of the circle. This unit is called a *radian* and its value for any circle is 57° 17′ 44″.8062470964

(to ten places of decimals) and is measured by the common or sexigesimal system. It is calculated from the formula: a radian $= \frac{180^\circ}{\pi}$.

circularly polarized waves

An elliptically polarized wave in which the direction of displacement at a point rotates with constant angular velocity about an axis in the direction of propagation, the magnitude of displacement being independent of its direction.

circumcircle

A synonym for *circumscribed circle.*

circumference

The boundary line of a circle or any closed curvilinear figure.

circumscribed

Said of a geometrical figure or solid into which another geometrical figure or solid has been placed so that they have the greatest possible number of points in contact.

circumscribed circles

A circle which *circumscribes* and touches a given figure at all of its vertices.

circumscribed polygon

If each side of a polygon is tangent to a circle and the circle lies inside the polygon, then the polygon is said to be circumscribed about the circle.

circumscribed quadrilateral

A quadrilateral whose sides are tangent to an inscribed circle.

circumscribed sphere

A sphere so situated about a polyhedron that every vertex of the latter lies on the sphere.

cis-

A prefix indicating the presence of *cis-trans isomerism* in a substance.

cissoid

The plane locus of the equation $y^2 = \frac{x^3}{2a - x}$. Also referred to as *cissoid of Diocles.*

cis-trans isomerism

A type of *stereoisomerism* in which two organic substances have the same constitutional formulas, but differ in all of their physical and most of their chemical properties, although they affect polarized light only if they contain an asymmetric grouping. In the *cis-* compounds certain atoms or atom groups are on the same side of a plane, in the *trans-* compounds they are on opposite sides of it. (Also called *axial stereoisomerism.*)

Cl

The symbol for the element *chlorine.*

Clapeyron equation

A mathematical expression of the rate of the change in the vapor pressure of a liquid with the temperature, or of the boiling points of liquids depending on the pressure.

class mark

In frequency tables and frequency distributions, the mid-point of a given class or group, which is used to represent the class or group.

Clausius-Clapeyron equation

An expression of the relationship between vapor pressure and heat of vaporization, or between change in volume, change in heat content and temperature.

cleavage

The tendency of a crystalline substance to split along definite planes (the *cleavage planes*), corresponding to the layers of the atoms composing the crystalline structure. Also the direct separation of the material along these planes.

cleavage plane

Any of the planes of easy fracture where *cleavage* usually occurs when the crystal is subjected to stress parallel to one or more of the faces of the system to which the crystal belongs. They are not necessarily related to the boundaries of the crystal, and occur in both minerals and metals.

climatology

The science which studies and describes the various types of climates and endeavors to explain the causes producing these climates, their variation, their effects on animal and plant life. It is divided, accordingly, into two general branches, *descriptive climatology* and *physical climatology.* It is closely related to and allied with meteorology and geography.

clockwise

In the same direction of rotation as that in which the hands of a clock move around the dial.

clockwise elliptically polarized wave

See *right-handed elliptically polarized wave.*

close-packed hexagonal structure

One of the two varieties of the *close-packed* crystal structure, i.e., a structure in which atoms are arranged according to one of the two possible systems which permits equal hard spheres to be packed so as to occupy the least possible space. (The other possible such arrangement is the *face-centered cubic* structure.) The close-packed hexagonal structure is based on a prismatic unit cell having a base in the form of a rhombus with 60° and 120° angles and a height that approaches, ideally, 1.63 times the length of an edge of the base. It has a lattice point at each corner. An atom lies at each corner and another atom, not on a lattice point, is located at half height directly above the center of one of the two equilateral triangles into which the base may be divided.

close-packed structure

A crystal structure in which atoms are arranged according to either of the two possible arrangements in which equal hard spheres can be packed so as to occupy the least total volume, namely: the *face-centered cubic* structure and the *close-packed hexagonal structure.*

closed circuit

In topology, any closed curve.

closed cycle

In thermodynamics, a process in which the working substance undergoes changes of heat into work or work into heat and returns periodically to its initial state.

cloud chamber

An apparatus, invented by C. T. R. Wilson (England) in 1911, and used to study the behavior of alpha particles, protons, electrons, positrons, mesons, photons, and the collision of these particles with atoms. The apparatus contains moist air, which is permitted to expand rapidly, producing condensation, and the trails of water droplets ("cloud tracks") reveal the paths of the rapidly moving sub-atomic particles.

cloud track

The trails of water droplets visible in a *cloud chamber* which indicate the paths of sub-atomic particles.

cluster of atoms

Atoms not in a permanent chemical union, held together by the electric attraction of a charged ion.

Cm

The symbol for the element *curium.*

cm

The abbreviation for *centimeter.*

CN
The chemical symbol for *cyan.*

Co
The symbol for the element *cobalt.*

coagulation
The formation of a solid or semisolid, gelatinous mass from a *sol* or a liquid.

coal
A native black or brownish organic rock, formed by the decomposition of accumulated plant material, consisting chiefly of carbon, but containing also other elements, including some inorganic matter. The principal grades of coal are: *lignite* (brown coal), *bituminous coal* (soft coal), and *anthracite* (hard coal).

coal gas
A gas obtained from the distillation of soft coal and used for illuminating purposes. It consists mainly of hydrogen, methane and carbon monoxide.

coalescence
The union or growing together of crystals or globules due to molecular attraction.

co-axial circles
A system of circles in the same plane, such that the radical axis of any pair of them is the same. The centers of these circles lie on the same line which is perpendicular to the radical axis.

cobalt
A metallic element in Group VIII of the periodic table, discovered by Brandt in 1739. Atomic number 27, atomic weight 58.94, density 8.9, melting-point 1490°C., boiling-point 2900° C., valence 2 or 3. Symbol: Co.

coefficient
A number or letter or symbol which has a fixed value; generally placed in front of a mathematical expression or combination of letters or symbols and used as a multiplier; thus $6x$; ax; $(b - c)x$ and $(au + bv + cw)$ $(x + y + z)$ in which x and $(x + y + z)$ are the variable or unknown quantities and 6, a, $(b - c)$ and $(au + bv + cw)$ are the coefficients or known quantities. (Cf. *detached, literal, numerical* and *undetermined coefficient.*)

coefficient of absorption
In the absorption of electromagnetic radiation, the ratio of the loss in intensity of the radiation produced by the absorp-

tion to the original radiation intensity prior to the absorption. Also called *absorption factor.*

coefficient of correlation

A number, between the limiting values of +1 and —1, which expresses the degree of linear relationship between two variables, and which can be determined from any one of several formulas.

coefficient of elasticity

See *elasticity, elastic modulus.*

coefficient of expansion

The degree of expansion of a body expressed as a proportion of its length at a standard temperature, usually 0°C. or 32°F.

coefficient of friction

In rolling, the nondimensional ratio between the force required in order to cause relative motion at constant speed between two surfaces and the normal force which presses the surfaces together.

coefficient of linear expansion

See *thermal expansion.*

coefficient of magnetic induction

See *magnetic permeability.*

coefficient of magnetization

The ratio of the intensity of magnetization produced in a substance to the magnetic field strength or intensity to which it is subjected; it is a measure of the attraction which a magnet exerts on a given substance. Also called *magnetic susceptibility.*

coefficient of restitution

In the collision of two solids, the ratio of the difference in velocity after the impact to the difference before the impact.

coefficient of rigidity

The ratio of shearing stress to shearing strain. (Also called *shear modulus.*)

coefficient of surface tension

The numerical magnitude of the force exerted by a surface on an adjacent surface per unit of length.

coefficient of viscosity

The value of the tangential force per unit area necessary in order to maintain unit relative velocity between two parallel planes which are a unit distance apart.

coefficient of volume elasticity

The ratio of volume stress to volume strain. (Also called *bulk modulus.*)

coefficient of volume expansion
See *thermal expansion.*

coenzyme
A nonprotein substance the presence of which is a prerequisite of the activity of an *enzyme.*

coercive force
The magnetizing force required for bringing the induction to zero in a magnetic substance which is in a symmetrically cyclically magnetized condition. Symbol: *Hc.*

coercivity
The property of a substance which is measured by the maximum magnitude of the *coercive force.*

cofunction
In trigonometry, the functions $sin\ \theta$ and $cos\ \theta$, the functions $tan\ \theta$ and $cot\ \theta$, and the functions $sec\ \theta$ and $csc\ \theta$ are said to be pairs of cofunctions of the angle θ. Any function of a given angle is always equal to the confunction of the complementary angle.

cohesion
That property of a substance which enables it to resist forces tending to tear it apart or separate it into parts. Cohesion is attraction between like molecules (cf. *adhesion*).

coincident lines
Two or more lines the equation of which is identical.

coincident plane figures
Two or more plane figures which occupy the same position.

coincident surfaces
Two or more surfaces the equation of which is identical.

coke
The gray, porous substance remaining after the distillation of coal. It is used as a fuel and in the manufacture of steel and iron products.

co-latitude
In geography, the complementary angle to the *latitude.*

cold junction
In a *thermocouple,* the unheated joint.

collective lens
See *convergent lens.*

colligative properties
Physical or chemical properties which can be related to each other by a mathematical function.

collimate
(1) With reference to rays of light or other radiation, to make

parallel to one another or to a chosen line or direction. (2) With reference to an optical instrument to adjust the line of sight to make it assume its proper position relative to the other parts; to align the axis of the optical elements to the mechanical axis.

collimator

An optical apparatus for producing parallel rays of light. It is used in spectroscopes.

collinear planes

Planes which have a common line.

collinear points

Points which are on the same line.

collision

(1) The encounter of two particles, atoms, molecules or bodies. (2) The interaction and mutual effect on one another of the colliding particles, atoms, molecules or bodies.

colloidal particles

Electrically charged particles, suspended in a medium, that are larger than atomic or molecular dimensions, but sufficiently small to exhibit the Brownian movement.

colloidal solution

A heterogeneous system (even though macroscopically it may appear to be homogeneous) consisting of two phases: the *dispersion medium* or *continuous phase,* which is a liquid, and in which suspended the *dispersed phase* which consists of minute particles of a solid. (If the dispersed phase consists of particles of a liquid, the system is called an *emulsion* or *emulsoid.*)

cologarithm

The cologarithm of a number is the logarithm of the reciprocal of that number.

color temperature

With reference to a given light source, the temperature at which a black-body must emit to give a color that matches the color of that source.

colorimeter

An optical instrument, used in colorimetric analysis to accurately measure color differences.

colorimetric analysis

A process by which the amount of an element present in a solution is determined by measuring the intrinsic color, or the color of the reaction product developed when selected reagents are added to the solution.

colorimetric purity

The percentage of monochromatic light in a mixture of monochromatic light and white light required to match a given specimen.

colorimetrics

A method of detection or determination based on a difference, qualitative or quantitative, of color.

columbium

See *niobium.*

coma

The spherical aberration of rays of light which pass through a lens in an oblique direction. It is present in a lens when the image of a bright point of light toward the margin of the field of view is distorted into a pear-shaped or coma-shaped blur. Sometimes known as *oblique spherical aberration* or *zonal aberration.*

combination

(1) Any group of a certain number, *r,* of things selected without regard for order of arrangement from among *n* such things. The number of combinations of *n* things taken *r* at a time is denoted by the symbol C (n, r). (2) The process of producing such combinations.

combining number

The number of atoms of hydrogen which will combine with or replace one atom of a given element.

combining weight

See *Dalton's law of definite proportions.*

combustion

Rapid oxidation of a substance, i.e., its rapid combination with atmospheric oxygen, accompanied by an emission of light and heat energy.

commensurable quantities

Two quantities for which there is a common measure, i.e., the ratio of which is expressed exactly by two integers.

common

(1) General, universal. (2) Ordinary. (3) Shared by two or more objects.

common difference

In an *arithmetic progression,* the constant difference between any two adjacent terms of the sequence. The common difference is regarded as *positive* if the terms of the progression are in an increasing order; if they are in a decreasing order, the common difference is said to be *negative.*

common elevation views

The front, right profile, left profile and rear views.

common factor

A number, quantity or expression that divides exactly into two or more numbers, quantities or expressions. Thus, 3 is a common factor of 9, 21 and 39; x^2 is a common factor of ax^3, bx^2 and cx^5.

common fraction

A fraction which has only a numerator and denominator without a whole number attached; e.g., ½; ⅔; 7/4. Also referred to occasionally as *vulgar fraction.*

common logarithms

See *Briggs' logarithms.*

common ratio

The quantity by which each term of a *geometric progression* is multiplied in order to obtain the succeeding term; thus r is the common ratio in the geometric progression: $-a + ar + ar^2 + ar^3 + \ldots$ or 3 is the common ratio in the geometric progression: $-4 + 12 + 36 + 108 + \ldots$

common salt

Sodium chloride $(Na^+ + Cl^-)$.

common scale of notation

The decimal system of notation, i.e., the system of numeration in which 10 is used as the *base* or *radix* of the system.

commutate

To exchange, interchange or substitute.

commutative law of addition

The order in which the terms are added does not alter their sum, i.e., $a + b = b + a$. Thus, the terms or parts of an expression which are connected by plus (+) or minus (−) signs can be written in any order. Thus: $3 - 6 + 10 = -6 + 3 + 10 = 10 + 3 - 6 = 7$ or $a + bc - \frac{d}{e} = bc + a - \frac{d}{e} = -\frac{d}{e} + bc + a$.

commutative law of multiplication

The order in which the terms are multiplied does not affect the product, i.e., $ab = ba$. Thus the factors of any product may be taken in any order. Thus $3b \times 4\ dc \times a = 3 \times b \times 4 \times d \times c \times a = 12\ abcd$. For division the total can be split up into any number of factors. Thus $12\ abcd = 2 \times 2 \times 3 \times a \times b \times c \times d = 2d \times 3b \times a \times 2c$.

compact *(noun)*
In metallurgy, an object produced by compression of single, mixed or alloyed metal powders, sometimes with nonmetallic substances. Also called *briquette.*

compact *(verb)*
To produce *compacts.*

compactibility
A measure of the plasticity of powder particles, i.e., the ability of a powder to form a compact of definite contours and structural stability at a given pressure.

compacting
Joining by thermal sintering.

comparator
An instrument for the accurate measurement of lengths or checking of scales by comparison with a standard scale.

comparison spectrum
A spectrum which contains a number of sharply defined, well identified lines of standard wavelength, which is used as a standard of comparison in studying other spectra, usually photographed on the same plate above and below of the spectrum studied.

compass
An instrument for drawing curves.

complement
See *complementary angle.*

complementary angle
The angle the addition of which to a given angle will produce a right angle.

complementary colors
Those pairs of colors which, when mixed, produce white or gray light: red and blue-green, green-yellow and violet, yellow and indigo-blue. In pigments, those opposite colors which, when mixed, produce a neutral gray; thus blue is complementary to yellow-red; yellow is the complement of purple-blue; red is the complement of blue-green.

complete diffusion
The process in which the diffusing medium scatters all the incident light, reflecting or transmitting no part of it.

complex
Not simple; complicated, intricate; having or composed of many parts.

complex fraction
A fraction that has a fraction or mixed number for one of its terms or for both terms. Thus $\frac{2\frac{3}{4}}{5/6}$; or $\frac{6}{4/7}$; or $\frac{4\frac{1}{3}}{7\ 2/5}$

complex ion
Any ion which consists of several atoms, except for the common radicals (as, e.g., SO_4^{--}).

complex number
A number having the form $a + ib$, or $a - bi$, where a and b are real numbers and $i = \sqrt{-1}$.

complex plane
See *Argand diagram.*

compliance
In physics, especially mechanics and acoustics, the reciprocal of stiffness.

component
An ingredient; any part that contributes to or goes to make up a compound, a structure or a complex unit.

component method of adding vectors
The method of adding vectors by resolving each of them into its *rectangular components* and adding these components algebraically, thus finding the resultant.

component parts
In mathematics, the factors into which a composite number or quantity can be divided. Thus 3 and 5 are component parts of 15; 2 and b and c are component parts of 2bc.

composite number
A number which is the product of any two or more numbers. It is a number which can be divided by another number without leaving a remainder. Thus 15 is a composite number, being the product of the numbers 3 and 5, which are called prime numbers.

composite reaction
A chemical transformation which takes place in the form of several reactions.

composition
The addition of *vectors.*

compound
See *chemical compound.*

compound addition
The addition of quantities of different kinds or denomina-

tions. Thus 5 miles, 6 furlongs, 15 yds. plus 8 miles, 3 furlongs, 21 yds. = 14 miles, 1 furlong, 36 yds.

compound amount

In computing compound interest, the original principal plus the compound interest.

compound division

The division of quantities of different kinds or denominations. Thus 5 tons, 4 cwt. divided by 11 = 9 cwt. 1 gr. 22 lbs. 14 6/11 ozs.

compound expression

A mathematical expression in which two or more terms are added to or subtracted from each other. E.g., $a + b - c$.

compound fraction

A fraction of another fraction or of a mixed number. Thus $\frac{2}{3}$ of $\frac{5}{8}$ (which is $\frac{2}{3} \times \frac{5}{8}$); or $\frac{3}{4}$ of 9 1/6 (which is $\frac{3}{4} \times$ 55/6).

compound interest

Interest obtained by the method according to which the interest on the original loan or principal is added to the principal at the end of a period (usually a year), and the interest obtained on this new principal at the end of the next period is added to the new principal, making a further increased principal and so on, until a chosen number of periods have elapsed or been reached.

compound lens

A lens composed of two or more separate pieces of glass which may or may not be cemented together.

compound microscope

A microscope which has two lenses or sets of lenses, the first being the eyepiece or eyepiece assembly, and the other the objective or objective assembly.

compound multiplication

The multiplication of quantities of different kinds or denominations. Thus £4. 6s. 8d. multiplied by 25 = £108. 6s. 8d. or 3 cwt. 2 gr. 7 lbs. multiplied by 18 = 3 tons, 4 cwt. 0 gr. 14 lbs.

compound quantity

A quantity composed of two or more simple quantities or terms, connected by the signs plus or minus. Thus a + bc − de.

compound ratios

Ratios formed by multiplying together the fractions which denote them. Or by multiplying together the antecedents (or numerators), to obtain a new antecedent, and multiplying the consequents (or denominators) to obtain a new consequent. Thus the compounded ratio of three ratios $4a : 3b^2$; $7\,ab : 5c$; $c : 6a = \frac{4a}{3b^2} \times \frac{7ab}{5c} \times \frac{c}{6a} = \frac{14a}{45b}$

compound subtraction

The subtraction of quantities of different kinds or denominations. Thus: £5. 2s. 6d. minus £2. 5s. 4d. = £2. 17s. 2d. or 6 tons 7 cwt. minus 3 tons 8 cwt.=2 tons 19 cwt.

compressibility

The reciprocal of the *bulk modulus* of a given object.

compression

(1) The resultant of forces or stresses tending to decrease the volume or length of an object. (2) The state or the measure of such diminishment in volume or length produced by such forces or stresses.

compression test

A sample of a material is placed under a hydraulic press and pressure is exerted until it "gives."

Compton effect

The decrease in frequency and increase in wavelength of x-rays or gamma rays when scattered by free electrons.

Compton electron

An electron ejected from its orbit in the extranuclear structure of an atom and possessing a momentum as a result of an impact with a photon or other high-energy quantum of radiation. (Also called Compton *recoil electron.*)

computation

The execution of a mathematical operation.

computer

A generic term for machines which can be used to solve mathematical problems. The two principal classes of computers are: (1) *Analog computers* which use physical magnitudes to solve the problems, and (2) *digital computers* which perform the calculations using numbers that represent the various quantities and are composed of digits represented according to a predetermined code.

concave

Hollow toward, or curving away from, the point of reference. E.g., a circle is concave toward its center or diameter.

concave lens

A lens which is thinner in the middle than at the periphery and acts as a *divergent lens.*

concave polygon
A closed plane figure one interior angle of which is greater than 180°.

concave quadrilateral
A four-sided plane figure one interior angle of which is greater than 180°

concentrate
In chemistry, to increase the strength of a solution, compound or mixture by reducing the quantity of undesired material and thus increasing the proportion per volume of the desired substance.

concentrate
In physics, to gather something diffused; to bring to a common center; (2) to bring about *concentration* in general.

concentrated
In chemistry: containing the minimum of water or other solute.

concentration
In chemistry, the number of molecules or ions present in a given volume, i.e., the mass per unit volume, or the strength of a solution.

concentration
In physics: (1) The process or act of increasing the quantity of a substance or energy in a given volume of space. (2) The result of such process or act, i.e., the number of units of mass of matter or units of energy present in a given volume of space.

concentric circles
Circles which have the same center.

concentric spheres
Spheres all the centers of which are the same identical point.

concrete number
A number referred to actual objects, as, e.g., *5 men, 8 houses,* where 5 and 8 are concrete numbers. But when used alone, without a qualifying noun, they are *abstract numbers.*

concurrence
The meeting of three or more lines in one point (called *point of concurrence.*)

concurrent
Said of three or more lines which meet in the same point (called *point of concurrence*).

concurrent forces
Forces (or *vectors*) lines of action of which intersect at one point.

concurrent reaction
Any one of two or more reactions which are related and usually interdependent.

concyclic points
Points which lie on the circumference of the same circle. Three points, if not in the same straight line, are always concyclic. Four or more points are not necessarily concyclic.

condensation
The process by which a vapor becomes a liquid or a solid.

condensation level
In meteorology, the level at which air becomes saturated when it is lifted adiabatically. Also called, more properly, *lifting condensation level.*

condensation pressure
The pressure to which air must be expanded dry-adiabatically in order that it may be cooled sufficiently to become saturated.

condenser
(1) See *capacitor.* (2) See *condensing lens.*

condensing lens
In optics, a lens or lens system which concentrates the illumination from a light source on a limited area. Also called *condenser.*

conditional equation
An equation that is true only for a particular value or values of the unknown quantity or quantities involved. E.g., $4x + 7 = 15$ is true only when $x = 2$.

conditional inequality
An *inequality* (q.v.) that holds true only for certain values of the terms involved.

conditionally convergent series
A series of terms which is convergent, whereas the corresponding series of absolute values is divergent.

conductance
See *electric conductance, thermal conductance.*

conductance water
Water of a high degree of purity, used for the solutions employed in measurements of *conductance.*

conduction
The flow of heat, sound, electricity, or some other physical

process or phenomenon through or along a substance (called a *conductor*).

conduction current

A current in which electricity is conducted or carried by electrons in motion.

conductivity

(1) The facility with which a substance permits a given form of radiant energy to flow through it.

(2) The *conductance,* i.e., the ability to conduct an electric current, through a cubic centimeter of the substance in question, measured and expressed in mhos. Also called *specific conductance.*

conductor

A substance which offers little resistance to the flow of heat, sound, electricity, or some other physical process or phenomenon, through itself or along itself.

cone

A solid formed by a *nappe* of a closed conical surface and a plane cutting all the elements. A cone the base of which is a circle is referred to as a *circular cone,* and if its axis is perpendicular to its base is a *right circular cone* (also designated as a *cone of revolution*).

cone of revolution

A cone, the *base* of which is a circle and the *axis* of which is perpendicular to the base; so called because it can be conceived of as formed by causing a right triangle to revolve about one of its legs or an isosceles triangle about its altitude. Cones produced by the revolution of similar right triangles about corresponding legs as axes are referred to as *similar cones of revolution.* (Cones of revolution are referred to also as *right circular cones.*)

confocal

Having the same focus or foci, real or imaginary.

conformal map

The representation of a sphere or other object with curved surfaces on a plane, in which the angle between any two arcs intersecting at a point on the surface of the object depicted is represented by an equal angle between the projections of the two arcs on the plane of the map.

conformal projection

The method of projection employed in producing a *conformal map;* this method does not preserve size, distance or area or scale of lengths, but it preserves the shapes of the objects depicted.

congruence

The exact similarity in every way of one thing to another. The mathematical sign for congruence is three parallel lines thus: $\equiv$. The things or equations are said to be identically equal for all values of unknown quantities involved.

congruent angles

Equal angles.

congruent figures

Two or more geometrical figures which are exactly alike in every respect and can be placed so as to coincide exactly.

congruent solids

Two or more solids which can be made to coincide in all their parts.

conic

A synonym for *conic section.*

conic section

Any section of a right circular cone cut by a plane. The conic sections produced, according to the angle of intersection, are: the *ellipse,* the *parabola* and the *hyperbola,* and in certain special cases the *circle,* the *straight line* and the *point* may also be regarded as such. (Also referred to simply as *conic.*)

conical surface

A surface generated by a line (the *generatrix*) which moves so that it always intersects a plane curve (the *directrix*) and always passes through a fixed point (the *vertex*). Any of the positions assumed by the moving line (generatrix) while forming the surface is referred to as an *element* of the latter.

conicoid

An *ellipsoid,* a *hyperboloid,* or a *paraboloid;* these quadric surfaces are so called because all of their plane sections are conics.

conjugate

Reciprocally related, as two points or axes capable of being interchanged without affecting the form of a statement or proposition.

conjugate algebraic numbers

The numbers which are roots of the same algebraic equation of the form $a_0x^n + a_1x^{n-1} + \ldots a_n = 0$.

conjugate angles

Two angles the sum of which is 360°.

conjugate axis of a hyperbola
The line segment which cuts the transverse axis perpendicularly at its midpoint and equals it in length.

conjugate complex numbers
Two complex numbers which differ only in the sign of the pure imaginary term. E.g., $a + bi$ and $a - bi$, when $b = 0$.

conjugate diameters of a conic
Two diameters each of which bisects the chords parallel to the other. E.g., the axes of an ellipse.

conjugate hyperbolas
Two hyperbolas which have the same asymptotes.

conjugate imaginaries
See *conjugate imaginary numbers.*

conjugate imaginary numbers
The imaginary numbers $a + bi$ and $a - bi$, when $b \neq 0$.

conjugate point
See *acnode.*

conjugate solutions
Two chemical solutions existing together in equilibrium at a given temperature in such a condition that any variation in the temperature results in a change of the relative proportions of the solutions and of their compositions.

conjugate triangles
Two triangles, the vertices of each of which are the poles of the other.

conjugated system
A chemical system characterized by a transmission of chemical reactivity from one atom to another.

consecutive mean
In meteorology, the mean of a small group of consecutive values, assigned to the middle time-term of the group, in the process of *smoothing.*

consecutive reaction
Two or more reactions in which a product or products of one are the reactants of another.

consequent
The second of the two terms of a ratio. Thus in the ratio A is to B, written A:B, the term B is the consequent and A is the *antecedent.*

conservation of areas
A general principle of the kinematics of central orbits, as demonstrated by Kepler's second law of motion.

conservation of energy
See *law of conservation of energy.*

conservation of mass
See *law of conservation of mass.*

conservation of momentum
See *law of conservation of momentum.*

consistency
That property of a body which makes it tend to resist deforming forces.

consonance
Resonance, acoustical or electrical, between bodies or systems not linked directly with each other.

constant
A property or quality that remains the same under the same conditions. In mathematics, a fixed, definite quantity which remains the same throughout the calculations of the problem or question.

constant acceleration
See *uniform acceleration.*

constant heat summation
See *Hess' law of constant heat summation.*

constant of gravitation
The acceleration that results from the attraction of a unit of mass at unit distance, determined by Heyl in 1942 to be 6.673×10^{-8} in C.G.S. units.

constituent
See *determinant.*

constitutive property
A chemical or physical property which depends on the quality of the substance involved, as on the boiling point, the molecular structure, etc.

constrain
To confine or limit to a determined position or path.

constringence
The *Abbe number* of a substance.

construction
In mathematics: (1) The process of drawing a figure that will satisfy certain given conditions. E.g., construction of a line perpendicular to another line and proving that it meets these requirements. (2) Construction in proving a theorem; drawing the figure indicated by the theorem and adding to the figure any additional parts that are needed in the proof.

Such additional lines, points, etc., are usually called "construction lines," "construction points," etc.

constructive interference

In physics, the mutual superposition of two waves which have the same period and the same phase.

contact force

A force whch acts by direct contact.

contamination

Impurities or extraneous material in a substance or compound or system.

continued fraction

A fraction the denominator of which contains a fraction, which in turn has a fraction in its denominator and so on. Thus:

$$\cfrac{a}{b+\cfrac{c}{d+\cfrac{e}{f+\dots}}}$$

continuous current

See *direct current.*

continuous function

A function $f(x)$ is said to be *continuous at the point a* if $\lim_{x\to a} f(x) = f(a)$, that is to say if the limiting value of the function, as x approaches a, actually equals the value of the function at the point a. A function is *continuous in an interval* $x_0 \le x \le x_1$ if it is continuous at every point in the interval.

continuous phase

The liquid phase of a *colloidal solution* in which the solid particles (the *dispersed phase*) are suspended.

continuous spectrum

A spectrum which shows all wavelengths in a continuous variation of intensity; e.g. the spectrum of an incandescent solid.

continuum

A continuous magnitude or extent. A perfectly connected set of elements or points, etc.

contraction

The shrinkage which occurs in solids as a result of the drawing together of the atoms within their molecules. The opposite process of *expansion.*

contravariant

A function of coefficients and variables (to be transformed by inverse substitution) that differs only by a multiplier (a power of a modulus) from the corresponding function of the transformed coefficients and variables.

control rod

The general designation of rods or bars made of a substance possessing a good neutron-absorbing capacity (such as cadmium or boron steel) which are inserted into uranium nuclear reactors to regulate the rate of propagation of the fission chain reaction by reducing the *multiplication factor* to unity.

convection

The transmission of heat by currents of the moving particles of a gas or liquid.

convection current

(1) A motion of matter which transfers heat from one point or region to another. (2) The transmission or conduction of electricity by masses in motion which are heavier than electrons.

convergence

The gradual decrease of the distance between two lines, paths, etc., with the result that they tend to meet at a point in a finite distance.

convergence angle

See *angle of convergence.*

convergent

An approximation to the exact value of a continued fraction, obtained by keeping only some of the first partial quotients. Thus: the convergents to $\pi=3.14159$ are 3/1; 22/7; 333/106; 355/113; etc. (They are gradually getting nearer the real value.)

convergent lens

A lens which is thicker in the middle than at the periphery and causes parallel light rays to converge, bringing the light beam to a point focus. Also called *converging lens, convex lens, collective lens* and *positive lens.*

convergent series

A series the sum of n of the terms of which approaches a limit as n increases in an unlimited fashion; that is, the series is said to converge when $\lim_{n\to\infty} S_n$ exists and is finite. If the corresponding series of absolute values converges, the series is said to *converge absolutely;* if the series itself is convergent, but the corresponding series of absolute values diverges, the series is said to *converge conditionally*.

converging lens

See *convergent lens*.

converse of a theorem

The theorem obtained by the process of interchanging the hypothesis and the conclusion of the given theorem. (Specifically bearing a reciprocal relation to another proposition or theorem. Example: The larger circle has the longer radius, or the longer radius describes the larger circle.)

conversion factor

A number by which a quantity expressed in one system of units may be converted to another system. (E.g., when converting meters per second to miles per hour, we multiply by the conversion factor 2.237.)

convex

Curving toward the point of reference. E.g., a circle with reference to a point outside it.

convex lens

A lens which is thicker in the middle than at the periphery and acts as a *convergent lens*.

convex polygon

A polygon none of the angles of which is greater than 180°, and consequently no side of which, when extended, will enter the polygon.

convex polyhedron

A polyhedron in which a cutting plane can produce only *convex polygons*.

convex quadrilateral

A four-sided figure each interior angle of which is less than 180°.

cooling lag

The difference between the critical points on heating and cooling due to the tendency of physical changes to lag behind temperature changes during heating and cooling operations. (Also called *hysteresis*.)

cooling power

The net rate at which heat can be taken up from any body by its environment.

coolometer

An instrument for measuring *cooling power*.

coordinate

One of a set of numbers which locate a point in space. If the point is known to be on a given line, only one coordinate is needed; if in a plane, two are required; if in space, three.

coordinate bond

In chemistry, the *bond* between atoms formed when an element lacking two electrons in its valence group acquires both of these by union with another atom.

coordinates

In graphs or analytical geometry, a system of straight lines springing from a common zero point or point of intersection. By measurement along or from these lines the position of a point or points on a surface or in space is obtained. The coordinates are straight lines, they are usually called the x, y or z axes, for Cartesian coordinates, and for polar coordinates a radius or vector and an angle are used. The axes are usually drawn at right angles to one another, though this is not always the case. For Cartesian coordinates, the distance of the point from the horizontal axis, measured along the vertical axis, is called the *ordinate,* the distance of the point from the vertical axis, measured along the horizontal axis, is called the *abscissa.*

coplanar

Situated in the same plane.

copper

A metallic element in Group I of the periodic table. Atomic number 29, atomic weight 63.54, density 8.95, melting-point 1083°C., boiling-point 2310°C., valence 1 or 2. Symbol Cu.

copunctal

Concurrent; meeting in a point.

core

(1) In a magnetic circuit, that part around which the winding is placed. (2) In a nuclear reactor, the region where the fissionable material is placed.

core loss

The energy expended in magnetizing a material with an alternating electric current. Also called *watt loss.*

Coriolis effect

The deflection of a moving object owing to the rotation of the earth, to the right in the northern hemisphere, and to the left in the southern hemisphere. (Also called *Coriolis acceleration*).

Coriolis force
The apparent force which produces the *Coriolis effect,* actually a result of the rotation of the earth.

corollary
A statement which follows from another with little or no proof. For example: Since we admit that all 90° angles are equal it follows also that all 45° angles are equal. Or two lines parallel to the same line are parallel to one another.

corona
In physics, a luminous discharge due to ionization of the air surrounding a conductor around which exists a voltage gradient exceeding a certain critical value.

corona
In meteorology, one or more rainbow-colored rings of small radii surrounding the sun or the moon when covered or partially obscured by a thin veil of cloud composed of water droplets.

coronagraph
An instrument for observing the corona of the sun.

coronavisor
An instrument for studying and photographing the corona of the sun, which makes use of television principles, and builds an image of the corona on a television tube.

corpuscle
A minute particle.

corpuscular
Relating to or having the nature of a corpuscle, or composed of corpuscles.

corpuscular theory of light
Newton's theory, that light is a stream of minute particles emitted by luminous bodies at very high velocities, and that the sensation of light is due to the bombardment of the retina of the eye by these particles. Although this theory was replaced by the wave theory of light, the concept of *photons* is reminiscent of Newton's corpuscular theory.

corrected lens
In optics, a compound lens, the component surfaces of which are so designed that the resulting lens is more or less free of an aberration or of certain aberrations.

correlation
Reciprocal relationship between two variables.

corresponding angles

If a straight line cuts two other straight lines, any two angles formed by these two straight lines and the cutting line, lying on the same side of the cutting line, and both angles lying either above or below the two straight lines being cut, are corresponding angles.

corresponding parts

In geometric figures, the sides and angles of one figure which are in the same relative position to sides and angles of the other.

corrosion

The slow, gradual destruction or disintegration of a substance, notably of a metal or an alloy, by the action of the atmosphere, water, or chemicals, or any other chemical or electrochemical agency. E.g., oxidation or the rusting of iron.

corrosive

Any agent that produces corrosion.

cos

Short for *cosine.*

cosec

Short for *cosecant.*

cosecant

One of the trigonometrical functions or ratios; in a right angle triangle, it is equal to the hypotenuse divided by the perpendicular or side opposite to the angle. (Written as *cosec.*)

cosecant curve

A graphic representation of the cosecant ratio, i.e., the graph of $y = \text{cosec } x$; it is one of the *trigonometric curves.*

cosine

One of the trigonometrical ratios or functions and in a right angle triangle is equal to the side adjacent to the angle divided by the hypotenuse, namely $\frac{b}{c}$. The cosine (written *cos*) of an angle equals the sine of the complementary angle.

cosine curve

A graphic representation of the cosine ratio, i.e., the graph of $y = \cos x$; it is one of the *trigonometric curves.*

cosine law

In any triangle, the square of any side is equal to the sum of the squares of the other sides minus twice their product times the cosine of their included angle. Thus: $a^2 = b^2 + c^2 - 2bc \cos a$.

cosine law for angles in spherical trigonometry

In any spherical triangle, $\cos A = -\cos B \cos C + \sin B \sin C \cos a$.

cosine law of spherical trigonometry

The cosine of a side of a spherical triangle equals the product of the cosines of the other two sides plus the products of the sines of the same two sides multiplied by the cosine of their included angle.

cosmic

Pertaining to or characteristic of the universe as a whole, or to celestial bodies (especially stars and galaxies) and their motions.

cosmic ray shower

The appearance of several light ionizing particles on downward paths, produced simultaneously by a phenomenon attributable to one single cosmic ray.

cosmic rays

Radiations which strike the earth, assumed to originate in interstellar space. These radiations are very penetrating (able to penetrate 18 inches of solid lead, or 200 feet of water); they are deflected by magnetic fields. They are stronger at the poles of the earth than at the Equator. First studied extensively by R. A. Millikan. Cosmic rays consist mainly of electrically charged particles which bombard the earth from all directions in space, with energies ranging from 2,000 to 17,000 Mev. While the origin of the *primary* cosmic rays (the radiation arriving directly at the atmospheric envelope of the earth) is still a point under debate, it is generally assumed that they are emitted by atoms in interstellar space and consist of high-velocity protons which collide with a great many atomic nuclei within the higher layers of the atmospheric envelope of the earth and cause these nuclei to split into *electrons, positrons* and *mesons,* which (especially the positrons) are imparted energies very closely approaching the energy of the primary proton ray, and form the *secondary* cosmic rays. (See also *plasma.*)

cosmic repulsion

A hypothetical force, postulated to be independent of mass and increasing in strength with the increase of the distance between the interacting objects; it was introduced by Einstein at one time to allow the formulaion of mathematical models of a static universe, but it was repudiated by him later.

cosmology

That branch of theoretical astronomy which deals with the universe as a whole, embracing the study of the structure, dimensions and interrelations of the galaxies, nebulae, etc.

cosmos

The totality of the observed and postulated physical whole, conceived as an orderly and harmonious system.

cosmotron

Designation of a proton synchrotron accelerator installed in the Brookhaven National Laboratory (Upton, L. I.), capable of imparting velocities of 2-3 bev.

cot

Short for *cotangent.*

cotangent

One of the trigonometrical ratios or functions and in a right angle triangle, equal to the adjacent side to the angle divided by the opposite (perpendicular) side, namely $\frac{b}{a}$. (Written as *cot.*)

cotangent curve

The graph of $y = \cot x$; it is one of the *trigonometric curves.*

coterminal angles

Angles which have identical initial sides and identical terminal sides, but are generated by different amounts of rotation of the terminal line.

coulomb

The unit used for measuring the quantity of an electric charge. It is equal to the charge of 6.28×10^{18} electrons. A coulomb is one-tenth of an abcoulomb and equals 3×10^9 statcoulombs.

Coulomb's law

The law which states that like charges repel, opposite charges attract each other, and the repulsion or attraction is proportional to the product of the magnitudes of the charges and inversely proportional to the square of the distance between them.

coulometer

An electrolytic cell arranged for the measurement of the quantity of electricity passing by the accurate measurement of one or more of the products of electrolysis.

count

In radiation measurement, the individual response of the

counting device to an ionizing event. Also the indication of the total of the events registered during a certain period or interval of time.

counter, counter tube

An apparatus for the detection and counting of ionizing events, processes and radiations. (Cf. *crystal counter, Geiger-Müller counter, radiation survey meters.*)

counterclockwise

In the direction of rotation opposite to that in which the hands of a clock move around the dial.

counterclockwise elliptically polarized wave

See *left-handed elliptically polarized wave.*

couple

In physics, two forces which are equal in magnitude, but opposite in direction, and do not act in the same line.

couple

In electrical terminology, two dissimilar conductors in electrical contact.

coupled circuits

Two circuits associated in such a way that power may be transferred from one to the other.

covalence

The *valency bond* which joins two atoms by the sharing of a pair of electrons, each of the two atoms providing one of these electrons; also, the number of electron pairs which an atom is capable of sharing with its neighbor atoms. Each such shared pair of electrons constitutes a *covalence bond* between the two atoms.

covalent bond

See *covalence.*

covariant

A function of both coefficients and variables of a quantic that retains its form when the quantic is transformed linearly. To change together with something else so as to preserve certain interrelations unchanged.

coversed sine

A trigonometric function of an acute angle, covers $\theta = 1 - \sin\theta$. It equals the hypotenuse of the right angle triangle minus the sine of the angle when the hypotenuse or radius of the circle = 1. Also referred to as *coversine* or *versed cosine.*

coversine

See *coversed sine.*

Cp

The symbol for *cassiopeium.*

Cr

The symbol for the element *chromium.*

cracking

Chemical decomposition of hydrocarbons, notably oils, while in a vapor condition by raising the temperature of the oil in the still above the normal point for the distillate being collected, and leaving the upper exterior surface of the still exposed to the atmosphere. Vapor condenses on the upper surface of the still and falls back into the fluid thereby reducing the hydrocarbons into products of lower density and boiling point. Cracking simply alters the molecules to an extent that produces an amount of low-boiling fractions that cannot be obtained by simple distillation.

creep

The slow, but continuous deformation of metals under steady load or prolonged stress.

crith

The weight of one liter of hydrogen: 0.089873 of a gram.

critical angle

The angle at which a ray of light which is just about to enter a medium of lesser density, from one of greater density, is refracted along the surface of the denser medium.

critical mass

The minimum mass, or size, which the fissionable material in a nuclear reactor or atomic bomb must have in order to be efficient in preventing the loss of neutrons by escape through the surface, i.e. non-fission capture, and thus in maintaining a spontaneous fission chain reaction. The larger the mass the less neutrons will escape fission capture.

critical potential

In atomic physics, the critical potential is used as a measure of the amount of energy necessary to raise an electron from a lower to a higher level.

critical pressure

The pressure which must be applied in order to condense a gas at the *critical temperature.*

critical state

A condition of a substance the liquid state and gaseous state of which are identical.

critical temperature

(1) The temperature above which a substance can exist only in its gaseous state and cannot be liquefied regardless of the magnitude of the pressure to which it is subjected. (2) In metallurgy, the temperature at which some change occurs in a metal or alloy during heating or cooling.

critical volume

The volume which one mole of a given substance occupies at the *critical temperature* and the *critical pressure.*

Crookes dark space

A synonym for *cathode dark space.*

Crookes tube

An early form of discharge tube, devised by Sir William Crookes for the study of *cathode rays,* containing a near-vacuum (roughly one one-millionth of an atmosphere) .

cross multiplication

The multiplication of the numerator on one side of an equation by the denominator on the other side of the equation. When this operation is carried out for both sides of an equation fractions are eliminated.

cross-ratio

If A, B, C, D, is a range of points, the ratio $\frac{AB \times CD}{AD \times CB}$ is called a cross-ratio of the four points. Sometimes called an anharmonic ratio.

cross-section

The surface exposed on a solid by a plane that cuts through that solid.

crucible

A thimble-shaped vessel made of clay, porcelain or other ceramic material, used in chemical experiments to hold a substance while it is being heated. (2) In metallurgy, a refractory vessel or pot used to melt down various grades of iron and steel to produce steel by the *crucible process.*

crucible process

A process for the production in steel, in which various grades of iron and steel are melted down in closed *crucibles,* or in special crucible furnaces, generally with the addition of carbon, manganese, and other materials, depending on the product desired.

crucible steel
Steel produced by the *crucible process.*

crunode
A point on a curve where the curve crosses itself and forms two branches each of which has distinct tangents.

cryogenic
Relating to, or producing low temperatures.

cryoscopy
Determination of the properties of substances by freezing points of mixtures.

crystal
A solid body, the atoms of which are arranged in a definite, regularly repeating, symmetrical pattern, and having faces at definite angles to each other; the faces of the crystal are the outward expression of the regular arrangement of the atoms, and thus present a geometrical form, when the crystal can grow freely without the interference of adjacent crystals. Nearly every pure element and compound is capable of forming crystals.

crystal counter
A *counter* that utilizes one of several known crystals which are rendered temporarily conducting by ionizing events.

crystal dynamics
The science and study of the movements of atoms and electrons in crystals.

crystal grating
A crystal used as a *diffraction grating* in the studies of x-rays and their spectra. (See *Laue method, Laue pattern.*)

crystal nucleus
The minute crystal particle deposited by a supersaturated solution which acts as a focal point for further crystal growth; and on which, according to its crystal form, the complete crystal is built up.

crystal optics
The science and the practical application of the transmission of light in crystals.

crystal system
Any one of the seven groups into which crystals are classified according to symmetry patterns. They are called: *cubic* (or *isometric,* or *regular*), *tetragonal, hexagonal, trigonal* (or *rhombohedral*), *orthorhombic* (or *rhombic*), *monoclinic, tri-*

clinic. These seven groups comprise a total of 32 possible types of crystal and 230 possible types of structure.

crystal water

See *water of crystallization.*

crystalline

Pertaining to, composed of, or resembling crystals.

crystallization

The formation of crystals, taking place when a substance changes from the gaseous or dissolved, or liquid or molten state to the solid state.

crystallogram

A photograph of the x-ray diffraction pattern produced by a crystal.

crystallographic axis

See *axis of a crystal.*

crystallographic planes

Any set of parallel and equally spaced planes that may be supposed to pass through the centers of atoms in crystals.

crystallographic system

Any of the major units of crystal classification, embracing one or more symmetry classes. (See *crystal system.*)

crystallography

That branch of physical science which treats of the geometrical forms and properties of crystals.

crystalloluminescence

(1) The emission of light during the precipitation of crystals from solutions. (2) The light so emitted.

Cs

The symbol for the element *cesium.*

Cu

The symbol for the element *copper.*

cu

An abbreviation for *cubic* in the names of units; e.g., *cu ft* for *cubic feet.*

cube

A rectangular solid bounded by six equal squares, and the three dimensions of which are equal. (Also called *regular hexahedron.*)

cube-centered

With reference to *space lattices,* a synonym for *body-centered cubic.*

cube root

The cube root of a given number, *a*, (written as $\sqrt[3]{a}$ is that value which used three times as a factor gives a product equal to *a*.

cubic

In mathematics, of the third degree or order.

cubic crystal

Any crystal which has three mutually perpendicular axes of equal length.

cubic curve

An *algebraic plane curve* of the third degree.

cubic equation

An equation which contains the cubic value (the third power) of the unknown quantity as the highest power of that quantity.

cubic system

In crystallography, the group which includes the crystals having three mutually perpendicular axes of equal length. (Also called *isometric* or *regular system.*)

cuboid

A solid shaped like a cube.

cumulative frequency

In frequency distributions, the sum of all measurements less than or equal to a given value.

cumulative relative frequency

In frequency distributions, the cumulative frequency of a specific value divided by the total number of counts.

cumulative symbols

The marks or signs (×) meaning multiplied by or (÷) meaning divided by.

cupro-nickel

An alloy containing about 70% copper and 30% nickel.

curie

The unit of the measurement of radioactivity. 1 curie = that quantity of any radioactive isotope which undergoes 3.7×10^{10} disintegrations per second. Formerly the curie was defined in terms of the radioactivity of *radon,* viz. 1 curie of radon = that quantity of radon which is in equilibrium with 1 gram of radium (3.7×10^{10} atoms disintegrating per second).

Curie point

The temperature above which a ferromagnetic substance becomes paramagnetic. The Curie point varies according to the substance.

curium

A transuranic element, not found in nature. Atomic number 96, mass number of the most stable known isotope 245, valence 3. First produced by G. T. Seaborg, R. A. James and A. Ghiorso in 1944, by cyclotronic alpha-particle bombardment of plutonium ($_{94}Pu^{239}$). Symbol: Cm.

current

(1) The flow of water or another fluid. (2) The flow of free electrons along a conductor, expressed in amperes. (See *electric current*.)

current density

The average current per unit area of an electrode, usually expressed in amperes per square foot or square decimeter. (Cf. *density of electricity*.)

current efficiency

In an electrochemical process, the ratio of the mass of the substance liberated or chemically changed, to the theoretical amount computed from *Faraday's law*.

curtate cycloid

A *trochoid* without loops.

curvature

The continual bending of a line from the straight line direction. It is the rate of change of inclination per unit change in arc length, i.e., the rate of change of direction of a curve.

curvature of a circle

A constant for a circle, equal to the reciprocal of the radius. The radius of the curvature is the absolute value of the reciprocal of the curvature.

curvature of field

In optics, an aberration owing to a lens defect which affects the longitudinal position of images off the axis so that objects in a plane perpendicular to the axis are shown as though in a curved or cup-shaped surface.

curve

(1) The resultant figure which can be imagined to be formed by the motion of a point in a plane or in space. (2) A bent line without angles; a line in which no three consecutive points are in the same straight line.

curve fitting
The ascertaining of the equation of a curve that satisfies certain established data.

curve of intersection
The curve which is common to two or more surfaces, i.e., the curve in which they meet, intercept or cut each other.

curve tracing
The process of plotting or drawing a curve by ascertaining the location of points that lie on it.

curved surface
A surface no portion of which is a *plane.*

curvilinear motion
Motion along a curve; motion which is not in a straight line. Curvilinear motion about a center of force is motion such as that of celestial bodies about the sun; the motion of a particle whose initial velocity was not directed toward the center of force at the given center. If this force is gravitation the path is a conic whose focus (or one of whose foci) is at the center of force.

cusp
A pointed end, apex or peak. A point at which the common tangent to two curves, coming from approximately the same direction, is drawn. (Cf. *tacnode.*)

cutting plane
With reference to *conics,* the plane which produces the conic by intersecting the real or imaginary cone.

cyan
The monovalent radical —CN, derived from hydrocyanic acid; symbol CN. (See *cyanide.*)

cyanide
Any compound containing the monovalent radical —CN (cyan). Cyanides are not considered organic compounds.

cybernetics
The science of automatic control.

cycle
(1) The complete series of values of a periodic quantity which occur during a period.— (2) A regularly recurring succession of events.

cyclic
(1) Concerning or relating to a circle or circles.— (2) Periodically recurrent.

cyclic compound

In chemistry, a compound in which the atoms, or at least some of them, are arranged in a ring.

cyclic polygon

A polygon all the apexes or vertices of which lie on the same circle.

cyclic quadrilateral

A four-sided figure all the four corners of which lie on the same circle.

cyclograph

An electronic instrument for testing and sorting metals.

cycloid

A curve which is the locus of a point on the circumference of a circle that rolls along a straight line. (Cf. *curtate cycloid, prolate cycloid, trochoid,* also *brachistochrone.*)

cyclotron

An apparatus invented (1931) and developed (1939–40) by E. O. Lawrence, for the nuclear bombardment of elements by highly accelerated positively charged particles (protons or deuterons), for the purpose of transmutation and the production of artificial radioactivity. The essential part of the cyclotron is a large electromagnet with a vacuum chamber between its poles. An electron arc is maintained in the chamber, to act as the source of particles, and voltage is applied to two semicircular electrodes (called *dees*) within the chamber. The electrical field between the electrodes accelerates the sub-atomic particles, and a magnetic field makes them move in semicircular paths; their acceleration and the radius of their orbit are increased every 180 degrees by an alternating electric field (which alternates just fast enough to have reversed when the particles reach the opposite end of their 180° orbit). The particles are imparted energies of many Mev, and are hurled against the nuclei of the target atoms, to produce the desired effect.

cylinder

A solid which is determined by a *closed cylindrical surface* and two parallel planes cutting all the elements of that surface. A cylinder the base of which is a circle is referred to as a *circular* cylinder, and if its axis is perpendicular to an *element,* as a *right circular cylinder* (also called a *cylinder of revolution*).

cylinder of revolution

A cylinder generated by causing a rectangle to revolve about one of its sides. The right sections of such a cylinder are circles, and its *bases* are perpendicular to an *element*. (Also called *right circular cylinder*.) Cylinders generated by the revolution of similar rectangles about corresponding sides are referred to as *similar cylinders of revolution*.

cylindrical surface

A surface generated by a line (the *generatrix*) which moves so that it always intersects a plane curve (the *directrix*) and always remains parallel to a fixed line that lies outside of the plane of the curve. Any of the positions assumed by the moving line (the generatrix) as it generates the surface is called an *element* of the latter.

cylindroid

(1) A cylindrical surface whose sections perpendicular to the elements are ellipses. (2) The surface which is the locus of a straight line moving so as to intersect two curves and remain always parallel to a given plane.

D

D

The symbol for *deuterium*.

d

The symbol for *deuteron*.

d-

The symbol for *dextro-*.

D layer

See *ionosphere*.

Δ (delta)

A symbol for a small quantity or a minor change in some large quantity.

Dalton's law of absorption of gases

The solubility of the various gases in a mixture is independent from and unaffected by the presence of other gases, providing that no chemical reaction takes place.

Dalton's law of definite proportions

In every sample of each compound substance the proportions by weight of the constituent elements are always the same; that is, when elements combine to form a compound, they do so in definite proportions by weight. These definite weights are called *combining weights*.

Dalton's law of partial pressures

The pressure of a mixture of several gases in a given space is equal to the sum of the partial pressures which each gas would exert if it were confined alone in the space.

damaging stress

The minimum stress which, if exceeded, would render a material unfit for service before the end of its normal expected life.

damped oscillation

The damped oscillation of a system is the oscillation when the amplitude of the oscillating quantity decreases with time. If the rate of decrease can be expressed by means of a mathematical function, the name of the function may be used to describe the damping. Thus if the rate of decrease is expressed as a negative exponential, the system is said to be an exponentially damped system. Also called *damped vibration*.

damped vibration
A vibration the amplitude of which decreases with time.

damping
Inhibition of motion through resistance. The damping of the motion of a body or system (including electric charges) which has those characteristics necessary to develop oscillations or vibrations, is the effect on the system of those continuously applied agencies, either external or internal, which hinder or prevent the execution of oscillations or vibrations. It is the progressive fading away of a motion or process until the equilibrium position is reached.

damping constant
The *natural logarithm* of the ratio of the first to the second of two values of an exponentially decreasing quantity separated by unit time. (It is the reciprocal of the *decay modulus*.)

daraf
The unit of *elastance;* the reciprocal of the *farad.*

dark current
The residual electric current present in a photoelectric cell when the cell is exposed to the effects of illumination.

dark-line spectrum
A spectrum that consists of an array of dark lines against a bright background, or which contains some lines that are darker than others. It may be described also as a continuous spectrum crossed by many dark lines.

dark space
A relatively nonluminous portion of an electrical discharge through a gas. (See *Aston dark space, cathode dark space, Faraday dark space.*)

dative bond
A type of *covalent bond* in which both electrons forming the bonding pair are supplied by one atom.

datum (*plural:* data)
Information supplied or ascertained in a study or investigation. Any fact, magnitude or relation given or admitted, from which other facts, magnitudes or relations can be deduced, plotted or calculated.

datum point
A basic, given or starting point.

daughter element
The product of the radioactive decay of a radioactive element (called the *parent*).

D-C

An abbreviation for *direct current.*

deactivate

To render inactive.

de Broglie equation

The equation defining the wavelength (λ) of a *de Broglie wave* as $\lambda = \frac{h}{mv}$, where h is *Planck's constant,* m is the mass and v the velocity (and thus mv the momentum) of the electron.

de Broglie wave

In wave mechanics, a wave or wave group associated with a moving elementary sub-atomic particle. Also referred to as *phase wave.*

Debye-Huckel theory of electrolytes

Recognizes that strong electrolytes are nearly completely ionized in water solution and that there are practically no un-ionized molecules present; instead it is considered that the free ions become bound by acquiring an ionic atmosphere which reduces their mobility.

Debye-Scherrer-Hull method

Technique of x-ray diffraction. A beam of x-rays is directed upon a powdered sample of material and the diffracted beams are received on a photographic plate.

decagon

A polygon having ten sides.

decameter

A unit of linear measure used in the metric system, equal to 10 meters or approximately 32.808 feet.

decantation

The separation of a liquid from a residue or a liquid of higher density by pouring off carefully the *supernatant liquid* without disturbing the heavier material which was previously allowed to settle at the bottom of the container.

decarburization

The loss or removal of carbon from iron, or steel, or any iron-based alloy, usually as a result of heating in a medium which reacts with the carbon. (The degree of decarburization decreases from surface to interior.)

decay

See *radioactive decay.*

decay constant

The ratio of the rate of the disintegration of a radioactive substance to the quantity of the substance still in its original state. It is the reciprocal of the *average life.* Also called *disintegration constant.*

decay modulus

(1) In a system exhibiting damped oscillations, the time at which the amplitude of the oscillation has diminished to $1/e$ of its original value (where e is the base of natural logarithms). Thus, this modulus is the reciprocal of the *damping constant.*—(2) With reference to radioactivity, a synonym for *average life.*

deceleration

Negative acceleration. The rate of decrease of speed, with respect to time.

deci-

A prefix meaning *one-tenth.*

decibel

The unit of noise intensity. One-tenth of a bel, the number of decibels denoting the ratio of two amounts of power being 10 times the logarithm to the base 10 of this ratio. The abbreviation *db* is commonly used for the term decibel. Expressed as a formula it is

$$N = 10 \log \frac{I_1}{I_2}$$

where I_1 and I_2 are the respective unit intensities (measured in watts) of two noises being compared, and N is the number of decibels by which one exceeds the other.

decimal

A fraction the denominator of which is 10 or a power of 10, as 100, 1000, etc. Thus $3/10$; $\frac{3615}{1000}$, etc. The denominator is usually not written, but a dot or point called the *decimal point,* is placed by counting from the extreme right-hand digit in the numerator toward the left a number of places equal to the number of zeros in the denominator. Thus $3/10$ is written .3; $\frac{3615}{1000}$ is written 3.615.

decimal logarithms

See *Briggs' logarithms.*

decimal point

The point (.) which is the symbol separating the digits representing integers from the digits representing fractions.

decimal system

(1) The *denary system,* i.e., a system which uses 10 as its *base.*—(2) Any system of decimal measurement; specifically, the *metric system* of weights and measures.

decimal system of angular measure

See *centesimal system of angular measure.*

decimetric waves

Electromagnetic radiation with wavelengths between 0.1 and 1 meter.

decinormal solution

A solution containing one-tenth equivalent moles per liter.

declination

In astronomy, the angular distance of a celestial body from the celestial equator (i.e., the arc of an *hour circle* included between the body and the celestial equator) measured through 90° and named "north" or "south" as the body is north or south of the celestial equator measured on an hour circle.

declination axis

In a telescope, the axis passing through the center of the *declination circle* around which the telescope moves to get the stars in the field.

declination circle of a telescope

A graduated circle fixed to the mounting of the telescope by which the instrument can be pointed to a star whose declination is known.

decomposition

The process of separating anything into its original parts or elements by chemical action or natural decay.

decrepitation

The shattering of crystals and mineral substances, usually accompanied by a loud crackling noise, when the internal stresses built up by heating have reached a certain magnitude.

deduce

To infer something from information obtained. To derive or draw a conclusion by reasoning from given principles.

deduct

To take away, to subtract.

deduction

A noun derived from either of the two verbs, "to deduct" or "to deduce." Derived from "to deduct" it means subtraction or taking away. Derived from "to deduce" it means inference, drawing conclusions by reasoning from given principles, stripping a problem of all its frills, leaving only the plain facts.

deductive reasoning

The reasoning which starts out from an assumption or premise and proceeds by logical steps to deduce a solution to the problem or question; this result is referred to as the *conclusion.*

deep space

A colloquial term for space beyond the outermost boundaries of our solar system.

defective equation

A derived equation that contains fewer solutions than the given equation from which it is obtained.

definite integral

The difference between the values of the *indefinite integral* for two distinct values of the independent variable; thus the definite integral of $f(x)$ is the difference between two values of the integral of $f(x)$ for two distinct values of the variable x. It is usually represented by the notation $\int_b^a f(x)dx$ and read as "the integral from a to b of $f(x)$ or "the integral of $f(x)$ with respect to x between the limits a and b."

definite proportions

See *law of definite proportions.*

definition

The explanation of the meaning of a word or phrase. A short description of something by pointing out its properties.

definition

In optics, the ability of a lens or lens system to record fine detail; also, the clearness and sharpness of the image produced.

deflocculation

The removal of *flocculation.* (See *dispersion.*)

deformation

Change of shape due to stress.

deformation curve

A curve in which the percentile deformation is plotted against time, under specified conditions.

degenerate conics

Limiting forms of conics, i.e., the intersections of a plane and a cone as the cutting plane moves into certain positions, viz.: a *pair of straight lines* (limiting case of the hyperbola, when the cutting plane occupies a position containing two generators of the cone), a *straight line* (limiting case of the parabola, when the cutting plane coincides with one generator), and a *point* (limiting case of an ellipse, when the cutting plane contains only the vertex).

degenerate quadric surfaces

A collective term for the following quadric surfaces; *cone* (real or imaginary), *elliptic cylinder* (real or imaginary), *hyperbolic cylinder, intersecting planes* (real or imaginary), *parabolic cylinder, pair of parallel planes* (real or imaginary), pair of *coincident plants* (real or imaginary). (Cf. *nondegenerate quadric surfaces.*)

degradation of energy

Any process as a result of which available energy becomes unavailable. Also, this unavailability of formerly available energy.

degradation of radiation

The transformation of a radiation into one lower in frequency and in quantum energy.

degree

Algebraically, a synonym of *power* to which a number or quantity is raised, i.e., the number of times that number or quantity is multiplied by itself.

degree

In mathematics, the unit of the *sexagesimal system of angular measure.* It is the largest division into which the circle is divided. In general practice there are 360 equal degrees in a circle. The smaller divisions are minutes, 60 to one degree, and seconds, 60 to one minute.

degree

In physics, the unit of the measure of heat.

degree of a polynomial

The exponent of the highest power of the variable that occurs in the given polynomial.

degree of a term

The sum of the exponents of the unknowns or variables contained in it. Thus x^2 is of the second degree, whereas xy^3 is of the third.

degree of an equation

The highest power of any term contained in the equation when in its simplest form, thus $y^4x + yx^3 = c^2$ is an equation of the "fourth degree."

degree of an expression

The highest number to which the powers of any term contained in it, add up: as $x^5 = 3x^4 + 5x - 7$ is an expression of the "fifth" degree; or $ax^6 + 5bx - 9y^3x^5$ is an expression of the "eighth" degree.

degree of angle

A small angle, measuring 1/360 of a complete rotation.

degree of arc

A small arc, measuring 1/360 of the circumference of a circle.

degree of freedom

In physics and chemistry, the number of independent variables, (as pressure, temperature and specific volume) which must be determined in order to define the state of a system or the position of every particle in a dynamical system. (See also *phase rule.*)

dehydration

(1) The process or act of removing water.—(2) The state produced by such process or act.

dehydrogenation

The removal of hydrogen from organic compounds by the use of an oxidizing or reducing catalyst.

deionization potential

The potential at which the gas within a tube ceases to be ionized.

deka-

A prefix indicating "ten."

dekametric waves

Electromagnetic radiation the wave length of which is between 10 and 100 meters.

deliquescence

(1) The property of absorbing moisture from the air and dissolving in the water thus received.—(2) The process or state of gradual dissolution as a result of absorbing moisture from the atmosphere.

deliquescent

(1) Able to absorb atmospheric moisture.—(2) In the process of absorbing atmospheric moisture.

delta

See Δ *(delta)*.

delta iron

The allotropic form of iron existing, in the case of pure iron, between 1405°C and the melting point, 1535°C. It is nonmagnetic, and atoms are arranged in the *body centered cubic structure,* as in *alpha iron.*

delta rays

Secondary radiation, composed of secondary electrons (electrons produced by ionization by primary radiations, such as alpha rays, beta rays, positrons, etc.) which are capable of producing ionization themselves (e.g. electrons emitted by a substance as a result of its bombardment by alpha particles.)

demagnetization

The reduction or elimination of the magnetism present in a magnetized body or in a space.

demagnetization curve

A curve which shows the relation between the magnetic flux density and the magnetizing force for values of the latter decreasing from saturated to zero induction.

de Moivre's theorem

A formula for raising complex numbers to an integer power when they are expressed in *polar form*, viz.: $[r(\cos \theta + i \sin \theta)]^n = r^n(\cos n\theta + i \sin n\theta)$.

denary system

The number system which uses 10 as its *base*.

denaturant

A harmful or inert substance added to another substance to make the latter unfit for human consumption. (E.g., methyl alcohol is added as a denaturant to ethyl alcohol.)

denaturation

The process of making a substance or product, notably alcohol or a salt, unfit for human consumption, by adding a harmful or inert substance (the *denaturant*).

denominator

The name given to the number below the line in a fraction. Thus in $\frac{1}{3}$ and a/b; *3* and *b* are the denominators or divisors.

densimeter

A device for measuring the specific gravity of liquids.

densitometer

An optical apparatus for measuring optical density, especially the density of spectrum lines, also of photographic plates, etc.

density

(1) In general, the compactness or matter or space distribution.—(2) Specifically, the ratio of mass to volume.—(3) See *photometric density*.—(4) See *density of electricity*.—(5) See *current density*.—(6) See *luminous density*.

density of electricity

The quantity of electricity per unit area at a given point on a surface, or per unit volume at a given point in space. (Cf. *current density*.)

density ratio
See *relative density (1)*.

deoxidation
(1) The removal of oxygen from a compound, also called *reduction*.— (2) In metallurgy, the lowering of the oxygen content of the molten steel by the addition of elements which have a high affinity for oxygen.— (3) A reaction in which the valence number becomes less positive or more negative.

departure
The horizontal distance on a plan or map of a given point from a chosen fixed point or vertical line.

dependent variable
A variable the value of which is a *function* of another (called the *independent variable*). E.g., in $x + 4 = y$, where the value of y depends on the value assigned to x, y is the dependent variable, and x is the independent variable.

depolarization
The process, act or result of counteracting, nullifying or preventing *polarization*.

depth of focus
In optical instruments, the allowable movement of the lens or lens system while still maintaining sharp focus and definition.

derivative
The instantaneous rate of change of a function referred to the variable, represented visually by the slope of the curve of the function with respect to the axis of the independent variable.

derived unit
Any unit which is based on a fundamental unit or on fundamental units.

Descartes' rule of signs
The number of positive roots of $f(x) = 0$ cannot exceed the number of *variations of sign* in $f(x) = 0$.

descriptive astronomy
That branch of *astronomy* which deals with the enumeration and description of celestial bodies, but without being concerned with physical or mathematical theories.

descriptive climatology
That branch of *climatology* which deals with the component

elements of weather and climate, and with their interaction and effect upon human life, health and activities.

descriptive geometry

The graphical method of solving solid (or space) analytic geometry problems. It can be defined also as the theory of engineering drawing.

desiccant

A substance which has as affinity for water (e.g., calcium chloride), used for removing moisture from the surrounding air.

desiccation

Drying, by making use of a substance which has a high affinity for water.

desiccator

A laboratory apparatus designed to keep residues or specimens free from moisture, by making use of a *desiccant.*

desorption

The removal of adsorbed gas from the surface of a solid; the reverse process of *sorption.*

destructive distillation

A process of the decomposition of an organic substance by heating it in the absence of air, and the condensation of the volatile products obtained.

destructive interference

In physics, the mutual superposition of two waves which have the same period but are 180° out of phase relative to each other.

detached coefficient

The coefficient separated from the unknown quantity of an expression. Thus 12 is the detached coefficient of the expression 12x.

detergent

(1) Possessing cleansing properties.— (2) As a *noun,* a cleansing agent (e.g., soap).

determinant

The sum of all possible (L^n) products of n^2 symbols, conventionally arranged in a square block of n columns and n rows, formed by taking one and only one symbol from each column and each row marked + or —, according as the number of arrangements in the order of the columns or rows is even or odd, the order of the columns or rows being held fixed or natural. Thus

$$\begin{matrix} a_1 & a_2 & a_3 \\ b_1 & b_2 & b_3 \\ c_1 & c_2 & c_3 \end{matrix}$$

is a determinant of the 3rd order or degree; and expanded in $a_1b_2c_3 - a_1b_3c_2 + a_2b_3c_1 - a_2b_1c_3 + a_3b_1c_2 - a_3b_2c_1$. The symbols are called *constituents,* the products *elements,* and the diagonals are *principal* or *secondary.*

determinate

Definite; having definite characteristics or limits.

determinate equation

An equation that is one of a set of as many equations as there are unknowns involved.

determinate problem

A problem in mathematics that admits only of a limited number of solutions.

determinate quantity

In mathematics, a quantity that has a limit in the number of values or solutions.

deuterium

The hydrogen isotope ${}_1H^2$, discovered by H. C. Urey, G. M. Murphy and F. G. Brickwedde in 1931, popularly also referred to as *heavy hydrogen.* Its atom number is 1, its atomic weight 2, its atomic mass 2.0147 (about twice that of the common hydrogen isotope H), its natural abundance in common gaseous hydrogen is 0.02% (about 1 part in 5000). The deuterium atom consists of a nucleus, called *deuteron,* with one orbital electron circling about it. Symbol: D.

deuteron

The nucleus of the *deuterium* atom, consisting of one proton and one neutron. Symbol: d.

developer

A solution containing an agent which makes a latent image on a photographic film or plate visible.

development of a solid

The process of "flattening out" the solid so as to make its surface form a plane figure.

deviation

In mathematics, the algebraic difference between the mean of a series of data and an individual member of the series. The deviation of a from b is $a - b$.

deviation

In physics, the bending of a light ray or other form of radiation from its original path, as measured by the angle between the directions of the incident ray and of the diffracted or refracted ray.

dew point

The temperature to which air must be cooled, at constant pressure and constant water vapor content, to enable *saturation* to occur.

Dewar flask

A vessel with a double wall in which the region between the walls has been evacuated, and in which the walls bordering this space have been silvered. Commonly used for the storage of liquefied gases.

dextro-

A prefix, abbreviated as *d-,* indicating that the substance the name of which it precedes is *dextrorotatory.*

dextro-compound

A *dextrorotatory compound.*

dextrorotatory

Turning to the right; producing a clockwise turn. (Abbreviation: *d-.*)

dextrorotatory compound

An *optically active* substance which turns the plane of polarization of light to the right.

di-

A prefix meaning *two* or *twice.*

diagonal

A straight line joining any two not consecutive vertices or corners of a polygon.

diagonal scale

A system of straight lines by which the hundredth parts of units may be drawn or measured.

diagonal triangle

See *harmonic triangle.*

diagram

Anything described or delineated by lines; a drawing that illustrates certain properties or data, possibly also conclusions based on given data.

dialysis

The separation of colloids from crystalloids, by utilizing the fact that the latter diffuse through a semipermeable membrane.

dialyzer

The semipermeable membrane used for *dialysis.*

dialyzate

The crystalloid substance which has been obtained by dialysis through a semipermeable membrane.

diamagnetic substance

Any substance which has magnetic properties lesser than unity and is therefore repelled by a magnet, or which tends to set its longest dimension at right angles to the lines of force when placed in a magnetic field.

diamagnetism

Negative magnetic susceptibility; the diamagnetic substance, of having magnetic properties lesser than unity (that of air) and being repelled by a magnet.

diameter of a circle

Any straight line passing through the centre of a circle and terminated on each side by the circumference. It divides the circle into two equal parts.

diameter of a conic

The locus of the midpoints of a system of parallel chords of the conic. It may be thought of as either a segment of a line or an entire line. Two diameters are said to be *conjugate* when each bisects the chords parallel to the other. (See also *diameter of a circle.*)

diameter of a sphere

Any straight line which passes through the centre of a sphere and terminates at the surface of the sphere.

diastrophism

In geology, (1) the processes resulting in the deformation of the earth's crust, producing the continents, mountains, ocean basins, etc.; (2) the results of such processes.

diathermancy

The property of being capable of transmitting heat; the opposite of being heat-insulating.

diathermometer

An instrument designed for examining the thermal resistance or the heat-conducting power of objects.

diatomic molecule

A molecule consisting of two atoms.

dibasic

Said of an acid in which the hydrogen atom is replaceable by two basic atoms or radicals.

dichroic

Possessing the property called *dichroism.*

dichroioscope

An instrument for exhibiting the two complementary colors of polarized light.

dichroism

The property which makes certain crystals show different colors when viewed from different directions, or by reflected or refracted light, owing to the unequal absorption of light rays by different planes.

dielectric *(noun)*

A medium having the property that the energy required to establish an electric field is recoverable, in whole or part, as electric energy, i.e., a medium in which an electric field can exist without creating a current; a non-conductor of electricity.

dielectric *(adj.)*

Capable of sustaining an electric field.

dielectric absorption

That property of an imperfect dielectric owing to which there is an accumulation of electric charges within the body of the material when it is placed in an electric field.

dielectric constant

That property which determines the electrostatic energy stored per unit volume for unit potential gradient. It is the ratio of the capacitance of a capacitor with a given dielectric between the electrodes to the capacitance with air between the electrodes to the capacitance with air between them. Thus, it is the term *d* in the equation $f = \frac{q_1 q_2}{rd}$, where *f* is the force of attraction between two charges q_1 and q_2 which are separated by a thickness *r* of the substance. The dielectric constant of a vacuum is unity by definition.

dielectric loss

The time rate at which electric energy is transformed into heat in a dielectric when it is subjected to a changing electric field.

difference of longitude

The difference of longitude of two places is the smaller arc of the equator intercepted between their meridians.

differential calculus

The study of the variation of a function with respect to changes in the independent variable or variables, by means of the concepts of derivative and differential; in particular, the study of slopes of curves, non-uniform velocities, accelerations, forces, approximations to the values of a function, maximum and minimum values of quantities, etc.

differential calorimeter

An instrument which utilizes the principle of the *thermocouple* for measuring the quantities of heat liberated or absorbed by bodies.

differential thermometer

An instrument based on the principle of the difference in the rates of expansion and contraction of two connected strips of different metals.

differentiation

An operation used in the calculus, of obtaining the differential coefficient; if $y = x^n$, the differential coefficient, $\frac{dy}{dx} = nx^{n-1}$. (It is the inverse process of *integration.*)

diffraction

The bending of radiation, specifically light, around the edges of opaque objects.

diffraction angle

See *angle of diffraction.*

diffraction fringes

The alternately light and dark bands seen at the edge of the shadow when light waves are diffracted at an angle.

diffraction grating

A device, consisting of a plate with parallel straight grooves in its surface, used to produce spectra. The grating employed to produce and study spectra of visible light consists basically of a flat glass plate on the surface of which parallel straight lines have been ruled, sometimes as close as 20,000 to the inch. A narrow light source seen through such a grating is accompanied on each side by one or more spectra, produced by diffraction.

diffraction spectrum

A spectrum produced by *diffraction,* as in a *diffraction grating.*

diffuse reflection

Reflection of light or other electromagnetic radiation from a *diffusing surface,* in all directions.

diffusing surface

A surface which is relatively rough in comparison with the wavelength of a radiation (especially light) incident on it, so that it breaks up the radiation and distributes it in all directions.

diffusion

(1) The gradual intermixture, without any outside agent or force, of contiguous gases, liquids or solids, due to the movement of ions, atoms, molecules or colloid particles. (2) The flow of molecules, usually, but not necessarily, through a membrane. (3) The spreading out of the atoms, ions or molecules in a fluid or porous medium or in an evacuated space.

diffusion coefficient

The number of grams of a substance which undergoes *diffusion* in one second across an area of one square centimeter through a unit gradient of concentration. (Also called *diffusivity*.)

diffusion of light

The scattering of light by reflection or by transmission.

diffusion potential

When the oppositely charged ions of an electrolyte, having different mobilities, diffuse through a solution or across a membrane because of concentration differences, the mobility difference results in a separation of charge. This in turn produces electric potential differences, known as diffusion potentials, which act to accelerate the slower and decelerate the more rapid ions so as to reduce the charge separation.

diffusionometer

An apparatus for ascertaining the rate at which diffusion of gases takes place.

diffusivity

A synonym for *diffusion coefficient.*

digestion

The maintenance of a fluid mixture for a length of time at a state of moderate heat until the mixture "ripens."

digit

Any of the integers 1, 2, 3, 4, 5, 6, 7, 8, 9, 0, as occurring in a number.

digital computer

See *computer.*

dihedral angle

An angle formed by two planes (the *faces* of the dihedral angle) which intersect in the *edge* of the dihedral angle.

dihedral angle of a polyhedral angle

Any of the dihedral angles formed by two consecutive *faces* of the polyhedral angle.

dilatometer
An instrument for measuring the expansion of contraction of a solid caused by changes in temperature or by chemical agencies.

diluent
An inert liquid or solid, used to increase the bulk of another substance of a solution.

dilute
Not concentrated.

dilute solution
A chemical solution which contains a relatively large quantity of the *solvent* and a small one of the *solute.*

dilution
(1) The reduction of the *concentration* of a solution. (2) The reciprocal of concentration; the volume of solvent in which a unit quantity of solute is dissolved.

dilution law
See *Ostwald's law.*

dilution ratio
The volume of diluent divided by the volume of solvent.

dimension
Extension in space. Any space is considered to have as many dimensions as the number of coordinates required to determine the position of a point in it. The three dimensions of Eucledian space are: *length, width* (or *breadth*) and *thickness* (or *height* or *depth*).

dimension of a term
The sum of the powers of each letter of the term, as, abc is said to be of three dimensions or of the third degree; or a^2bc^3 is said to be of six dimensions or of the sixth degree. A numeral coefficient is not counted, as: $-6a^3b^2$ is of five dimensions.

dimensional formula
Any formula expressing the relation of physical quantities and units of measurement to the units of length, mass and time.

dimer
A combination of two identical molecules; a *polymer* of two molecules.

dimolecular
Pertaining to, consisting of or involving two molecules. Also called *bimolecular.*

dimolecular reaction
A reaction in which two molecules participate and undergo a change. Also called *bimolecular reaction* and *reaction of the second order.*

dimorphic
Possessing the property of *dimorphism;* said, for instance, of substances which may exist in either of two different crystal systems.

dimorphism
The condition of having two different forms.

dimorphous
See *polymorphism.*

diode
A two-element vacuum tube containing an anode and a cathode.

diopter
A unit used for expressing the refractive power of a lens or prism. In a lens or lens system, the diopter is the reciprocal of the focal length stated in meters. In a prism, one diopter (called one *prism* diopter) is the power required to deviate a ray of light by one centimeter at a distance of one meter from the prism.

dip of horizon
The angle of depression of an object below the horizon due to the elevated position of the observer.

dipole
(1) A system consisting of two electric charges equal in magnitude, but opposite in sign. (2) A system consisting of two opposite magnetic poles of equal magnitude. (3) Short for dipole antenna.

dipole ion
See *zwitterion.*

dipole moment
The product of one of the charges of a dipole unit by the distance between them.

direct current (D-C)
A unidirectional current in which the changes in magnitude are either zero or so small that they may be neglected. Also called *continuous current.*

direct proportion
The mutual relationship of two variables the quotient of which is always a constant.

direct variation
See *directly proportional.*

direct-vision spectroscope
A *spectroscope* that employs a train of prisms so arranged as to make the collimator, prism and view telescope lie along the axis of a single tube housing the entire system. A diffraction grating may also be used.

directed angle
An angle stated to be positive or negative.

directed line
A line the direction from one end of which to its other end is specified to be positive, the reverse direction being negative. Also referred to as *sensed line.*

directed line segment
See *directed line.*

directed number
A number that has a sign, indicating that it is positive or negative, thus indicating the direction in which it is to be measured from the zero point. Also called *signed number.*

directing curve
The fixed line with reference to which another line moves to generate a space locus (cylinder, cone, etc.). Also called *directrix.*

direction angles
The smallest angles formed by a line with the positive directions of the axes of the coordinate system.

direction cosines
The cosines of the *direction angles.*

direction numbers
Any two numbers proportional to the *direction cosines* of a line.

direction of an electric field
At any point of the field, the direction of the force on a positive charge placed at that particular point.

directly proportional
Said of two variables, as *A* and *B,* which depend on one another in such a way that, when the value of A changes, the value of B changes in the same ratio. This means that the one quantity divided by the other quantity is always equal to the same constant; or $\frac{A}{B} = m$, so that if A changes then B changes in such a manner that $\frac{A}{B}$ always equals m.

director circle

The circle which is the locus of points from which the tangents of a central conic are at right angles. It is concentric with the ellipse and hyperbola and has a radius $\sqrt{a^2 + b^2}$. In the parabola it becomes the *directrix.*

directrix of a cone

The closed line, or base line, which determines the motion of the *generatrix.*

directrix of a conic section

The fixed straight line the distance from which bears a constant ratio to the distance from a fixed point, called the *focus;* these two distances give points on the conic sections. The directrix and focus are polar and pole for the conic.

directrix of a cylinder

The closed line, or base line, which determines the motion of the *generatrix.*

directrix of a surface

The curve or broken line which directs or guides the generation of a conical, cylindrical, prismatic or pyramidal surface. (See also *directing curve.*)

disaccharides

A group of organic compounds which yield two *monosaccharides* through hydrolysis.

discharge

The release of accumulated electric charges.

discharge tube

A generic term for all apparatus in which electric discharge occurs in a vacuum, near vacuum or gas at a low pressure, and which permits the passage of electricity through the vacuum of gas upon application of sufficient voltage.

discontinuous phase

See *dispersed phase.*

discrepancy

A difference between measures obtained through separate, but identical or comparable operations, or through different processes based on data from the same source.

discrete

Remaining separate, isolated or distinct.

discriminant

In the formula for the solution of a quadratic equation,

$$x = \frac{-b \pm \sqrt{b^2 - 4ac}}{2a},$$

the expression $b^2 - 4ac$ is called the discriminant.

disequilibrium

Lack of equilibrium between opposing forces or reactions.

disinfectant

(1) Possessing the property of killing bacterial life. (2) As a *noun,* a substance which kills bacteria.

disintegration

A process resulting in the change of a radioactive nucleus through the emission of alpha or beta particles (in certain cases, of positrons), producing a change in the atomic number of the element.

disintegration constant

A synonym for *decay constant.*

dispersed phase

In a colloidal solution, the phase consisting of the particles suspended in the liquid called the *continuous phase.*

dispersion

In chemistry, (1) the separation or scattering of fine particles in a liquid medium; (2) any mixture where one substance is very intimately intermingled with another. Most frequently a dispersion refers to a colloidal suspension.

dispersion

In physics, (1) scattering in general; (2) the separation of a radiation into component parts according to frequency, wavelength, or some other characteristic.

dispersion medium

The continuous phase of a *colloidal solution.*

dispersive lens

See *divergent lens.*

dispersive power

The ratio $\frac{\mu_1 - \mu_2}{\mu - 1}$ where μ_1 and μ_2 are the *indices of refraction* of the medium in question for wavelengths 1 and 2, respectively, and μ is the average of μ_1 and μ_2.

dispersoid

A colloidal or finely dispersed substance. In metallurgy, finely scattered constituents, relatively insoluble, present in the microstructure of an alloy.

displacement

In physics, a vector quantity indicating the change of position of a moving body (especially one in a periodic motion) from its position of equilibrium.

displacement

(1) In chemistry, the removal of an element, molecule or

radical by another element, molecule or radical. (2) See *ionic displacement.*

displacement law

See (1) *Soddy Fajans displacement law,* (2) *Wien's displacement law.*

displacement series

A synonym for *electrochemical series.*

display

In *telemetry,* an occasionally used synonym for *recorder.*

dissociation

The breakdown of a substance into two or more others. The breaking up of the constituent molecules of a substance into ions when the substance goes into solution.

dissociation constant

For a dissociated compound at equilibrium, the product of the activities of anion by the activity of cation, divided by the activity of undissociated compound, is a constant for a definite temperature, called the *dissociation constant, ionization constant* or *equilibrium constant.*

dissociation point

The temperature at which the breaking up of a substance into two or more other substances takes place.

dissonance

(1) Discord. (2) A combination of sounds so unrelated harmonically that they produce beats.

distance formula

The formula defining the length of a straight line joining two given points in a plane or in space. For any two points, (x_1, y_1) and $(x_2\ y_2)$ in the xy plane, the distance

$d = \sqrt{(x_2 - x_1)^2 + (y_2 - y_1)^2}.$

For any two points, (x_1, y_1, z_1) and (x_2, y_2, z_2), in space, the distance $d = \sqrt{(x_2 - x_1)^2 + (y_2 - y_1)^2 + (z_2 - z_1)^2}.$

distillate

Liquid obtained by a condensation of the vapor in the process of *distillation.*

distillation

The process of removing impurities from a liquid by heating it to its boiling point and then causing its vapors to condense again.

distortion

In optics, a term synonymous with *lens distortion.* Frequently used, loosely, for any aberration or lens defect in general.

distribution coefficient

The constant ratio mentioned in the *distribution law.* Also called *partition coefficient.*

distribution graph

A graph showing the relation of parts to a whole and to each other. (A good example is the circular graph in which relative percentages of quantities are represented proportionally by sectors of the circle.)

distribution law

If a substance is added to a system of two mutually totally or nearly immiscible liquids, in both of which it is soluble, it distributes itself between the two by dissolving in each in fixed proportions, independent of the quantity of the solute, the ratio of its concentrations in the two solvents being constant at a constant temperature; this ratio (called the *distribution coefficient* or *partition coefficient*) is constant and equals the ratio of the values of the solubility of the substance in each of the two solvents. (Also called *partition law* and *Nernst theorem* or *law of distribution.*)

distributive law of multiplication

The multiplication of a compound expression by a factor is the sum of the partial product of each term of the expression by the factor. Thus $(a + b + c)d = ad + bd + cd$.

divalent

In chemistry, having a *valence* of two, i.e., capable of combining with two atoms of hydrogen or their equivalent.

divergent

Receding progressively farther and farther.

divergent lens

A lens which is thinner in the middle than at the periphery and causes parallel light rays to diverge, i.e., a beam or parallel rays to spread out, so that no real focus is obtained. Also called *divergent lens, concave lens, dispersive lens* and *negative lens.*

divergent series

A series the sum of *n* of the terms of which does not approach a definite limiting value. That is, the series is said to diverge, or to be divergent when $\lim_{n \to \infty} S_n$ fails to exist or is infinite.

diverging lens

See *divergent lens.*

divide

To separate a number or quantity or thing into two or more quantities or pieces.

dividend

The number which is divided in the process of division. In a fraction it is the numerator.

division

The process of finding one of two numbers when their product and the other number are given. Division is usually indicated by the symbol ÷ or by a fraction whose numerator is the *dividend* and whose denominator is the *divisor.* (Cf. *compound division.*)

divisor

The number or quantity by which another number or quantity is divided. It is the denominator in a fraction. Thus 5 is the divisor of 20, which gives an answer or quotient of 4; or $20/5 = 4$.

dodecagon

A polygon bounded by twelve straight lines.

dodecahedron

A polyhedron with twelve faces. (In a *regular dodecahedron,* the twelve faces are regular pentagons, and the polyhedral angles are congruent.)

domain

A bounded region of uniform or unidirectional magnetization or electric or magnetic polarization within a ferroelectric or ferromagnetic body.

domain of a relation or function

The set of all first elements of the set of ordered pairs forming the relation or function. (Cf. *range of a relation or function.*)

Doppler effect

The apparent or observed frequency of a wave motion changes as the distance between its source and the observer changes, regardless of whether the source is stationary and the observer approaches it or recedes from it or the source approaches or recedes from the stationary observer. If the distance between source and observer decreases, the frequency of the wave increase, and if the distance increases, the frequency of the wave decreases. This effect applies to sound, light, as well as the other electromagnetic radiations.

Doppler-Fizeau principle

The principle of the *Doppler effect* as applied by Fizeau to explain the shifting of the spectral lines.

Doppler radio

An apparatus that utilizes the *Doppler shift* manifested in the shift in the frequency of radio waves for tracking missiles and rockets in flight.

Doppler shift

The change in the observed frequency of a wave produced by the *Doppler effect,* expressed in cycles per second.

dosemeter

An instrument for measuring the quantity or dosage of radiation. Also called *dose meter* and *dosimeter.*

dosimeter

See *dosemeter.*

dot product

See *scalar product.*

double-angle cosine formula

$\cos 2\theta = \cos^2\theta - \sin^2\theta$.

double-angle formulas

Formulas expressing the trigonometric functions of twice a given angle in terms of functions of that angle itself. The most important of these formulas are: the *double-angle sine formula, the double-angle cosine formula,* and the *double-angle tangent formula.*

double-angle sine formula

$\sin 2\theta = 2 \sin \theta \cos \theta$.

double-angle tangent formula

$$\tan 2\theta = \frac{2 \tan\theta}{1 - \tan^2\theta}$$

double cusp

See *tacnode.*

double law of the mean

See *extended law of the mean.*

double point

A point on the common base of two homographic ranges which corresponds to itself for the two ranges. There may be two or more double points which may be real, coincident or imaginary. When a point corresponds to itself on the base of an involution.

double point of a curve

A point where the curve crosses itself or becomes tangent to itself.

double ray

In mathematics, a line through the vertex which corresponds to itself when two homogeneous pencils have a common vertex.

double refraction

A phenomenon, also called *birefringence,* observed when light passes through certain crystals; the light is split into two component beams, called the *ordinary ray* and the *extraordinary ray,* which undergo polarization at right angles to each other, propagate at different velocities within the crystal, and generally in different directions.

double replacement

A chemical reaction between two compounds, involving the *displacement* and *replacement* of two elements, molecules or radicals and resulting in the formation of two new compounds.

double salt

A molecular combination of two salts which crystallize together from solutions, but decompose into the two constituent salts when dissolved: the combination of two metal atoms and an acid radical or of one metal and two acid radicals.

double vision

A fault in a binocular instrument, caused by the non-parallelism of the optical axes of the two telescopes, resulting in the formation of two separate images instead of one sharply defined image.

doublet

A pair of spectral lines which differ so little in wavelength than they are likely to appear as one line to the casual observer.

doublet

In chemistry, two electrons shared by two atoms, and representing a single nonpolar valence bond.

drawing

The process of pulling metal, generally through diamond dies, in order to make wire.

dry analysis

Chemical analysis by *spectrography,* i.e., without the use of reagents in solutions.

dry cell

A cell in which the electrolyte is in the form of a jelly or absorbed in a porous medium or otherwise restrained from flowing from its intended position, so that it is non-spillable and the cell is completely portable.

dry ice

Solidified carbon dioxide (CO_2).

duality
The principle that all theorems relating to points or lines have a reciprocal theorem by substituting line for point, and point for line. Thus: Two lines determine a point, and two points determine a line. (Also referred to as *reciprocity*.)

ductility
A measure of that property of a solid which permits it to be drawn into a wire.

duodecagon
A regular polygon bounded by twelve straight lines of equal length; all the interior angles are 150°.

duodecimal system
The number system using 12 as its *base*. Also called *duodenary system*.

duodenary system of scale of notation
See *duodecimal system*.

duplicate ratio
The result of compounding a ratio with itself once. E.g., the ratio $a:b$ multiplied by itself becomes the duplicate ratio $a^2:b^2$.

Dy
The symbol for the element *dysprosium*.

dyad
A chemical element having a *valence* of two.

dye
A substance possessing color and capable of changing the color of fibers when absorbed by them from solutions.

dynactinometer
An instrument for measuring the intensity of photogenic (light-producing) rays and computing the power of object-glasses.

dynamic balance
The balance of a body which is rotating or in motion.

dynamic equilibrium
The state produced in a system when two opposing forces are operating in it or acting on it simultaneously.

dynamic geology
The science of the causes and processes of geological change.

dynamic load
An alternating or variable load.

dynamic strength
Resistance to impact or vibratory stress.

dynamic stress

In the testing of materials, stress which is suddenly applied and thus tends to produce motion in the part tested.

dynamics

The study of masses and forces. The study of the effect of forces in causing or modifying the motions of masses and producing strains in elastic bodies. It is usually treated under two heads: *statics* and *kinematics*.

dynamo

A machine which converts mechanical energy into electrical energy.

dynatron

An obsolete name for *meson* (q.v.).

dyne

The unit of force in the CGS system; 1 dyne is the force which imparts an acceleration of 1 cm. per second per second to a mass of 1 gram; 1 dyne is equal to a weight of approximately $\frac{1}{981}$ gram.

dysprosium

A trivalent metallic element, one of the rare earths, discovered by Boisbaudran in 1886. Atomic number 66, atomic weight 162.46. Symbol: Dy.

E

E

The symbol for the element *einsteinium.*

e

The base of the *natural* or *Napierian logarithms.* Its numerical value is 2.7182818284 . . .

e, e^-, $_{-1}e^\circ$

Symbols for *electron.*

e^+, $+e^\circ$

Symbols for *positron.*

e-center

See *excenter.*

E layer

See *ionosphere.*

e/m

The symbol of *specific electronic mass.*

ebullioscope

An instrument for observing the boiling points of liquids, and in particular for determining the alcoholic strength of a mixture by the temperature at which it boils.

eccentricity

In a conic, the ratio of the distance of any of its points from a fixed point (the *focus*) to the distance of the same point from a fixed line (the *directrix*). This ratio is always less than 1 in an ellipse, always equal to 1 in a parabola, and always greater than 1 in a hyperbola.

echelon

See *echelon grating.*

echelon grating

A diffraction grating, consisting of 20-40 parallel plates of equal thickness arranged in optical contact, staggered to form a row of steps about one millimeter wide each. The gratings of this type have very high powers of resolution, ranging as high as 1,000,000.

echo

(1) In general, a wave which has been reflected at one or more

points in the transmission medium, with sufficient magnitude and time difference to be perceived in some manner as a wave distinct from that of the main transmission. (2) In acoustics, the phenomenon produced by the return of a reflected sound wave to the observer 1/10 of a second or more after the original, unreflected wave, resulting in his perceiving a distinct repetition of the original sound.

echo depth-sounding
See *sonar*.

echo ranging
The determination of the distance and direction of underwater objects by the same process used in *sonar*.

eclipse
The obscuration of one celestial body by another. It is usually restricted to the sun, moon and satellites of other planets. (The hiding of a star by the moon or a planet is called occultation.)

eclipse of the moon
An obscuration of the moon by the shadow of the earth when the earth is between the sun and the moon and on a straight line with both.

eclipse of the sun
A partial or complete obscuration of the sun by the shadow of the moon when the moon is between the sun and the earth and on a straight line with both.

ecliptic
The apparent path of the sun among the stars during a year; it is the projection of the plane of the earth's orbit on the celestial sphere. It is a great circle inclined $23\frac{1}{2}°$ to the celestial equator. It intersects the equator at two points, the *equinoxes;* the points where it is farthest from the equator are the *solstices.*

economic geology
The branch of geology which deals with the application of the science in industrial relations and operation; the application of geological methods to problems in civil and mining engineering and the production of mineral resources.

ecosphere
In geophysics, the layer inhabited by or suitable for the existence of living organisms.

eddy
The more or less circular motion produced by an obstruction in the path of a moving fluid.

eddy conductivity
The transfer of a physical property, such as heat by eddies, from one layer to another of a fluid in turbulent motion.

eddy currents
Electric currents induced in surrounding masses of conducting material by circuits carrying alternating currents, or by a varying magnetic field. (Referred to also as *Foucault currents.*)

edge of a dihedral angle
The line of intersection of the *faces* of the dihedral angle.

edges of a polyhedral angle
The edges of the *dihedral angles of the polyhedral angle.*

edges of a polyhedron
The lines of intersection of the *faces.*

Edison effect
The fundamental principle of the operation of the *vacuum tube,* namely the flow of a current through a vacuum tube containing a metal plate and a heated filament when and only when the plate is connected to the positive side of the battery. (See *thermionic emission.*)

effective atmosphere
That part of the atmosphere which effectively influences a particular process or motion, its outer limits varying according to the terms of the process or motion considered. (Cf. *sensible atmosphere.*)

effervescence
The formation of froth or bubbles on the surface of a liquid owing to the escape of gases (not as a result of boiling).

effervescent
Emitting bubbles of escaping gas.

efficiency
The efficiency of a device with respect to a physical quantity which may be stored, transferred or transformed by the device is the ratio of the useful output of the quantity to its total input. Unless specifically stated otherwise, the term "efficiency" means efficiency with respect to power. (See also *current efficiency.*)

efflorescence
The property possessed by certain crystals of giving up their water of crystallization to the air.

efflorescent
(1) Capable of *efflorescence.* (2) In the state of efflorescence.

effusion

The passage of gases through porous bodies, i.e., the flow or escape of gases through holes larger than those to the passage through which the term *diffusion* would be properly applicable.

eikonometer

An optical apparatus for determining the magnifying power of a microscope or for the measurement of microscopic objects.

Einstein shift

A shift toward the red in the lines of the spectrum of light which according to the theory of relativity has its frequencies reduced slightly upon emerging from a strong gravitational field, such as that of a dense star.

einsteinium

A transuranic element, not found in nature. Atomic number 99, mass number of the most stable isotope known 254. Discovered in 1952; produced by bombardment of uranium ($_{92}U^{238}$) by nitrogen nuclei. Symbol: E.

Einstein's law of the equivalence of mass and energy

This fundamental principle was announced by Einstein in 1905, as follows: $E = mc^2$ (i.e. $m = \frac{E}{c^2}$), where E is the energy, m is the mass, and c is the speed of light in vacuum. From this equation it follows that mass and energy are interconvertible aspects of the same property, and when a body gives up energy, it also loses mass, i.e. weight. This principle holds true in all cases, but in ordinary chemical or molecular processes the change in mass is too microscopically small to be measured or even observed. In the case of nuclear energy, however, it is of an order sufficiently great to be measured.

Einstein's law of the photochemical equivalent

In a chemical reaction induced by exposure to light, every molecule which undergoes the reaction absorbs one *quantum* of radiation.

elastance

The measure of the difficulty of producing an *electric displacement* in a capacitor. It is the reciprocal of *capacitance,* and is measured in *darafs* (the daraf is the reciprocal of the *farad*).

elastic

Said of a substance capable of regaining its original, normal shape altered by a deforming force upon removal of that force.

elastic collision

A term applied to a collision of two particles resulting in a mutual deflection with no loss of energy, i.e. a collision in which none of the translational kinetic energy is transformed into internal energy of either particle.

elastic deformation

A deformation in which the body does not acquire a permanent set, but regains its original shape or volume when the deforming force is removed.

elastic failure

The permanent distortion of a body or piece of apparatus to a degree which prevents it from functioning properly.

elastic fatigue

The temporary loss of perfect elasticity shown by certain materials which having been subjected to deforming stress for long periods, do not return to their original form on the removal of the stress, but slowly recover their elasticity.

elastic lag

The failure of a solid body to regain its original shape and size upon removal of the deforming which produced a change in its shape and size.

elastic limit

The smallest stress that produces permanent deformation of a body, i.e., after exposure to which the body is no longer able to regain its original shape and size (and is often said to have acquired a *permanent set*).

elastic modulus

The ratio, within the limit of elasticity, of the stress to the corresponding strain. For stresses in tension and compression, it is called *Young's modulus;* for stresses involving change of shape with no change in volume, it is referred to as *shear modulus* or *modulus of rigidity;* for stresses involving change of volume without change of shape, it is designated *bulk modulus.*

elastic range

See *range of proportionality.*

elastic stiffness

Resistance to deformation within the elastic range.

elasticity

That property of a body which enables it to resist deforming forces and to regain its original shape after those forces are removed.

electric

Containing, producing, arising from, actuated by, or carrying electricity, or designed to carry electricity and capable of so doing. Examples: electric energy, motor, vehicle, wave. Compare "electrical." Some dictionaries indicate these terms as synonymous but usage in the electrical engineering field has in general been restricted to the meaning given in the definition above. It is recognized that there are borderline cases wherein the usage determines the selection. (From "American Standard Definitions of Electrical Terms"—1941).

electric cell

An apparatus which transforms chemical energy into electrical energy (or electrical energy into chemical energy). It is a combination of two electrodes immersed in an electrolyte; an electromotive force is set up, and if the two electrodes are connected externally, an electric current will flow.

electric charge

An excess or shortage in the number of electrons. An excess number of electrons is a *negative* electric charge, the shortage of an electron is a *positive* electric charge. Substances are said to possess electric charge when (1) forces exist between them; (2) they are acted upon by forces when moving in a magnetic field which possesses a component at right angles to their direction of motion.

electric circuit

A conductor or a system of conductors so arranged and connected as to permit an electric current to flow through it.

electric conductance

Ability to conduct an electric current. The conductance of a substance is the reciprocal of its resistance. It is expressed in *mhos*.

electric conductivity

The facility with which a substance permits electricity to be transmitted through it, measured in terms of the current density per unit potential gradient in the direction of the flow.

electric current

The flow of charges between two points between which there is a difference in potential. An electric current through a surface is the time rate at which positive or negative electricity passes through the surface. If both positive and negative electricity simultaneously pass through the surface, it is the time rate of passage of the algebraic sum of the two.

electric displacement
Electric flux density or electric induction.

electric eye
The popular name for a photoelectric cell.

electric field
(1) The area of space surrounding an electrically charged body in which the forces due to the charge are detectable. (2) The electric component of the electromagnetic field associated with electrons in motion and with radio waves.

electric field intensity
The property of a radio wave expressed in terms of the electromotive force which it produces. (Also called *electric field strength,* or simply *electric intensity*.)

electric flux
The quantity of electricity displaced across a given area.

electric force field
See *electric field*.

electric generator
A machine which transforms mechanical power into electric power.

electric induction
The process which electrifies a body as a result of its being merely placed in an electric field. (Also called *electrostatic induction*.)

electric intensity
See *electric field intensity*.

electric motor
A machine which transforms electric energy into mechanical energy.

electric potential
The electric potential of a point is the work done in moving a unit charge against the forces of the field from a point of zero potential to the point in question. A point which has a potential higher than a zero surface is said to have a positive potential, one having a lower potential is said to have a negative potential.

electric resistance
Resistance to the flow of electricity; it is the value of the ratio of potential difference at the two ends of a conductor to the current which flows through it.

electric steel
Steel made by one of the electric processes where the heat re-

quired for melting the steel is provided by an electric arc made to pass between the metal itself and a carbon electrode, or by eddy currents induced by a high-frequency current. In both processes the furnace lining many be either acid or basic.

electrical

Relating to or associated with electricity, but not having its properties or characteristics. (Cf. *electric*.)

electrical capacity

See *capacitance.*

electrical degree

The 360th part of the angle subtended, at the axis of a machine, by two consecutive field poles of like polarity. One mechanical degree is thus equal to as many electrical degrees as there are pairs of poles in the machine.

electrical resistivity

See *specific resistance.*

electricity

A physical agent pervading the atomic structure of matter and characterized by being separable, by the expenditure of energy, into two components designated as positive and negative electricity, in which state the electricity possesses recoverable energy.

electrochemical

Relating to or involving electrochemistry; relating to or involving both electricity and chemistry.

electrochemical cell

A vessel containing two electrodes in an electrolyte, capable of delivering an electric current created by the chemical reaction taking place in it.

electrochemical equivalent

An electrochemical equivalent of an element, compound, radical, or ion is the weight of that substance involved in a specified electrochemical reaction during the passage of a specified quantity of electricity, such as a faraday, ampere-hour, or coulomb (usually one coulomb).

electrochemical equivalent of an ion

The weight of the ion which is liberated or deposited by one coulomb of electricity; numerically equal to 1/96490 of the *chemical equivalent.*

electrochemical series

A tabular arrangement of the elements in the order of the *electrode potential* developed when an element is immersed in a solution of normal ionic concentration. (Also called *electromotive, e.m.f., galvanic* or *potential series.*) The sign is positive for elements the potentials of which are cathodic to hydrogen, and negative for those with potentials anodic to hydrogen. The order of the elements varies slightly under different conditions of measurement.

Metal at 25°C	Normal Electrode Potential (potential in salt solution of normal ionic activity relative to the normal hydrogen electrode as arbitrary zero point) *Volts*	
"Noble" or Cathodic End		
Gold	−1.42	
Platinum	−1.2	
Silver	−0.7995	
Mercury	−0.7986	
Copper	−0.3448	
Hydrogen (1 atm)	±0.0000	
Lead	+0.126	
Tin	+0.136	
Nickel	+0.250	
Cadmium	+0.4020	
Iron	+0.440	
Chromium	+.71	Potential almost independent of original concentration of metal ions (Chromium to Lithium)
Zinc	+0.7620	
Aluminum	+1.67	
Magnesium	+2.34	
Sodium	+2.712	
Calcium	+2.87	
Potassium	+2.922	
Lithium	+3.02	
"Base" or Anodic End		

electrochemistry
That branch of physical chemistry which deals with the interrelations of electricity and chemistry and with the reciprocal transformations of electrical and chemical energy.

electrode

A conductor belonging to the class of metallic conductors, but not necessarily a metal, through which a current enters or leaves an electrolytic cell, arc, furnace, vacuum tube, gaseous discharge tube, or any conductor of the non-metallic class. Specifically, in an electrolytic cell, an electrode is a conductor of the metallic-conductor class, at which there is a change from conduction by electrons to conduction by ions or colloidal particles.

electrode holder

A device for holding the electrode and conducting current to it.

electrode potential

The potential developed on an inert metal, such as platinum, when it is inserted in a reversible oxidation-reduction system.

electrodeposition

The process of depositing a substance upon an electrode by electrolysis. Electrodeposition includes electroplating, electroforming, electrorefining, and electrowinning.

electrodissolution

The process of dissolving a substance from an electrode by electrolysis.

electrodynamic instrument

An instrument which depends for its operation on the reaction between the current in one or more moving coils and the current in one or more fixed coils.

electrodynamics

That branch of physics which treats of the forces and energy transformations related to electric currents and the magnetic fields associated with them.

electrodynamometer

An instrument in which the mechanical reactions between two parts of the same circuit are used for detecting or measuring an electric current.

electroencephalograph

An apparatus for detecting, measuring and recording the wave forms of voltages developed in the brain.

electro-endosmosis

Endosmosis produced by the application of an electric potential.

electroextraction

The direct recovery by electrolysis of metal from a solution of its salts.

electrogalvanizing

The *electrodeposition* of a protective coating of zinc on metal objects.

electrogravity

A proposed designation of the suggested field of research on the possibility of a direct control of gravity.

electrojet

A concentrated stream of electricity of limited width, believed to be produced by an intensification of atmospheric electric sheet currents usually present over the magnetic poles, and they form the boundaries of the zones of the *aurorae*. There is a strong electrojet flowing westward on the morning side of the earth, and a weaker one flowing eastward on the evening side, with a zero-current area where the two meet.

electrokinetics

That branch of science which deals with the laws of electricity in motion. The study of motion set up by an applied electromotive force, and of the production of an e.m.f. by motion such as the flow of liquids.

electrolysis

The production of chemical changes, in particlar the separation of the components of liquefied or molten compounds, by a current passing through an *electrolyte* (q.v.). The positively charged ions migrate to the negatively charged electrode, the ions carrying a negative charge migrate to the positively charged electrode.

electrolyte

(1) A conducting medium in which the flow of electric current is accompanied by the movement of matter. (2) A substance which when dissolved in a specified solvent (usually water) produces a conducting medium. (3) An acid, base, or salt, which when dissolved in water, dissociates into electrically charged particles called ions. The solution can conduct an electric current. (See also *Debye-Huckel theory of electrolytes.*)

electrolytic cathode

An electrode at which positive ions are discharged or negative ions are formed, or at which other reducing reactions occur.

electrolytic cell

A unit apparatus designed for carrying out an electrochemical reaction and including a vessel, two or more electrodes, and one or more electrolytes, in which two electrodes, the *anode* and the *cathode,* are immersed for the purpose of producing an electrochemical action or enabling it to be transmitted.

electrolytic conduction

The flow of electricity in the form of the movement of ions in solution, involving a chemical reaction at the electrodes.

electrolytic deposition

The process of producing a metal from a solution of its salts by passing an electric current through it.

electrolytic dissociation

The phenomenon which occurs when an acid, a base, or a salt is dissolved in water or any other electrolyte. Some or all of the molecules of the dissolved substance are broken up into an equal number of *cations* (ions charged with positive electricity) and *anions* (ions with negative electric charges).

electrolytic reduction

The removal of positive charges from, or the addition of electrons to, an atom or ion; this process occurs in an electrolytic cell at *cathode* only.

electrolyzer

An electrolytic cell for the production of alkali, chlorine, hypochlorites or other allied products.

electromagnet

A temporary magnet formed by winding a coil of wire around a piece of soft iron (called the *core*), the iron becoming magnetic when an electric current flows through the wire.

electromagnetic constant

A constant denoted in expressions of relations between electric and magnetic quantities as *c*; its numerical value is 2.99776×10^{10} cm. per second, the speed of propagation of electromagnetic waves.

electromagnetic energy

Energy created by a rapidly moving electric field which sets up a coincident, moving magnetic field at right angles to the lines of electric force and to their direction of motion. (This is the condition which is believed to exist in the space traversed by electromagnetic radiation.)

electromagnetic force

Equal to $N \frac{d\varphi}{dt}$ where N equals the number of turns and $\frac{d\varphi}{dt}$ equals the change in flux per unit of time.

electromagnetic induction

The production of a voltage in a coil due to a change in the number of magnetic lines of force (flux linkages) passing through the coil.

electromagnetic radiation

Gamma (γ) rays (wavelength between 10^{-10} and 10^{-9} cm.), x-rays (wavelength between 10^{-9} and 10^{-7} cm.), ultraviolet radiation (wavelength between 10^{-7} and 4×10^{-5} cm.), visible light (wavelength between 4×10^{-5} and 7.7×10^{-5} cm.), infra-red waves (wavelength between 7.7×10^{-5} and 0.4 cm.), and Hertzian waves (wavelength between 0.4 and 3×10^{6} cm.).

electromagnetic spectrum

A chart or graph showing the relation of all known electromagnetic wave forms classified by wavelength. The visible light spectrum occupies an extremely minute portion of the electromagnetic spectrum. (See *electromagnetic radiation.*)

electromagnetic wave

A wave in which there are both electric and magnetic displacements. (See *electromagnetic radiation.*)

electromagnetics

That branch of physics which treats of the mutual relationships existing between electric currents and the attendant magnetic fields.

electromechanics

That branch of electrical engineering which deals with machines producing or operated by electric currents.

electrometallurgy

The application of the principles, methods and processes of *electrochemistry* and *electrothermics* to metallurgical operations.

electrometer

An instrument for detecting or measuring potential difference by means of the mechanical or electrostatic forces exerted between electrically charged bodies.

electromotive force

(1) That which causes current to flow. (2) The energy added per unit charge due to the mechanical or chemical action producing the current. (3) The open circuit difference in potential between the terminals of a cell or generator. (4) The property of a physical device which tends to produce an electric current in a circuit. Measured in *volts.*

electromotive series

See *electrochemical series.*

electron (e, e^-, $_{-1}e^0$, β)

A fundamental particle, discovered by J. J. Thomson in 1897. It is the smallest known particle having a negative charge capable of isolation and measurement, and all charges are exact multiples of this "unit of electric charge" which has a magnitude of 4.8025×10^{-10} e.s.u. (equal to 1.6023×10^{-20} e.m.u.). Atoms are composed of a nucleus and a certain number of orbital electrons circling about it. A shortage of electrons is called "positive electricity" or "positive electric charge," whereas an excess of electrons is called "negative electricity" or "negative electric charge." Electric currents are produced by mass movements of electrons in electric conductors.

electron emission

The liberation of electrons from an electrode into the surrounding space. Quantitatively, it is the rate at which electrons are emitted from an electrode.

electron gun

An electrode structure which produces and may control, focus, and deflect an electron beam.

electron microscope

A microscope using electrons instead of light rays and electromagnetic fields instead of glass lenses. It sends a beam of electrons through a thin sample of the material examined, magnifies the resulting shadow caused by denser portions of the sample, and makes this shadow visible on a fluorescent screen or records it on photographic film. Magnifications up to 100,000 diameters are obtainable, which makes the electron microscope from 50 to 100 times more powerful than the strongest optical microscope. It has important applications in chemistry, biology, and metallurgy.

electron optics

That branch of electronics which deals with the control of electron beams in a vacuum by means of electric or magnetic fields.

electron shell

See *electron structure, shell.*

electron structure of atoms

According to the currently accepted views and theories of atomic structure, the atom consists of a central nucleus with a certain number of electrons circling around it. The number of these electrons (called *orbital* or *planetary* electrons) equals the atomic number of the element. They do not circle around the nucleus all in orbits of the same radius, but are arranged in various *shells,* which are designated by the letters *K, L, M, N, O, P* and *Q,* in this order, the *K*-shell being the shell nearest to the nucleus, the *L*-shell the one immediately next to the *K*-shell, etc. The electrons in the same shell may be arranged in various *subshells,* which are designated by the letters *s, p, d, f, g* and *h,* in this order.

electron telescope

An apparatus, operating on the same basic principle as the *electron microscope;* it focuses the infra-red image of an object on a photosensitive cathode, to produce an enlarged image on a fluorescent screen.

electron volt (e.v., ev)

The basic energy unit in nuclear physics: 1 electron volt is the kinetic energy acquired by an electron in falling freely through a potential drop of 1 volt; it is the equivalent of 1.6020×10^{-12} erg. Since the electron volt represents such a minute amount of energy, it is customary to calculate in units of million electron volts; 10^6 ev = 1 mev = 1.6020×10^{-6} erg. 10^9 ev = 1000 mev. = 1 bev.

electronate

To add one or more electrons to an atom; a synonym for *reduce.*

electronegative

Having the tendency to form negative ions, or carrying a negative electric charge.

electronegative potential

A potential corresponding in sign to potentials of the "base" or anodic members of the *electrochemical series.*

electronic

Pertaining to or functioning through electrons.

electronic charge (e)

The natural unit or quantum of electric charge, also called *elementary electric charge* and *elementary quantum of electricity;* its exact magnitude was ascertained by R. A. Millikan, through his oil-drop experiments, as 4.8025×10^{-10} e.s.u. (equal to 1.60203×10^{-20} e.m.u. or 1.60×10^{-19} ampere-second) .

electronic instrument

An instrument which utilizes for its operation the action of an electronic (thermionic) tube. Also called "thermionic instrument."

electronics

That branch of science and technology which relates to the conduction of electricity through gases or in vacuo. The science of the behavior of free electrons.

electroosmosis

Osmosis in an electric field.

electrophoresis

The movement, produced by the application of an electric potential, of fine particles suspended in a liquid, toward the anode (*anaphoresis*) or toward the cathode (*cataphoresis*).

electropositive

Having the tendency to form positive ions, or carrying a positive electric charge.

electropositive potential

A potential corresponding in sign to potentials of the "noble" or cathodic members of the *electrochemical series.*

electroscope

An instrument for detecting or measuring small electric charges by utilizing the electrostatic repulsion between charged bodies to indicate a charge or potential difference. The simplest of the many types of electroscopes is the *gold-leaf electroscope.*

electrostatic induction

The process of charging an object electrically by bringing it into the electric field of a charged object. (Also called *electric induction.*)

electrostatic unit (e.s.u.)

The fundamental unit of electric charge in the C-G-S system; 1 e.s.u. is the electric charge which will repel, in a vacuum, a similar charge at a distance of 1 cm. with a force of 1 dyne, and it equals one three-billionth $\left\{\frac{1}{3 \times 10^9}\right\}$ coulomb.

electrostatic units

A system of units based on the reciprocal force of two electric charges; the fundamental unit of this system is the *electrostatic unit* of charge.

electrostatics

That branch of physics which deals with the laws and properties of electricity at rest.

electrostenolysis

The discharge of ions or colloidal particles in capillaries through the application of an electric potential.

electrostriction

In chemistry, the contraction of the solvent of a solution as a result of the attraction exerted by the ions of the solute on the water dipoles.

electrostriction

In physics: (1) the deformation of a solid when subjected to the effect of an electric field; (2) a synonym for *inverse piezoelectric effect*.

electrothermics

That branch of science and technology which deals with the direct transformations of electric energy and heat.

electrotropism

Reaction to the electric current.

electrovalence, electrovalency bond

"Electrovalence" is a term which embraces both "positive valence" and "negative valence." "Electrovalency bond" is the valency bond formed by the transfer of orbital electrons from one atom to another; the atom which gives up the electron (*positive valence*) becomes a positive ion, the one which acquires the electron (*negative valence*) becomes a negative ion, and the two ions are held together by electrostatic attraction.

electrovalent

Made up of ions.

electrowinning

The electrodeposition of metals or compounds from solutions derived from ores or other materials, using insoluble anodes.

electrum

An alloy of gold (80%) and silver (20%).

element

A substance which cannot be broken down into simpler substances by chemical means. There are 102 known elements, of which 88 are found in nature, whereas the other 14 (technetium, promethium, astatine, francium, and the 10 *transuranic* elements) have been synthethized in the laboratory or produced in nuclear reactors. Eleven elements (argon, chlorine, fluorine, helium, hydrogen, krypton, neon, nitrogen, oxygen, radon and xenon) are gases, two (bromine and mercury) are liquids, and the others are solid, at normal temperature and pressure.

List of the Elements and their International Atomic Weights

Element	Symbol	Atomic No.	Atomic Weight
Actinium	Ac	89	227
Aluminum	Al	13	26.98
Americium	Am	95	(243)
Antimony	Sb	51	121.76
Argon	A	18	39.944
Arsenic	As	33	74.91
Astatine	At	85	(211)
Barium	Ba	56	137.36
Berkelium	Bk	97	(245)
Beryllium	Be	4	9.013
Bismuth	Bi	83	209.00
Boron	B	5	10.82
Bromine	Br	35	79.916
Cadmium	Cd	48	112.41
Calcium	Ca	20	40.08
Californium	Cf	98	(248)
Carbon	C	6	12.011
Cerium	Ce	58	140.13
Cesium	Cs	55	132.91
Chlorine	Cl	17	35.457
Chromium	Cr	24	52.01
Cobalt	Co	27	58.94
Copper	Cu	29	63.54
Curium	Cm	96	(245)
Dysprosium	Dy	66	162.46
Einsteinium	E	99	(255)
Erbium	Er	68	167.2
Europium	Eu	63	152.0
Fermium	Fm	100	(252)
Fluorine	F	9	19.00
Francium	Fr	87	(223)
Gadolinium	Gd	64	156.9
Gallium	Ga	31	69.72
Germanium	Ge	32	72.60
Gold	Au	79	197.0
Hafnium	Hf	72	178.58
Helium	He	2	4.003

Element	Symbol	Atomic No.	Atomic Weight
Holmium	Ho	67	164.94
Hydrogen	H	1	1.0080
Indium	In	49	114.82
Iodine	I	53	126.91
Iridium	Ir	77	193.1
Iron	Fe	26	55.85
Krypton	Kr	36	83.80
Lanthanum	La	57	138.92
Lead	Pb	82	207.21
Lithium	Li	3	6.940
Lutetium	Lu	71	174.99
Magnesium	Mg	12	24.32
Manganese	Mn	25	54.94
Mendelevium	Mv	101	(256)
Mercury	Hg	80	200.61
Molybdenum	Mo	42	95.95
Neodymium	Nd	60	144.27
Neon	Ne	10	20.183
Neptunium	Np	93	(237)
Nickel	Ni	28	58.69
Niobium	Nb	41	92.91
Nitrogen	N	7	14.008
Nobelium	No	102	(251-253?)
Osmium	Os	76	190.2
Oxygen	O	8	16.0000
Palladium	Pd	46	106.7
Phosphorus	P	15	30.975
Platinum	Pt	78	195.23
Plutonium	Pu	94	(242)
Polonium	Po	84	210
Potassium	K	19	39.100
Praseodymium	Pr	59	140.92
Promethium	Pm	61	(145)
Protactinium	Pa	91	231
Radium	Ra	88	226.05
Radon	Rn	86	222
Rhenium	Re	75	186.31
Rhodium	Rh	45	102.91
Rubidium	Rb	37	85.48
Ruthenium	Ru	44	101.1

Element	Symbol	Atomic No.	Atomic Weight
Samarium	Sm	62	150.43
Scandium	Sc	21	44.96
Selenium	Se	34	78.96
Silicon	Si	14	28.09
Silver	Ag	47	107.880
Sodium	Na	11	22.991
Strontium	Sr	38	87.63
Sulfur	S	16	32.066
Tantalum	Ta	73	180.95
Technetium	Tc	43	(99)
Tellurium	Te	52	127.61
Terbium	Tb	65	158.93
Thallium	Tl	81	204.39
Thorium	Th	90	232.05
Thulium	Tm	69	168.94
Tin	Sn	50	118.70
Titanium	Ti	22	47.90
Tungsten	W	74	183.92
Uranium	U	92	238.07
Vanadium	V	23	50.95
Xenon	Xe	54	131.3
Ytterbium	Yb	70	173.04
Yttrium	Y	39	88.92
Zinc	Zn	30	65.38
Zirconium	Zr	40	91.22

NOTE: The figures in parentheses indicate the atomic weight of the most stable (i.e., longest-lived) isotope.

element of a determinant

See *determinant.*

element of a surface

With reference to conical, cylindrical, prismatic and pyramidal surfaces, any position assumed by the *generatrix* in the process of generating the surface.

element of an algebraic expression

One of the terms in the algebraical expression. A part of the same nature as the entire magnitude. Part of a determinant. Thus in 4ab, 4 and a and b are each elements.

elementary mathematics

That part of mathematics which includes ordinary arithmetic, algebra, geometry and trigonometry.

elementary particles

The particles which at the present time it is assumed to be impossible to resolve into further components, as the proton, neutron, electron, positron, neutrino, meson and photon or γ-ray. Although the x-particle and the deuteron are composite particles, they are usually included under this collective term, due to the important part which they play in nuclear reactions.

elements of a triangle

The three angles and the three sides.

elevation view

Any view, i.e., representation on a plane surface of a point or object in space, in which the lines of sight of the observer are horizontal or level. The front, right profile, left profile and rear views are the *common elevation views,* all others are referred to as *auxiliary elevation views.*

eliminant

In mathematics, the resultant equation or answer obtained after eliminating certain unknown values or variables from two or more equations. Also called *resultant.*

ellipse

A *conic* section formed by a plane cutting a cone obliquely to its base. It is the locus of a moving point the sum of the distances of which from two fixed points (the *foci*) is always constant. Defined also as the locus of a moving point the ratio of the distance of which from a fixed point (one of the *foci*) and from a fixed line (one of the *directrices*) is a constant (called the *eccentrity* of the ellipse) that is less than unity.

ellipsoid

The surface of a solid, all the cross-sections of which are ellipses or circles.

ellipsoid of revolution

An ellipsoid resulting when an ellipse is rotated about one of its axes.

elliptic conical surface

A conical surface the *directrix* of which is an ellipse.

elliptic cylinder

A cylinder the bases of which are ellipses.

elliptic involution

See *involution.*

elliptic paraboloid

A *paraboloid* that can be held so that those of its sections which lie parallel to one coordinate plane are ellipses, whereas those lying parallel to the other coordinate plane are parabolas.

elliptic polarization

Polarization which produces an *elliptically polarized wave.*

elliptically polarized wave

A polarized wave in which the direction of displacement at a point rotates about a point in a plane and the magnitude of displacement varies as the radius of an ellipse.

ellipticity

In an ellipse the ratio of the difference between the two semi-diameters to the major semi-diameter; it is a measure of the oblateness of the ellipse.

elution

The extraction of a solid from a mixture of solids with the aid of a liquid.

emanation

Radon and its isotopes, *actinon* and *thoron.*

emanium

An early name for *actinium.*

e.m.f.

The abbreviation for *electromotive force.*

e.m.f. series

See *electrochemical series.*

emission

The radiation of energy, or the liberation or sending out of particles.

emission spectrum

Spectrum showing the radiations emitted by a substance.

emissive power

The time rate, at a given temperature, of the total emission of radiant energy per unit area of the radiating surface. (Also called *intrinsic radiance.*)

emissivity

The ratio of the *emissive power* of a radiating surface to the emissive power of a black body at the same temperature.

empirical curve

A curve determined and drawn to conform fully or nearly with certain empirical or statistical data.

empirical equation

The equation of an *empirical curve.*

empirical formula

In chemistry, a formula which shows the species of atoms present in a molecule as well as their numerical ratio, but not their grouping.

emulsifier

See *emulsifying agent.*

emulsifying agent

A substance, also called *emulsifier,* the admixture of which to an emulsion makes the emulsion more stable and thus produces a *permanent emulsion.*

emulsion

A dispersion of one immiscible liquid in another, i.e., a suspension of oil-in-water or water-in-oil.

emulsoid

A colloidal suspension of a liquid in another liquid. Certain solids, such as gelatin, show all the properties of this class, so the term is not strictly accurate.

enantiotropic form

See *enantiotropy.*

enantiotropy

The property which enables certain substances to exist in two crystal forms which are in a reversible equilibrium with each other, one of these *enantiotropic* forms being stable above a certain temperature (the *transition point*), the other below it.

end point

That stage in *titration* at which the reaction is complete.

end product

In a radioactive series, the final element which remains stable.

end-on array of antennas

A system of antennas coupled together for directional effects, arranged in the direction of the transmission. Also called *alignment array.*

endosmosis

Osmosis into a vessel or container.

endothermic

Relating to, accompanied or characterized by the absorption of heat.

endothermic reaction

A reaction in which heat must be supplied continually to keep the reaction going.

energy

The capacity to do work. Energy takes any of several forms,

such as mechanical energy (which is either kinetic energy, i.e. energy due to motion, or potential energy, i.e. energy due to position), electrical energy, heat or thermal energy, etc. Mechanical energy is a product of the force at action and of the distance over which it acts.

energy level

One of the different states that may exist in an atom, i.e., the energy of the atom in a given state of excitation. The transition of the orbital electrons from one energy level to another produce the emission of different radiation frequencies and quanta.

engine

A machine which produces power to do work.

enhanced line

A spectral line of extraordinary or exceptional intensity, produced by a spark or some other source of exceptionally high temperature; as more and more atoms of a gas or gaseous body are ionized owing to a constant rise in temperature, more and more of these lines appear or become strengthened in its spectrum.

enthalpy

The heat content per unit mass expressed in B.t.u. per pound. When a change occurs at constant pressure, as when water is boiled, he change in enthalpy is equal to the heat added, i.e., *latent heat.*

entropy

An index of the relative amount of the unavailable energy in a physicochemical system. The concept was introduced by Clausius. The entropy of a physical system approaches its maximum as the system approaches its final state of equilibrium. Thus, entropy may be regarded as the measure of thermodynamic degeneration.

entropy function

An expression of entropy as a function of temperature.

envelope

A curve marked out by a succession of straight lines or curves drawn in a regular way. Example: All the lines drawn perpendicular to the radii of a circle which forms the circumference of the circle.

enzyme

A generic term for a group of organic, biochemical catalysts, usually of a colloid nature, endowed with specific powers of reaction, formed by living cells.

eolian

In geological terminology, this expression indicates "caused or produced by the wind." (E.g., eolian deposits are wind-blown deposits, sediments accumulated by the wind.)

epicycloid

A curve which is the locus of a point on a circle that rolls around the outside of a fixed circle while remaining coplanar with it. (Cf. *hypocycloid.*)

epitrochoid

A curve which is the locus of a point on the radius (or the extension of the radius) of a circle that rolls around the outside of a fixed circle while remaining coplanar with it.

equal

In mathematics, this term is used to indicate that two or more terms are alike in magnitude, quantity, number or degree. In geometry, it denotes exact agreement with respect to some specific property, but without necessarily implying actual *congruence.*

equal figures

Two or more geometric figures which are exactly identical in some or most respects, without being *actually congruent.* (Also called *equivalent* figures.)

equal solids

Two or more solids which have equal volumes. (Also called *equivalent solids.*)

equation

A statement of the equality between two expressions which appear on the two sides of the *sign of equation* (=).

equation of state

In chemistry, any equation which relates the pressure, volume and temperature of a gas, as well as the *ideal-gas constant.* (E.g., the *ideal-gas law, van der Waals's equation,* etc.)

equation of the first degree

See *simple equation.*

equation of the second degree

See *quadratic equation.*

equation of the third degree

See *cubic equation.*

equator

The great circle of the earth, or any heavenly body, the plane of which is at right angles to the axis.

equi-

A prefix indicating equality.

equiangular

With reference to a configuration, this adjective means "having all angles equal." When applied to two figures, it indicates that the angles of one of them are equal to the angles of the other when taken in the same order respectively.

equiangular spiral

See *logarithmic spiral.*

equiconjugate diameters

The pair of conjugate diameters of an ellipse which are equal to one another—namely those which lie along the diagonals of the rectangle formed by the tangents parallel to the major and minor axes.

equidistant

Situated at equal distance.

equilateral

Having sides of equal length.

equilateral hyperbola

See *rectangular hyperbola.*

equilateral triangle

A triangle all three sides of which are of equal length, and all three angles of which are exactly 60°.

equilibrant

The force which is necessary to balance or keep in equilibrium two or more concurrent forces. The equilibrant is equal in value to the *resultant,* but its direction is opposite to that of the latter.

equilibrium

(1) The property of a number of forces having their resultant equal to zero. (2) The property of a particle or body when the resultant of all forces acting on it is zero. (3) *Radioactive equilibrium* of a radioactive element or isotope is the state in which the rate of the formation of atoms is equal to the rate of their disintegration by radioactive decay, so that the amount of the element or isotope is constant.

equilibrium constant

See *dissociation constant.*

equimolecular mixture

A mixture which contains substances in equal molecular proportions, i.e., in the ratio of their molecular weights.

equimultiple

One of the products of two or more quantities multiplied by the same quantity. Thus 8 × 3 or 24 and 8 × 5 or 40 are equimultiples of 3 and 5.

equinoctial

That great circle of the *celestial sphere* which is everywhere 90° from the poles. It is the earth's equator projected on the celestial sphere.

equinoxes

Points at which the ecliptic intersects the equator.

equipartition of energy

A principle formulated by Boltzmann, stating that the mean kinetic energy of molecules is divided equally among the various degrees of freedom which they possess.

equipotential surface

A surface all points of which have the same electric potential.

equitangential curve

A curve the tangent of which is equal to a constant line.

equivalent

See *chemical equivalent.*

equivalent conductance

For a solution, the *specific conductance* multiplied by the volume of solution (expressed in milliliters) containing one *equivalent weight* of solute.

equivalent equations

Two or more equations in one variable that have exactly the same solution.

equivalent fractions

Two fractions, one of which is obtained from the other one by the multiplication of both the numerator and denominator of the latter by the same number. (E.g., $\frac{x}{y}$ and $\frac{nx}{ny}$.)

equivalent polygons

Polygons having the same area. The term is often used as a synonym for *congruent polygons,* and also for *equal polygons* (cf. *equal figures*).

equivalent solids

See *equal solids.*

equivalent weight

That mass of a given substance which will combine with or replace 8 grams of oxygen in a specified chemical reaction. (Also called *chemical equivalent.*)

Er

The symbol for the element *erbium.*

erbium

A trivalent metallic element, one of the group of rare earths of the periodic system, discovered by Mosander in 1843. Atomic number 68, atomic weight 167.2. Symbol: Er.

erg

The unit of work or energy in the C-G-S- system; 1 erg = 1 centimeter-dyne, i.e. the work done by a force of 1 dyne when its point of application is moved 1 cm. in the direction of the force; 1 erg = 10^{-7} joule = 7.3756×10^{-8} foot pounds.

erosion

(1) The abrasion of metal or other material by liquid or gas, usually accelerated by the pressure of solid particles of matter in suspension. (2) The wearing away of rocks and other substances by atmospheric effects or by the action of chemicals.

escape velocity

The minimum velocity which a mass or particle must possess in order to escape the gravitational attraction of a body.

escribed circle

The intersections of the bisectors of the exterior angles of a triangle are the centers of three circles each touching one side of the triangle and the prolongation of the other two sides. These circles are called *escribed circles* and their centers are excenters.

essential oils

Oils of vegetable origin, containing the odoriferous principles of vegetable organisms. Being highly volatile, they are also called *volatile oils.*

ester

An organic salt formed by the action of an alcohol on an organic acid, resulting in the dehydration of the acid.

esterification

The formation of an *ester.*

e.s.u., E.S.U.

Abbreviations for *electrostatic unit* of charge.

Et

The chemical symbol for *ethyl.*

etching

In metallurgy, a process of revealing the structure of metal by removal of the cold worked metal produced on the surface during the polishing operation, followed by selective chemical attack on the structure. The method is based on the principle that the differently oriented crystals have different rates of solution in the *etching reagents*. In some cases, etching may be brought out by a differential oxidation produced by *heat tinting*.

etching reagent

A means of revealing the structure of a metal or alloy by differential corrosion effects, thus the different constituent parts of the various grains can be distinguished. This substance is usually a solution of the reagent in water, acid or alkali.

ethanoic acid

See *acetic acid*.

ethanol

Chemists' name for *ethyl alcohol*.

ether

In physics, the medium assumed to permeate the entire space of the universe, including both the spaces between bodies and the bodies themselves, and believed to be the vehicle of propagation of light and other forms of radiation.

ether

An alkyl or aryl oxide, obtained by the replacement of the hydrogen of the hydroxyl group in alcohol by another radical, i.e., by the elimination of water.

ethyl

The movement C_2H_5—radical, derived from ethane; symbol: Et.

ethyl alcohol

Ordinary *alcohol*.

ethyl group

The monovalent alkyl radical $—C_2H_5$.

Eu

The symbol for the element *europium*.

eu-colloids

Colloids whose colloid-particles are identical with single molecules of a substance; if many molecules coalesce the colloids are "association colloids."

eudiometer

A graduated glass tube which is closed at one end and which is used for work with gases. The displacement of the liquid indicates the volume of the gas which is present.

Euler's formula (or theorem) of polyhedrons

In simple polyhedrons $E + 2 = V + F$, where E is the number of edges, V the number of vertices, and F the number of faces.

europium

A metallic element, one of the group of rare earths in the periodic table, discovered by Demarcay in 1896. Atomic number 63, atomic weight 152.0. Symbol: Eu.

eutectic

(1) A mixture of definite composition, consisting of two or more constituents, which solidifies out of the liquid at a minimum freezing point. (The essential difference between a eutectic and a *eutectoid* is that the eutectic is formed from the melt, whereas the eutectoid is formed from a solid.) (2) See *eutectic alloy, eutectic mixture.*

eutectic alloy

A mixture of two or more metals which solidifies at the *eutectic temperature.*

eutectic change

The transformation from the liquid to the solid state in a *eutectic alloy.* It involves the simultaneous crystallization of two constituents in a binary system, and of three in a ternary system.

eutectic mixture

A mixture of two or more substances which has the lowest possible melting point of any mixture of its constituents.

eutectic reaction

A reversible reaction in which a liquid solution solidifies or transforms at constant temperature to form a solid mass made up of two kinds of crystals.

eutectic structure

The characteristic arrangement of the constituents in an *eutectic* resulting from their simultaneous crystallization from the melt.

eutectic system

An alloy system in which one alloy crystallizes out at a lower temperature than that of the other constituents of the system.

eutectic temperature

The melting point of a *eutectic mixture.*

eutectoid

The structure obtained by the breakdown of a *solid solution* during cooling. A mixture of two or more constituents which forms on cooling from a *solid solution* and transforms again on heating (e.g., pearlite). The essential difference between a *eutectic* and a eutectoid is that the eutectic is formed from the melt, whereas the eutectoid is formed from a solid.

ev, e.v.
Abbreviations for *electron volt.*
evaluate
To find the exact value of the given expression.
evaporation
The process by which a substance changes from the liquid to the gaseous state, without necessarily reaching the boiling point.
evolute of a curve
The locus of the centers of the curvature of the curve. It is the curve the *involute* of which is the given curve.
evolute of a surface
The locus of centers of curvature of the surface.
excenter
The center of a circle touching the exterior side of a triangle and the other two sides prolonged beyond their intersection with that side. This center is obtained by bisecting the exterior angles formed by the side and the prolongations of the other two sides. (Also called *e-center.*)
excentral triangle
A triangle the vertices of which are the *excenters* of a given triangle.
excess of nine
The remainder after the division of a positive integer by 9; it equals the remainder obtained in the division of the sum of the digits by 9.
excitation
In electricity, the electrical energy which, when applied to a device, causes that device to produce an effect.
excitation
In nuclear physics, the raising or transferring of an atom from its normal quantum state or energy level to a higher quantum state or energy level.
excited atom
An atom raised to an *excited state.*
excited state
A quantum state or energy level of an atom which is higher than its normal or ground state or energy level.
exciter
An auxiliary generator which supplies energy for the field excitation of another electric machine.
excluded region
In preparing graphs, that region of the plane of the graph

lying between a and b when there are no real values for the *function,* y of the *independent variable,* x, for any of the values from $x = a$ to $x = b$.

excluded value

A value for an *independent variable* for which its *function* has no real value.

exclusion principle

See *Pauli exclusion principle.*

exosmosis

Osmosis out of a vessel or container.

exosphere

The outermost fringe or layer of the atmosphere, above the *ionosphere,* where collisions between molecular particles are so rare that only the force of gravity will return escaping molecules to the upper atmosphere.

exothermic

Relating to, accompanied or characterized by the development or release of heat.

exothermic reaction

A chemical or physical reaction in which heat is released.

expanding universe

The theory that the universe is constantly expanding, based on the fact that the mean distances between celestial bodies and configurations is apparently constantly increasing (as though they were on the surface of a balloon which was being blown up steadily to an ever larger size), their recession from us being indicated by the shift of the lines of their spectra toward the red end (see *Doppler effect*). The galaxies appear to be flying away from each other at tremendous speeds, and the farther they are the faster they seem to be moving. The original theory of Lemaître postulated that the origin of our universe had been a giant "primordial atom" that exploded as a result of violent radioactive decay processes. This theory has, however, been discarded as inconsistent with more recent findings and conclusions. One current view is that until 5 billion years ago our universe was collapsing, shrinking or condensing at a uniform rate from a state of a maximum, infinite rarefaction toward a state of maximum compression, which it ultimately reached 5 billion years ago; when that state of maximum compression had been reached, the process reversed itself, and the present expansion toward an infinite rarefaction began again.

expansion

The developed result of an indicated mathematical operation. It is the expression of a function in the form of a converging series; as the expansion of $(a + b)^2$ is $a^2 + 2ab + b^2$.

explement

Either of two *conjugate angles* with reference to each other.

explicit

Distinctly stated or expressed, not merely implied.

explosive

A generic term for a class of substances which upon being heated or struck undergo a rapid, violent, exothermic chemical change, in which a great deal of energy is released instantaneously, in the form of a great deal of heat, light and sound.

exponent

A number which indicates how many times its base is to be used as a factor, or a number which indicates the power to which another number is to be raised. Example: in x^2, 2 is the exponent of x. (See *law of exponents.*)

exponential curve

The graph of $y = a^x$, i.e., of $x = \log_a y$.

exponential equation

An equation in which one of the unknowns occurs as an exponent. (e.g., $4^x = 64$) .

exponential function

(1) A function in which the variable or variables appear as exponents. (2) Specifically, e^x where $e = 2.7182818 \ldots$

exponential increase

The increase of a quantity at a given time at a rate proportional to the amount present at that time.

exponential law

(1) Any law according to which exponents can be combined, divided, etc. (2) The increase or diminishment of a physical quantity in conformity with the expression $y = Ae^{ax}$, where A and a are constants, and x is a variable denoting a time, a distance or some other coordinate.

expression

A collection of numbers, algebraical symbols (or letters) and signs. The various parts of the expression which are connected by plus (+) or minus (−) signs, are called terms; as $4 + ab - 3cd + \frac{b}{c}$.

exradius

The radius of a circle which passes through all the apexes of a triangle or other polygon.

exsecant

A trigonometric function of an acute angle, exsec $\theta = (\sec \theta) - 1$.

extended law of the mean

If both $f(x)$ and $g(x)$ have continuous first derivatives on the interval (a,b), there exists at least one value of x, e.g., x_1 such that

$$\frac{f(b) - f(a)}{g(b) - g(a)} = \frac{f'(x_1)}{g'(x_1)}, \ a < x_1 < b$$

(Also called *double law of the mean, second mean value theorem* and *Cauchy's mean value formula.*)

extended mean value theorem

See *extended law of the mean.*

extended source of light

Any source of light that is regarded as not a point source.

exterior angle

The angle formed by a side of triangle or polygon and the extension of the adjacent side. Also the outside angles made by one straight line cutting two other straight lines.

exterior-angle theorem

The statement that any *exterior angle* of a plane triangle is equal to the sum of the two opposite interior angles.

external angle

See *exterior angle.*

extraction of a root

The operation of finding a root of a given number or quantity.

extraneous root

In the process of solving of an equation, a number arrived at which does not satisfy the equation.

extranuclear electron

One of the *orbital* or *planetary* electrons circling around the nucleus of an atom, arranged in shells and subshells. The arrangement of the extranuclear electrons determines the chemical properties of the element, and disturbances of this extranuclear electron structure produce optical spectra.

extraordinary ray

In *double refraction,* that one of the two refracted rays which has the lesser *index of refraction.*

extrapolation

The approximation of the value of a quantity or function for a value of the argument lesser or greater than all the values of the same argument employed for this operation of approximation.

extremes

The first and fourth terms of a proportion.

eye lens

The smaller lens of the *eyepiece,* of a telescope, placed opposite the field lens.

eyepiece

In a telescope, the lens or lens system through which the observer views the image formed, and which magnifies that image for better and more efficient viewing. Modern eyepieces consist of a system of at least two lenses, the larger of which is called the *field lens,* and the smaller one the *eye lens.* The simplest systems can be classified broadly in two groups: *positive* and *negative eyepieces* according to the shape and arrangement of the two lenses.

F

F.

(1) The symbol for the element *fluorine;* (2) the symbol for *faraday;* (3) the symbol for *Fahrenheit.*

F layer

See *ionosphere.*

f number

See *f/ratio.*

f/ratio

The ratio between the focal distance and the diameter of the *lens.*

face angle of a polyhedral angle

Any of the angles formed by the consecutive *edges of the polyhedral angle.*

face-centered cubic lattice

One of the two varieties of the *close-packed* crystal structure, i.e., a lattice in which atoms are arranged according to the one of the two possible systems which permits equal hard spheres to be packed so as to occupy the least possible space. (The other possible such arrangement is the *close-packed hexagonal* structure.) In the face-centered cubic lattice the atomic centers are disposed in space in such a way that they may be supposed to be situated at the corners and the middle of the faces of a set of cubic cells, i.e. that the lattice possesses half an atom in the middle of each face of the cubic lattice, the corners each sharing one atom with seven other cubes. The facial atom is shared with the adjacent cube and thus each unit cell contains four atoms.

face of a cathode-ray tube

The front or viewing surface.

face of a crystal

Any plane surface on the crystal which is in a particular and invariable relation to the faces and planes of reference and to other such plane surfaces.

face of a dihedral angle

One of the planes forming the angle.

face of a polyhedral angle
One of the planes forming the angle.

face of a polyhedron
One of the polygons which bound the polyhedron.

facet
Any flat surface of a crystal.

facies
In geological terminology, the totality of the features and characteristics of a geological unit or formation.

factor
Any of the numbers used in multiplying. Every factor of a product is also an exact divisor of the product, i.e., 5 and 8 are factors of 40.

factor
To resolve into factors. E.g., factoring 10 means writing it as 2×5.

factor theorem
If r is a root of the polynomial $f(x) = 0$, the polynomial is divisible without remainder by $(x - r)$, i.e., $(x - r)$ is a factor of $f(x)$. Conversely, if $(x - r)$ is a factor of $f(x)$, then r is a root of the polynomial $f(x) = 0$.

factorial
Pertaining to or concerning factors.

factorial *n*
$n!$ representing the product $n(n - 1)\ (n - 2)\ (n - 3) \ldots 5,4,3,2,1$.

factorial notation
The method of expressing the products of n consecutive numbers as $n!$, called *factorial n* (q.v.).

factorize
To factor (q.v.).

Fahrenheit scale
A temperature scale according to which the freezing point of water is marked 32° and its boiling-point 212°. The abbreviation of Fahrenheit is F. To convert Fahrenheit degrees into centigrade, subtract 32, multiply by 5, and divide by 9.

family of elements
A group of chemical elements which have allied properties, as, e.g., the *halogens.* (See also *radioactive series.*)

family of lines
All those lines in a plane which are obtained by assigning

values to a single parameter, and which possess a certain common geometric property.

farad

A unit of capacitance equal to the amount of capacitance present when 1 volt can store 1 coulomb of electricity. A farad is one-billionth part (10^{-9}) of an abfarad.

faraday (F)

The quantity of electricity (96,500 coulombs) required for an electrochemical reaction involving one chemical equivalent, i.e. number of coulombs per gram equivalent of an ion; equal to 9.651.94 e.m.u. gm mole^{-1}. Also called *Faraday constant* or *Faraday's electrolytic constant.*

Faraday constant

See *faraday.*

Faraday dark space

The non-luminous zone which separates the negative glow from the positive column in a Crookes tube or other gaseous discharge tube.

Faraday effect

See *magnetic rotation.*

Faraday's electrolytic constant

See *faraday.*

Faraday's law of electromagnetic induction

The induced electromotive force in any circuit is proportional to the rate of change of the number of lines of force linked with the circuit.

Faraday's laws of electrolysis

(1) The amount of chemical change produced by a current is proportional to the quantity of electricity passed. (2) The amounts of different substances liberated or deposited by a given quantity of electricity are proportional to the chemical equivalent weights of those substances.

faradmeter

An instrument for measuring electric capacitance. Faradmeters are provided with a scale usually graduated in microfarads.

Farmer's reducer

A mixture of a solution of potassium ferricyanide added to a solution of sodium hyposulphite, used for the purpose of reducing the density of a photographic negative.

fat

A generic term for a group of organic compounds which consists of the glyceryl esters of the higher *fatty acids.*

fata morgana

A complex form of mirage, characterized by a marked distortion of images.

fatigue

In metals, machine parts, etc., a progressive loss of strength and a steady deterioration owing to repeated stresses.

fatigue

In optics, the gradual decrease of some chacteristic property due to external causes; e.g., the decrease of the photoelectric sensitivity of certain surfaces upon prolonged exposure to light.

fatigue test

Test involving placing a material under tension until it yields.

fatty acid

In chemistry, a monobasic aliphatic acid. Some authors restrict the use of the term to saturated acids. Cyclic acids, such as chaulmoogric acid, are sometimes termed fatty acids when they occur naturally as constituents of fats.

Fe

The symbol for the element *iron.*

feedback

In an electrical system, the returning of a fraction of the energy output into the input circuit.

Fermat's principle

A light ray passing between two points, the course of which includes reflections and refractions, will always follow the path of least time, i.e., the time which it takes to travel from one of the two points to the other will be either a minimum or a maximum.

ferment

An organic catalyst which produces *fermentation.*

fermentation

The process of chemical change in organic matter produced by an *enzyme* or a *ferment,* and tending toward a simplification of the complex compounds present.

fermentation tube

A glass tube used for observing and studying fermentation processes.

fermium

A transuranic element, atomic number 100, the discovery of which was announced, simultaneously, by the Argonne National Laboratories and the University of California. It was produced in the nuclear reactor from plutonium (Pu^{239}) by the absorption of 15 neutrons, resulting in fermium of mass number 254. (In a separate process, fermium was produced from einsteinium.) Symbol: Fm.

ferri-
A prefix indicating the presence of a ferric ion in the compound.

ferric
Containing trivalent iron.

ferrite
A term once restricted to pure *alpha iron,* but now extended to include any solid solutions of which alpha iron, as distinct from *gamma iron,* is the solvent.

ferro-
A prefix meaning *iron.* In the names of chemical compounds, it indicates the presence of metallic iron or of divalent iron.

ferroalloy
A generic term for alloys of iron with another metal, as ferroaluminum, ferrochromium, ferromanganese, etc.

ferroelectric
A dielectric substance which shows a spontaneous polarization.

ferromagnesian mineral
In geological terminology, a mineral in which iron and magnesium predominate.

ferromagnetic
Possessing a *magnetic permeability* considerably higher than unity, and being therefore attracted by a magnet. (Frequently referred to simply as *magnetic.*)

ferromagnetism
The property of being *ferromagnetic.*

ferrous
Containing divalent iron.

ferruginous
Containing iron.

fertilizer
A substance which can be used to enrich the soil or to replace plant food consumed by the existing vegetation.

field
See *force field.*

field coil
The coil which, when carrying a current, magnetizes a *field magnet* of an electrical machine.

field force
A force which appears to act at a distance, without direct contact.

field lens
The larger one of the two lenses of the *eyepiece* of a telescope, placed opposite the *eye lens*.

field magnet
A magnet (usually an electromagnet) which supplies the magnetic field in an electrical machine.

15° calorie
A synonym for *normal calorie*.

figurate number
Any number or series of numbers formed from an arithmetical progression of which the first term is unity and the difference a whole number, by taking the first two, first three, first four, etc. terms and forming a new series of progressions, and so on, by being formed in the same way. The numbers in each sequence, when represented by points, can be systematically arranged in various geometrical figures, such as triangles, pentagons, tetrahedrons, etc.

figure
A character denoting a number, e.g., 3, 14, etc. (See *geometric figure, plane figure*.)

filar micrometer
An instrument for measuring very small distances in the field of an *eyepiece*. It consists essentially of two spider threads placed parallelly; one of them is fixed, the other movable and controlled by a fine screw with a graduated head and a counter of its revolutions.

film
A thin, not necessary visible, layer of a substance.

filter
In general, a sheet or plate of a suitable substance that is interposed between the source of a radiation and the observer or some exposed object to absorb selectively a part of the radiation.

filtrate
A solution separated by filtration.

finite
Having limits or bounds; having an end.

finite graph
A term used in topology. (See *network*.)

finite series

A series that terminates at some assigned term. The indicated sum of a finite number of terms.

finite straight line

A straight line which is limited in length.

firebrick

A brick made from highly heat-resistant (*"refractory"*) clays, which in the fired state consists essentially of silica and alumina.

fireclay

In general, any clay which has a fusion point higher than approximately 1600°C and can therefore well be used for making fire-brick.

first condition of equilibrium

In physics, the principle stating that a body is in equilibrium if the vector sum of the forces which act upon it is zero.

fissile material

Any element or isotope, the nuclei of which can be caused by nuclear bombardment to undergo nuclear fission and to produce a fission chain reaction. (According to this definition, this term can be applied to uranium, plutonium, thorium and protactinium; however, protactinium is too rare to be put to much practical use and thorium fails to produce a satisfactory fission chain reaction, wherefore it is used only as an admixture to U^{235} or plutonium.)

fission chain reaction

See *reactor.*

fission (nuclear)

The breakdown of the nucleus of an atom by bombardment with the proper projectiles. Slow neutrons cause fission of uranium and other atoms. This discovery led to the atomic bomb.

fission product

One of the nuclei into which nuclear bombardment, usually by neutrons, causes the nucleus of a fissile heavy atom (uranium, plutonium, thorium, protactinium) to split. Fission products are nuclei of isotopes in the middle of the periodic table, ranging approximately from atomic number 34 to 58.

Fitzgerald-Lorentz contraction

A phenomenon postulated independently by Fitzgerald (in 1893) and Lorentz (in 1895), consisting in a contraction in length of a body moving through the ether with high velocity, occurring in the direction of the motion. (Such contraction of a moving body was later confirmed by the theory of *relativity* which, however, explains it as a result of the relative motion of the body and the observer.)

fixation

(1) The process of making permanent. (2) In photography, the process of dissolving light-sensitive silver salts from films, plates, or paper, thus making them insensitive to the further action of light.

fixed oils

Oils consisting of fatty substances of animal and vegetable organisms; they contain the esters of fatty acids.

flame

The gaseous reaction product of the burning of a gas in air or another gas.

flammable

A synonym, preferred in industrial terminology, for *inflamable;* able to be set on fire, readily combustible.

flash point

The temperature of a liquid fuel, below the ignition point, at which it gives off a combustible vapor.

flash spectrum

The solar spectrum as seen an instant before a total eclipse of the sun.

fleximeter

An instrument for determining the state of stress due to bending moments alone.

flexural strength

Resistance to bending.

flexural stress

The stress set up in a body on bending.

flexure

The capacity of a solid object to bend without fracture.

flocculation

The coalescence of a finely divided precipitate.

flotation

A process in which air is blown into a mixture of very finely powdered materials suspended in water, the air bubbles being used to float one kind of material and thus separate it from another.

flow

The motion or passage of a fluid.

fluid ***(noun)***
A collective term for liquids and gases.

fluid ***(adj.)***
Said of a substance the particles of which change their relative positions readily without causing a breaking up of the mass of the substance.

fluid friction
Viscosity.

fluid pressure
The force per unit area exerted by a fluid perpendicularly against a surface in contact with it.

fluidity
The degree of the ability of a substance to flow, i.e., of its freedom from *viscosity;* mathematically, it is the reciprocal of viscosity.

fluorescence
The absorption of radiant energy by a substance followed immediately by its re-emission in the form of visible light of a greater wavelength. Fluorescence persists only so long as the exciting radiant energy is acting on the absorbing and re-emitting substance.

fluorescent lamp
An electric discharge lamp in which the radiant energy from the electric discharge is transformed by suitable materials into wavelengths giving higher luminosity.

fluorescent screen
A sheet of suitable material coated with substance which fluoresces visibly when x-rays, radium rays or electrons impinge upon it. Also used in cathode-ray tubes to form the visible images.

fluorine
A monovalent gaseous element in Group VII of the periodic table, one of the halogens, discovered by Scheele in 1771. Atomic number 9, atomic weight 19.00, melting-point −223°C., boiling-point −187°C., density 1.697 grams per liter at normal temperature and pressure. Symbol: F.

fluoroscope
A device consisting of a fluorescent screen suitably mounted either separately or in conjunction with an x-ray tube, by means of which x-ray shadows of objects between the tube and the screen are made visible.

flux
(1) In physics, in general, the quantity of energy or power

passing through a surface in unit time. (2) In nuclear physics, the product of the number of particles passing through in unit time and of their mean velocity. (3) See *magnetic flux*.

flux

In metallurgy, (1) a substance used to facilitate the smelting of metals and the removal of impurities from them; (2) material used to dissolve or exclude the formation of oxides from the surface of the metal during welding, brazing or soldering, and/or to add ingredients to improve the properties of the weld metal; (3) material which lowers the melting point of a refractory material.

flux line

A term often applied to *magnetic lines of force*.

fluxmeter

An instrument for measuring the intensity of a magnetic field.

Fm

The symbol for the element *fermium*.

focal chord

With reference to conics, a chord that passes through a focus.

focal length of a lens

The distance of the *principal focus* from the lens, i.e., the distance at which the lens must be placed from a screen in order to have the projected image of an object at infinity be in focus on the screen.

focal ratio (f/ratio)

The ratio between the *focal length* and the diameter of a lens. (Also called *f-number*.)

foci

The plural of *focus* (q.v.).

focus (*plural:* foci)

With reference to a conic, one of two points the sum or difference of the distance of which to the circumference of the curve is constant. In the ellipse, the sum of the distances is constant, in the hyperbola their difference; in the parabola, one focus is in infinity, the other is a point whose distance from any point on the curve is equal to the perpendicular distance of the latter point from the directrix. The focus and directrix are "pole and polar" for the conic.

focus (*plural:* foci)

In optics, the point where converging rays meet and produce a desired effect, as the intensification of illumination, heat, etc., or a visible image.

focus
In optics, to adjust lenses or mirrors in order to produce a clearly defined image.

fog
Finely distributed droplets of a liquid suspended in a gas.

foil
A thin, flexible sheet of metal, formed by rolling and hammering.

folding line
In *orthographic projection,* the line of intersection between two mutually perpendicular projection planes. Also called *reference line.*

folium of Descartes
An algebraic curve of the third degree, the Cartesian equation of which is $x^3 + y^3 = 3axy$.

foot
A unit of length equal to 12 inches. Used in the *FPS system of units.*

foot-candle (ft-c)
A unit of illumination; 1 ft-c = the illumination on a surface all points of which are at a distance of 1 foot from a uniform point source of 1 candle, i.e. 1 ft-c = 1 lumen per square foot. (For *apparent foot-candle,* see *foot-lambert.*)

foot-lambert (ft-L)
A unit of brightness; 1 ft-L = $1/\pi$ candle per square foot, and it equals the uniform brightness of a perfectly diffusing surface emitting or reflecting light at the rate of 1 lumen per square foot, or the average brightness of any surface which emits or reflects light at that rate. Called also *apparent foot-candle.*

foot of a line
The point in which a straight line passes through a plane.

foot-pound (ft-lb)
A unit of work; 1 ft-lb equals the work done in raising a mass of 1 pound vertically through a distance of 1 foot, against the force of gravity.

foot-poundal
The absolute unit of work in the FPS system, equal to 1/32.2 foot-pound.

foot-pound-second system
See *FPS system of units.*

forbidden lines
The designation of certain spectral lines which are produced

by certain extremely rarely occurring, improbable transitions (called *forbidden transitions*), so that their occurrence under ordinary laboratory conditions is extremely unlikely.

force

Any physical cause which is capable of modifying the motion of a body. The vector sum of the forces acting on a body at rest or in uniform rectilinear frictionless motion is zero.

force field

An area or region or space in which a certain force is active and exerts its effect.

force of inertia

The force with which a body offers resistance to any change in its state of rest or motion.

formaldehyde

(CH_2O) The simplest aldehyde derived from methanol.

formula

A general expression for solving certain problems or cases. It is a relation established amongst certain quantities any one of which may be taken as the unknown if the other quantities are known or can be ascertained. Thus $M = \frac{SI}{C}$ is a formula for steel beams. It is an equation which expresses some common rule. (See also *chemical formula.*)

formula weight

The molecular weight of a substance expressed in grams.

fossil fuel

Fuels formed underground in the geologic past (e.g., coal and oil).

Foucault current

See *eddy current.*

4° calorie

A synonym for *standard calorie.*

(4n), (4n + 1), (4n + 2), (4n + 3) series

See *radioactive series.*

four-leafed rose

The graph of the equation $r = a \sin 2\theta$ or $r = a \cos 2\theta$. (Cf. n-*leafed rose.*)

fourth dimension

A hypothetical spatial dimension which would be at right angles to all three known spatial dimensions (length, breadth and thickness).

fourth proportional

The fourth number that will form a proportion when three given numbers are used in the order stated.

FPS (fps) system of units

A system of units of measurement, most widely used in the United States, in which the fundamental units are the *foot* (unit of length), the *pound* (unit of weight or force), and the *second* (unit of time).

Fr

The symbol for the element *francium.*

fraction

An indicated quotient of two quantities. If unity (one) is divided into any number of equal parts, then one or more of these parts is called a fraction. Thus if we divide unity into seven equal parts (of 1/7 each) and take four of these parts, we have a fraction equal to four-sevenths or 4/7.

fractional

Non-integer, involving or relating to a fraction or fractions.

fractional combustion

A process of separating gas mixtures, by producing the combustion of one component of the mixture under conditions which do not affect the other components.

fractional condensation

A process of separating gases in a mixture by partial condensation, varying the temperature or pressure until one of the components of the mixture liquefies.

fractional crystallization

A process of separating a mixture in solution by a successive slow crystallization of the components of different solubilities.

fractional distillation

The process or operation of separating liquids having different boiling points; a mixture of two or more such liquids is heated, and the portions of the *distillate* are collected separately.

fractional equation

An equation containing fractions. Thus $\frac{x}{y} = \frac{a}{b}$; or $\frac{x}{x} - \frac{15}{16} = \frac{x-6}{x-7}$.

fractional expression

An expression containing fractions. Thus

$$\frac{a}{(a-b)\ (a-c)} + \frac{b}{(b-c)\ (b-a)} + \frac{c}{(c-a)\ (c-b)}$$

fractionation

Separation by successive operations, each removing from a liquid some proportion of one of the substances. The operation may be one of precipitation, or of crystallization, or of distillation.

frame of reference

A set of points, lines or planes used as a system of reference for defining space co-ordinates.

francium

A very rare radioactive element, in Group I of the periodic table, one of the alkali metals, formerly also called *alabamine* (Ab) . It does not occur in nature except for minute traces of Fr^{223}. This isotope was discovered by Mlle. Perey, in Paris, in 1939, who named it first *actinium K* and later *francium,* the name which has been officially adopted. Symbol: Fr.

Fraunhofer diffraction

Diffraction of light when the light source as well as the screen on which the phenomena are observed are effectively infinite distances from the diffracting apparatus. (Cf. *Fresnel diffraction.*)

Fraunhofer lines

Fine black absorption lines dividing the various bands of color in the solar spectrum.

free

Said of an element or substance which occurs in natural form unmixed or uncombined with other elements or substances.

free electron

An electron within a substance, not attached permanently to any atom, or an electron liberated from an atom by, e.g., ionization.

free energy

That portion of the total energy content of a system which is aavilable for doing work. All forms of potential energy (mechanical, electrical, and chemical) constitute free energy in the presence of appropriate structural arrangements. Heat is convertible into work only in a fraction dependent on the difference of temperatures between which it is transferred, being completely convertible only with an infinite temperature drop.

free fall

The condition of a body or system when it moves acted upon by no force of any appreciable magnitude other than the force of gravity. The condition of a rocket or missile when it coasts in outer space with its driving power off, producing an effect equal to that of *zero-gravity*. The acceleration of a freely falling body at sea level, at a latitude of 45°, is 980.6 centimeters (i.e., 32.17 feet) per second per second.

free radical

An atom or group of atoms broken away from a stable compound by application of external energy, and, although containing unpaired electrons, remaining free for transitory or longer periods.

freezing point

The temperature at which a liquid solidifies: identical with the point at which the solid melts (*melting point*).

Freon

Trade name for a group of halogenated hydrocarbons containing one or more fluorine atoms. Widely used as refrigerants and propellants.

frequency

The number of completed cycles, waves, vibrations, etc., of a periodic phenomenon per second. It is the reciprocal of the *period* (the time required for the completion of one cycle, etc.).

frequency band

See *band of frequencies.*

frequency distribution

A term used in the study of statistics. It represents a method of presenting accumulated data by reclassifying the data into subgroups and then re-presenting the original facts in terms of these subgroups.

frequency modulation

A type of radio and television broadcast in which the frequency of the carrier wave is modulated. It provides less distortion but it can be received only over a comparatively small distance.

frequency polygon

In the graphical representation of frequency distributions, a broken-line graph which connects the points the coordinates of which are class marks and class frequencies.

Fresnel diffraction

Diffraction of light when the light source as well as the screen on which the phenomena are observed are at finite distances from the diffracting apparatus. (Cf. *Fraunhofer diffraction.*)

friction
The action between two bodies at the surfaces of contact which opposes their movement.

frictional electricity
See *triboelectricity.*

front surface mirror
A mirror produced by applying a thin coating of silver or aluminum to the front face of a plate of glass to serve as a reflecting surface.

frontal line
In orthographic projection, a line that is parallel to the vertical projection plane.

fructose
The sugar $C_6H_{12}O_6$, also called *fruit sugar,* found in fruits and in honey.

frustum of a cone, prism or pyramid
That part of a cone, prism or pyramid which lies between its base (or a plane parallel to the base) and another parallel plane cutting through the solid. Both parallel planes are referred to as the *bases* of the frustum.

fuel
Any matter used to produce energy by *combustion.*

fuel cell
A source of energy by oxidation by electrochemical means at more or less ordinary temperatures.

fugacity
The tendency of a substance to escape or disappear from the phase in which it is present. The tendency of a gas to escape from an enclosed space.

function
A magnitude or variable quantity so related to another magnitude or variable quantity, that to values of the latter there correspond values of the former. Either magnitude or variable quantity is a function of the other; but the one the values of which are observed or arbitrarily fixed or assigned is called the *argument* or *independent variable;* the other is called a *dependent variable* or a *function* of the first. It is a quantity so connected to another that any change in the value of one correspondingly affects the value of the other.

function in *n* variables
A quantity (referred to as the *dependent variable*) that assumes a value or values according to each set of values of the *n*

variables called the *independent variable* or the *argument.* A common notation of a function in *n* variables is f $(x_1, x_2 \ldots x_n)$.

function in one variable

A function involving one variable; e.g., the area of the circle is a function of the radius. A common notation for a function in one variable, x, is $f(x)$.

fundamental

In music, the lowest single note or tone in a compound tone (chord or dischord), all the others being the *upper partials.*

fundamental particles

This term is applied in general to the *elementary particles,* i.e. those particles which are at the present time assumed to be the ultimate building blocks of matter, i.e. the proton, the neutron, the electron, the positron, the neutrino, the mesons and the γ-ray or photon, and others, which according to our present state of knowledge cannot be resolved into components. Although the x-particle and the deuteron are composite particles, they are often included in this group of elementary particles, since they play an important part in nuclear reactions. The fundamental particles which serve as building blocks of the atomic nucleus are the nucleons (i.e. the proton and the neutron) only, although it has been suggested that the mesons, too, should be included.

fundamental theorem of algebra

Every polynomial equation with coefficients which are real or complex numbers has at least one root which may be a real or a complex number.

fundamental units

The units of length, mass and time.

furnace spectrum

Spectrum of material heated to incandescence or vaporized in an electric furnace.

fuse

To melt.

fuse

A device for protecting electrical apparatus from damage by an overcurrent or a short-circuit, consisting essentially of a circuit-opening fusible member which is directly heated and destroyed by the passage of overcurrent through it.

fusible alloy

Any alloy of low melting point, usually consisting of bismuth, lead, cadmium and tin.

fusion

Change of a substance from the liquid to the solid state at a definite temperature, called *melting point,* and accompanied by the absorption of latent heat of fusion.

fusion (nuclear)

Isotopes of light elements may unite forming heavier nuclei. Temperatures of millions of degrees are necessary to bring about this fusion. The hydrogen bomb is a fusion bomb.

fusion point

See *melting point.*

fusion welding

The welding of metals in a molten state.

G

g

(1) The abbreviation for *gravity*. (2) The abbreviation for *gram*.

γ

(1) The symbol for *γ-ray* or *γ-ray photon;* (2) the symbol for *gamma* in all its meanings.

Ga

The symbol for the element *gallium*.

gadolinium

A rare trivalent metallic element, one of the group of rare earth elements of the periodic table, discovered by Marignac in 1886. Atomic number 64, atomic weight 156.9. Symbol Gd.

gallium

A trivalent metallic element in Group III of the periodic table, discovered by Boisbaudran in 1875. Atomic number 31, atomic weight 69.72, melting-point 29.75°C., boiling-point 1700°C., specific gravity 5.885. Symbol: Ga.

galvanic cell

Two plates of dissimilar metals immersed in an electrolyte; more frequently called a *voltaic cell*.

galvanic series

A synonym for *electrochemical series*.

galvanism

An obsolete term for the science of electric currents.

galvanize

To coat a metal with a thin layer of zinc by dipping it into a bath of molten zinc (referred to commercially as *spelter*).

galvanometer

An instrument for detecting or measuring a small electric current or a function of the current by means of a mechanical motion derived from electromagnetic or electrodynamic forces which are set up as a result of the current.

galvanoscope

An instrument capable of detecting, but not measuring, electric currents.

gamma (γ)

(1) 10^{-6} gram; (2) a unit of magnetic field intensity ($1\gamma = 10^{-5}$ oersted); (3) a unit for measuring density in photographic negatives.

gamma iron

An allotropic form of iron which has a *face centered cubic lattice* and is nonmagnetic.

gamma-ray (γ**-ray**)

Electromagnetic radiation of nuclear origin, of very short wavelength, emitted by the nuclei of certain atoms in the course of their radioactive decay. Gamma rays are not composed of particles, but consist of high-energy photons; they have the speed of light, have no electric charge and no rest mass, but do have a very slight mass due to their motion and velocity (0.00107 a.m.u. at 1 Mev). Their range, both in air and matter, is much longer than that of either α or β particles or rays, but their ionizing capacity is much weaker than that of either of the latter. The wavelength of the gamma ray is characteristic of the element emitting it, and varies between 0.0466 and 3.9 Angstrom units. Their penetration power is inversely proportional to the density of the substance penetrated, and is far higher than that of both alpha and beta rays; even low-frequency ("soft") gamma rays will penetrate up to about 12 mm. of lead, and high-frequency ("hard") gamma rays can be stopped by several centimeters of lead only.

gangue

The earthy part of an ore, i.e., that part which contains no metal.

gas

Matter in fluid state, low in density, of high compressibility, which owing to the negligible internal cohesion of its molecules can fill a container completely and to a uniform density, and has neither a definite shape nor volume, nor a free surface.

gas constant

The *ideal-gas constant* or *molar gas constant.* (See *ideal-gas law.*)

gas-law equation

The *ideal-gas law.*

gas laws

The laws which express the effects of temperature and pressure on the volume of a given mass of gas. (See *Boyle's law, Charles' law, perfect-gas law.*)

gas producer
See *producer.*

gas tube
A vacuum tube in which the pressure of the contained gas or vapor is such as to affect substantially the electrical characteristics of the tube.

gaseous
Gas-like, pertaining to or involving gases.

gaseous pressure
The pressure exerted by a gas on the walls of the container in which it is confined, resulting from the impacts of the molecules on those walls.

gaseous solution
A homogeneous mixture of two or more gases.

gauge
An instrument for measuring dimensions, volumes or pressures.

gauss
The practical CGS electromagnetic unit of magnetic induction of flux density; 1 gauss = 1 line (or maxwell) per cm^2. (Prior to 1932, the term *gauss* was the name of the practical CGS unit of magnetic intensity, now called *oersted.*)

Gay-Lussac's law
The volumes of inter-reacting gases and the volume of the gaseous product of their reaction bear simple proportions to each other which can always be expressed numerically in small whole numbers.

Gay-Lussac's law of gas volumes
An alternative name for *Charles' law.*

Gd
The symbol for the element *gadolinium.*

Ge
The symbol for the element *germanium.*

geiger
A newly coined technical term, derived from the name, *Geiger counter,* meaning: to subject to an examination by a Geiger counter (or Geiger-Müller counter).

Geiger-Müller counter, Geiger counter
A counting tube, i.e. an apparatus for detecting and counting ionizing particles, perfected by Hans Geiger, with the collaboration of Müller, in 1928. The modern version of this counter

is an oblong box with headphones attached to it. When a fast alpha or beta particle enters the apparatus, it produces ions in the gas contained in it, and the ions cause a loud tick to be heard in the headphone; the ticks are counted automatically by a mechanical device.

Geissler tube

A special form of *discharge tube,* for demonstrating the luminous effects of discharges through rarefied gases; it is a two-electrode discharge tube, exhausted to at least 1/100 atmosphere, which gives a flow discharge.

gel

The solid phase of a colloidal solution, obtained by partial drying or cooling of the solution.

general gas law

The *ideal-gas law.*

generating function

A fraction which can be developed indefinitely in ascending powers of the unknown quantity to form a series. Thus

$$\frac{1}{(1-x)^2} = 1 + 2x + 3x^2 + 4x^4 \ldots\ldots\ldots\ldots$$

generator

(1) See *generatrix.* (2) See *electric generator.* (3) See *producer.*

generator gas

See *producer gas.*

generatrix

The moving line which generates a conical, cylindrical, prismatic or pyramidal surface; it is the line parallel to the axis of a prism or cylinder which if moved along the boundary of the base makes the surface of the prism or cylinder.

geochemistry

The study and evaluation of the chemical composition of the earth and of the chemical changes within it, past and present.

geodesic

Short for *geodesic line;* the line of least distance between any two points on a mathematically defined surface.

geodesy

The science which deals with the determination of the size and shape of the earth by direct measurements and observations.

geodetic line

The line of the shortest distance between any two points on the surface of a spheroid.

geoid

The earth considered as a geometric solid the surface of which coincides with the mean level of the oceans.

geology

The science which treats of the history of the earth and its life, especially as recorded in the rocks. Three principal branches or phases are usually distinguished: 1. *structural,* or *geotectonic* geology, treating of the form, arrangement, and internal structure of the rocks. 2. *dynamic* geology, dealing with the causes and processes of geological change. 3. *historical* geology which, aided by other branches, aims to give a chronological account of the events in the earth's history. Other subdivisions are *economic* geology, that branch of geology which deals with the application of the science in industrial relations and operations; *stratigraphic* geology, a study of the succession of the beds of rock laid down during the progress of geologic ages; and *petroleum* or *oil* geology, which treats of the geologic factors involved in the origin, accumulation and production of petroleum.

geomagnetic

Relating to the magnetic field of the earth.

geomagnetism

The *terrestrial magnetism.*

geometric figure

Any configuration on a plane surface that is bounded or enclosed by or consists of curved or straight lines.

geometric mean

Any number of numbers or quantities is that root of their product which is represented by their number. Thus the geometrical mean of two numbers is the square root of the product of the two numbers; or the geometrical mean of three numbers is the cube root of the product of the three numbers.

geometric progression

A sequence of terms such that the ratio of each term to the immediately preceding one is the same throughout the sequence. The general form is usually written: a, ar, ar^2, . . . ar^{n-1}, where a is the first term, r the *common ratio* (or simply *ratio*) and ar^{n-1} the last or nth term. The sum of such terms is sometimes called a *geometric progression,* although *geometric series* is more common.

geometric series

A series the terms of which form a *geometric progression.*

geometric surface
A closed surface enclosing a portion of space.

geometry
The branch of mathematics which investigates the relations, properties and measurements of solids, surfaces, lines and angles.

geophysical constant
A quantity that expresses a fixed value for a law or magnitude that applies to the physics of the earth. (The constant of gravitation and the constant of nutation are examples of a geophysical constant.)

geophysics
A branch of experimental physics which deals with the earth's structure, composition, physical properties and development.

geotectonic geology
That branch of geology treating of the form, arrangement and internal structure of rocks. (Also called *structural geology* or *tectonics*.)

geothermal energy
Power derived from the heat of the interior of the earth.

germ
A microbe or spore; the popular term for all species of bacteria.

germanium
A rare metallic element in Group IV of the periodic table, discovered by Winkler in 1886. Atomic number 32, atomic weight 72.60, melting-point 958°C., specific gravity 5.465, valence 4. Symbol: Ge.

germicide
A substance which kills germs.

getter
An *alkali metal* introduced into a vacuum tube during manufacture and fired after the tube has been evacuated, to react chemically with any gases which may have been left in manufacture.

ghost
A line seen in a *line spectrum* which is not a part of the spectrum, but is caused by imperfections in the ruling of the *diffraction grating* used.

Gibbs' phase rule
See *phase rule*.

gilbert

The electromagnetic unit of magnetomotive force in the C-G-S system; a gilbert is the equivalent of $5/2\pi$ (i.e. 0.795) ampere-turn.

glaciation

Changes produced in the features of the solid crust of the earth owing to the deposition and erosion of glacier ice.

glancing angle

With reference to radiation impinging on a surface, the complement of the *angle of incidence.* (The term is often used to indicate merely a very small, unspecified angle of incidence.)

glow discharge

A luminous sparkling discharge of electricity in a gas, usually not accompanied by sound.

glucose

The sugar $C_6H_{12}O_6$, also called *grape sugar,* contained in fruits and honey, and in small amounts in the blood.

glycogen

$(C_6H_{10}O_5)_x$ Animal starch. A carbohydrate occurring in large amounts in the animal organism, especially the liver.

glycol

The generic designation of a class of organic compounds derived from aliphatic hydrocarbons by the replacement of two hydrogen atoms by hydroxyl groups.

gold

A metallic element in Group I of the periodic table. Atomic number 79, atomic weight 197.0, melting-point 1063°C., boiling-point 2600°C., specific gravity 19.3, valence 1 or 3. Symbol: Au.

gold family

The elements osmium, iridium, platinum and gold (atomic numbers 76-79). Also called *gold period.*

gold-leaf electroscope

An *electroscope,* consisting of a glass jar containing two pieces of gold leaf, fastened to a metal rod which passes through the cover of the jar and ends in a knob. If the metal rod is charged, the two gold leaves fly apart. (See also *Lauritsen electroscope.*)

gold period

The elements found in the *gold family* in the periodic chart.

Goldschmidt process
See *aluminothermic process.*

goniometer
An instrument for measuring angles.

grade
In mathematics, the unit of the *centesimal system of angular measure.*

gradient
In mathematics, the incline of a straight line, it is the distance traveled vertically divided by the distance traveled horizontally and is usually given for the unit vertical distance. Example: $1/300 =$ one unit (vertically) in three hundred units (horizontally). It is the tangent of the angle which the line makes with the positive direction of the horizontal axis.

gradient
In physics, the rate at which a variable quantity, such as temperature or pressure, changes in value; in these instances, called "thermometric gradient," and "barometric gradient," respectively. (See also *thermal gradient.*)

gradient of a curve
The gradient of the tangent to the curve at any particular point.

gradient of a straight line
In mathematics, the increase of y divided by the increase of x. Thus if the coordinates of one point on a line are x and y and the coordinates of another point on the line are $(x + 3)$ and $(y + 2)$, then the gradient of the line between the two points is $2/3$ or three units measured horizontally and two units measured vertically.

graduate
A container with marks on the side graduating the capacity of the vessel in terms of a system for measuring fluids.

Graham's law
The relative rates of the diffusion of gases under the same conditions are inversely proportional to the square roots of their densities.

grain
In metallurgical terminology, an *allotriomorphic crystal* present in metals and in one-component alloys.

gram (gramme) (g)
The unit of mass or weight in the CGS system, originally de-

fined as the mass or weight of 1 cm^3 of pure water at its maximum density (4°C.), now defined for practical purposes as 1/1000 of the International Standard Kilogram, a platinum-iridium bar, kept at Sèvres, France. 1 g $= \frac{1}{28.35}$ ounce avoirdupois.

gram-atom

The mass of an element the weight of which in grams equals numerically the atomic weight of the element. The number of the atoms contained in 1 gram-atom of any element is 6.0234×10^{23} (i.e. *Avogadro's number*).

gram atomic weight

A synonym for *gram-atom*.

gram calorie

See *calorie*.

gram-equivalent

The chemical equivalent weight of a substance expressed in grams.

gram formula weight

The mass in grams equal numerically to the *formula weight*.

gram-ion

The sum total of the atomic weights of the atom composing an ion, in grams.

gram mole

See *gram molecular weight*.

gram-molecular volume

The volume occupied by one mole of a substance at the temperature of 0°C and the pressure of 760 millimeters of mercury. For an ideal gas, this volume is 22.4 liters. (Also called *molal volume*.)

gram-molecular weight

The weight of a *gram-molecule*.

gram molecule

The mass of a pure substance the weight of which in grams equals numerically the molecular weight of the substance, i.e., a quantity of the substance that weighs *a* grams where *a* equals the *molecular weight* of the substance. The number of the molecules contained in 1 gram-molecule of any pure substance is 6.0234×10^{23} (i.e. *Avogadro's number*). Also called *mole* or *mol*.

granular
Grain-like, containing or composed of grains.

granulation
The process or act of changing a substance into grains or granular particles.

graph
The representation of varying relationships as between quantity and time by visual means. There are *bar graphs, broken* and *curved line graphs,* and *ideographs* in which quantities are represented by tiny figures of people, trains, factories, etc. Graphs can also be used to solve algebraic equations.

graphical geometry
That branch of geometry which treats of relative positions in space of points, lines and surfaces.

graphical representation
Showing on a graph the relation between two variables connected by a formula or equation.

graphing
Drawing the graph of an equation or one that represents visually a set of data.

graphite
One of the two naturally occurring forms of crystalline carbon (the other being diamond). Specific gravity 2.0-2.5, hardness 0.5-1.0.

grating
See *diffraction grating.*

grating spectroscope
A *spectroscope* in which the prism is replaced by a *grating* or by a piece of glass into which a series of parallel lines have been ruled.

gravimetric analysis
In chemistry, a form of quantitative analysis by weighing the substances and precipitates obtained.

gravimetry
The science or technique of measurement by weight.

gravisphere
The spherical extent in which the force of a given celestial body's gravity is predominant.

gravitation
The force with which, according to the law of Newton, any

two particles of matter attract each other, directed along the straight line which joins the two particles. It is directly proportional to the product of the masses of the particles and inversely proportional to the square of the distance between them.

gravitational astronomy

The study of astronomy from the point of view of gravitational forces.

gravity

The apparent force per unit mass with which the earth attracts bodies near its surface, as measured by the acceleration of a body in free fall relative to the surface of the earth. This apparent force acting on a body of mass m at a point where the acceleration of gravity is g is mg and is called the weight of the body.

gray body

A body which emits radiant energy, having the same relative spectral energy distribution as a black body at the same temperature, but in a smaller amount.

great calorie

See *kilocalorie.*

great circle charts

Charts on which the earth's surface is projected on a plane tangent to the earth's surface by rays from the earth's center. On such a chart all great circles appear as straight lines, while meridians appear as straight lines converging toward the pole. The distortion is least at the point of tangency and increases as that point is left.

great circle of a sphere

Any section produced by a plane that passes through the center of the sphere.

Greek alphabet

The Greek letters, which are extensively used as physical and mathematical symbols, are given here for purposes of reference:

Cap.	Lower case	Greek name	English sound
Α	α	Alpha	A
Β	β	Beta	B
Γ	γ	Gamma	G
Δ	δ	Delta	D

Cap.	Lower case	Greek name	English sound
Ε	ε	Epsilon	E
Ζ	ζ	Zeta	Z
Η	η	Eta	E
Θ	θ	Theta	Th
Ι	ι	Iota	I
Κ	κ	Kappa	K
Λ	λ	Lambda	L
Μ	μ	Mu	M
Ν	υ	Nu	N
Ξ	ξ	Xi	X
Ο	ο	Omicron	O short
Π	π	Pi	P
Ρ	ρ	Rho	R
Σ	σ	Sigma	S
Τ	τ	Tau	T
Υ	υ	Upsilon	U
Φ	φ	Phi	F
Χ	χ	Chi	Ch
Ψ	ψ	Psi	Ps
Ω	ω	Omega	O long

grenz rays

A form of electromagnetic radiation similar to ordinary white light, but having a wavelength between that of ultraviolet light and x-rays (about 2 Angstrom units). They may considered as very soft x-rays with the ability to penetrate small thicknesses of material opaque to white light. (The term is an adaptation of the German expression *Grenzstrahlen*), literally meaning "border rays.")

Grignard's reagent

Any one of a group of organic magnesium compounds of the general type R.Mg.X where R is an organic radical and X is a halogen. It must be absolutely dry and so ether is used as a dispersing agent for the reagent.

Grotthus-Draper law

The only radiation which can produce any chemical change in a system exposed to it is that radiation which is absorbed by the system.

ground line
In descriptive geometry, the line of intersection between the vertical and horizontal planes.

ground state
The state of lowest energy level of an atom or electron; it is also the most stable state, all other states being *excited* states.

group
In chemistry (1) a number of elements having similar chemical properties, or which exhibit similar chemical reactions; (2) a number of atoms which behave as a distinct unit and go through a chemical reaction unseparated.

Grüneisen's law
The ratio of the coefficient of expansion of a metal to its specific heat is a constant regardless of the temperature, providing that the pressure is constant.

Guldberg-Waage law
See *law of mass action.*

gyrocompass
A compass consisting of a continuously driven *gyroscope* whose supporting ring confines the spinning axis to a horizontal plane, so that the earth's rotation causes it to assume a position parallel to the earth's axis, and thus to point to the true north. This mechanical compass is highly insensitive to magnetic disturbances.

gyrodynamics
The dynamics of rotating bodies, especially those affected by precession.

gyromagnetic
Pertaining to the magnetic properties of rotating electric charges, especially of electrons within atoms.

gyroscope
An instrument which utilizes a rotating mass to maintain a stable equilibrium. A wheel or disk is mounted so as to spin rapidly about an axis, and also free to rotate about one or both of two axes perpendicular to each other and to the axis of the spin. If when the wheel is spinning, a twisting force is applied about one of the two axes, it will produce a rotation, called *precession,* about the other axis. Used as a stabilizer of ships or airplanes or as a steering apparatus.

H

H

The symbol for the element *hydrogen.*

h

The symbol for *Planck's constant.*

habit

A term used to cover the varying development of the crystal forms possessed by any one mineral. Thus, calcite may occur as crystals showing the faces of hexagonal prism, basal pinacoid, scalenohedron, and rhombohedron; according to the relative development or dominance of one or other of these forms, the habit may be prismatic, tabular, scalenohedral, or rhombohedral.

hafnium

A metallic element in Group IV of the periodic table, discovered by Coster and Hevesy in 1923. Atomic number 72, atomic weight 178.58, melting-point 1700°C., specific gravity 13.3, valence 4. Symbol: Hf.

half-angle cosine formula

$$\cos\frac{\theta}{2} = \sqrt{\frac{1 + \cos\theta}{2}}$$

half-angle formulas

Formulas expressing the trigonometric functions of half a given angle in terms of functions of the angle itself. The most important of these formulas are: the *half-angle sine formula,* the *half-angle cosine formula,* and the *half-angle tangent formula.*

half-angle formulas for the solution of plane triangles

Expressions of the relations between the sides of a triangle and any of its angles:

$$\tan\frac{A}{2} = \frac{r}{s - a}, \quad \tan\frac{B}{2} = \frac{r}{s - b}, \quad \tan\frac{C}{2} = \frac{r}{s - c},$$

where A, B and C are the angles of the triangle, a, b and c the respective opposite sides, and r is the radius of the circle inscribed in the triangle.

half-angle sine formula

$$\sin\frac{\theta}{2} = \pm\sqrt{\frac{1-\cos\theta}{2}}$$

half-angle tangent formula

$$\tan\frac{\theta}{2} = \pm\sqrt{\frac{1-\cos\theta}{1+\cos\theta}}$$

half-life

The period of time required by the radioactivity of a substance to drop to one-half its original value, i.e. the time which it takes for one-half of the atoms of a radioactive substance to disintegrate. After a period of time corresponding approximately to ten times the half-life, the radioactivity of the substance, i.e., the number of the atoms in it, drops to about 0.1% of the original value. The half-life value is a physical constant for each individual isotope, and is absolutely independent of temperature, pressure, etc. The half-life of an element or isotope equals 0.6931 times its *average life* (or *decay modulus*).

half-period

A synonym for *half-life*.

half-sectional view

Such a view may be drawn for a symmetrical object. This type of section is accomplished by passing two cutting planes at right angles to each other along the center lines or symmetrical axes. Thus one-quarter of the object is considered removed and the interior exposed to view. Invisible outlines are generally omitted from the sectioned and unsectioned portions of the object but are shown in the unsectioned portion if the object is thus more clearly described. The half section provides a view which shows the internal and external features of an object.

half-thickness

The thickness of a uniform layer of an absorbent substance which is required for reducing the intensity of a radiation to one-half of its original value; its numerical value is calculated by dividing 0.6931 by the *absorption coefficient*.

half-value thickness

See *half-thickness*.

half-wave plate

A thin plate of quartz or mica cut to the proper thickness to produce a phase difference of 180° between the *ordinary* and *extraordinary rays* into which it splits an incident beam of light.

halide

Any of the binary salts (*fluorides, chlorides, bromides* and *iodides*) formed by the union of a *halogen* with a metal.

halogens

The non-metallic elements fluorine, chlorine, bromine, iodine and astatine, in Group VII of the periodic table. They are usually monovalent, electronegative, and their atoms have the extranuclear electronic structure of the rare gas atoms minus 1 extranuclear electron.

hard coal

See *anthracite.*

hard direction

Any of the directions along which it is most difficult to induce magnetization in an *anisotropic crystal,* in contradistinction to the *directions of easy magnetization.*

hard radiation

A term applied to radiation of high penetration power, i.e. of high frequency and short wavelength.

hard solder

A generic term for *solders* consisting essentially of alloys of copper and silver and zinc.

hard soldering

See *soldering.*

hard water

Water which contains sulfates, bicarbonates, and other calcium and magnesium salts and which therefore forms insoluble compounds with soap, with the result that little or no lather is formed. (Cf. *permanent hardness.*)

hardness

In solids, the property of resistance to deformation, cutting, grinding, scratching, and to impacts. Determined by measuring the depth of indentation of a standard steel ball at a standard pressure. The value obtained is directly proportional to the applied pressure and inversely proportional to the depth of penetration. (See also *Mohs' scale.*)

hardness of radiation

The penetrating power of the radiation.

hardness of water

See *hard water.*

harmonic

Having relations or properties bearing some resemblance to those of musical consonances; said of certain numbers, ratios, points, lines, motions, etc.

harmonic conjugates

The division of a line segment at two points, internally and externally, in the same ratio.

harmonic mean (H.M.)

The mean found by inverting the *harmonic progression* and thereby turning it into an *arithmetic progression*. Then find the required number of means by the method for *arithmetic means;* invert these and they become the *harmonic means* required.

harmonic pencil

Four coplanar, copunctual rays that cut every transversal in a harmonic row.

harmonic progression

The relationship to each other of three quantities, a, b and c when $\frac{a}{c} = \frac{a - b}{b - c}$. The reciprocals of quantities in *arithmetic progression* are in harmonic progression; e.g., if 7 and 1/6 are terms of a harmonic progression, then 1/7 and 6 are the terms of the corresponding arithmetic progression.

harmonic range

Four collinear points forming two pairs of harmonic conjugates. Also called *harmonic row.*

harmonic row

See *harmonic range.*

harmonic series

A series the terms of which are in *harmonic progression* (e.g., the series $1 + \frac{1}{2} + \frac{1}{3} + \frac{1}{4} + 1/5 + \ldots$); the reciprocals of these terms form an *arithmetic series.*

harmonic triangle

A triangle formed by joining the points of intersection of the sides of a quadrilateral and the point of the intersection of the diagonals of the quadrilateral. Also called *diagonal triangle.* (Cf. *self-conjugate triangle.*)

harmonics

In music, those tones which sound above any actual musical note. Each basic frequency in a harmonic series is an integral multiple of a fundamental frequency.

haversine

A trigonometric function of an acute angle, equal to one-half the *versed sine,* i.e.: $\text{hav}\ \theta = \frac{1 - \cos\theta}{2}$. (The word "haversine" is a contraction of the words "half-versed sine.")

haversine law in spherical trigonometry

In any spherical triangle, hav a = hav (b — c) + sin b • sin c • hav A; etc.

H-bomb

See *hydrogen bomb.*

H.C.F.

Highest common factor (q.v.).

He

The symbol for the element *helium.*

heat

The total energy of the random molecular motion of a body. It is defined also as the average kinetic energy of translation of the random molecular motion.

heat

In metallurgy, (1) the steel melting operation from the charging of raw materials, to the tapping of the molten metal; (2) the batch of steel produced by a single melting operation; (3) in forging, the period of working before reheating becomes necessary; (4) the amount of forging stock heated up at one time.

heat capacity

See *thermal capacity.*

heat conductance

The time rate of heat flow through unit area of a body, of given size and shape, per unit temperature difference. Also called *thermal conductance.*

heat conduction

Transmission of heat by molecular collisions.

heat conductor

A material which readily transmits heat by means of conduction. Also called *thermal conductor.*

heat conductivity

The rate of transfer of heat along a body by conduction, i.e., the facility with which a substance permits heat to flow through it, measured by the quantity of heat transmitted per unit time, per unit temperature gradient along the direction of the flow, per unit cross-sectional area, e.g., in calories flowing per second across a cubic centimeter of the substance having a temperature difference of 1 °C on opposite faces. Also called *thermal conductivity*.

heat engine

A machine which transforms heat into mechanical energy.

heat of absorption

The quantity of heat released when a gaseous substance is dissolved.

heat of adsorption

The quantity of heat evolved in the absorption of a definite quantity of gas on a bare surface.

heat of aggregation

A collective term for the quantities of heat absorbed or released in forming aggregates, as, e.g. the *heat of condensation, heat of crystallization, heat of fusion, heat of vaporization,* etc.

heat of combustion

The heat, measured in calories, released per gram-atom or gram-molecule when an element, respectively a compound, is completely oxidized or burned in oxygen at constant volume.

heat of condensation

The amount of heat given up by a unit mass of a substance when it passes from the state of vapor into the liquid state.

heat of crystallization

The heat which is released when a unicomponent system crystallizes, or which is absorbed when such a component melts, under isothermal conditions.

heat of dilution

The quantity of heat absorbed or released by diluting a solution with water or another diluent, or by mixing a strong solution with a weaker solution of the same substance.

heat of dissolution

See *heat of solution.*

heat of formation

The quantity of heat absorbed or released when a compound is formed from its component elements.

heat of fusion

The heat absorbed by a unit mass of a solid when it melts; the quantity of heat required to change a unit mass of a substance from the solid to the liquid state at its melting point.

heat of hydration

The quantity of heat absorbed or released when a substance takes up water.

heat of ionization

The quantity of heat absorbed or released when one gram-equivalent of a substance is broken up completely into positive and negative ions.

heat of linkage

The quantity of heat required for forming or disrupting a chemical bond.

heat of melting

See *heat of crystallization, heat of fusion.*

heat of mixture

The quantity of heat released when two liquids which do not react together are mixed. Also called *heat of mixing.*

heat of neutralization

The quantity of heat released when one gram-equivalent of an acid is neutralized by a base.

heat of reaction

The quantity of heat absorbed or released in a chemical reaction when the reaction takes place at constant volume, in the direction indicated by, and between the amounts of substances represented in a chemical equation. (This term subsumes a number of expressions, as *heat of combustion, heat of linkage, heat of neutralization,* etc.)

heat of solution

The heat released or absorbed as a solid is dissolved in a liquid. In a unicomponent system, it equals the heat of fusion.

heat of sublimation

The amount of heat used when a solid changes directly to its gaseous state in doing external work (expansion) and internal work (to increase the molecular velocity).

heat of transition

A general term for the quantity of heat absorbed or released, at the *transition temperature,* as a substance passes from one alotropic form into another. Also called heat of *transformation.*

heat of vaporization

The heat absorbed by a unit mass of a liquid when it vaporizes; the quantity of heat required at a given temperature to convert a unit mass of liquid into vapor.

heat summation

See *Hess' law of constant heat summation.*

heat tinting

A method depending upon differential oxidation which involves heating a *metallographic* specimen, with a suitably polished surface, in air, for the purpose of developing the microstructure by oxidizing or otherwise affecting the different constituents.

heat transmission coefficient

Any one of a number of coefficients used in the calculation of heat transmission by conduction, convection, and radiation, through various materials and structures.

heat treatable alloy

An alloy the properties of which can be modified by heat treatment.

heat treatment

A process in which steel in the solid state is taken through one or more temperature cycles for the purpose of obtaining certain desired properties. Heating for the sole purpose of hot working is excluded from the meaning of this definition.

Heaviside layer

The ionized layer above the earth. Known also as the *Appleton layer.*

heavy hydrogen

See *deuterium.*

heavy mesons

See *K-mesons.*

heavy metal

A metal which has a *specific gravity* greater than 4.

heavy water

Deuterium oxide (D_2O) ; water, in the molecules of which the place of the ordinary hydrogen atoms (${}_1H^1$) is taken by two atoms of *deuterium.* Pure D_2O is a colorless, odorless liquid, identical in appearance with ordinary water (H_2O) , but has a density of 1.105 and a molecular weight of 20, as against the density of 1.000 and molecular weight of 18 of H_2O; its freezing-point is 3.82°C. and its boiling-point 101.42°. (as against the freezing-point of 0°C. and boiling-point of 100°C. of H_2O) . Ordinary water as occurring in nature contains 1 part of D_2O in 5000; the D_2O can be isolated from ordinary natural water by electrolysis (20 liters of ordinary water will yield, after successive electrolytic treatments, 12 $cm.^3$ of 99.9% pure D_2O) . Heavy water has adverse effects on many living organisms, acting especially as a retarder of growth; organisms on a lower level of the evolutionary scale die in pure D_2O. It has been found to be very suitable for use as *moderator* to slow down neutrons in nuclear reactors. First isolated by H. C. Urey and collaborators.

hecto-

A prefix meaning *one hundred.*

heft

In geological terminology, the specific gravity or weight of a mineral in comparison with the weight of an equal volume of water.

height of a plane figure

See *altitude*.

helical

Shaped like a *helix*.

heliography

The study and description of the sun and its properties and characteristics.

helium

An inert, zerovalent element in Group O of the periodic table, lightest of the rare gases, discovered by Lockyer, spectrographically, in 1868, and by Ramsay in the mineral clevite in 1895. Atomic number 2, atomic weight 4.003, melting-point −272.2° C., boiling-point −268.9°C., density 0.17848 gram per liter. Symbol: He. A He^4 atom stripped of its two extranuclear electrons, and therefore carrying two positive charges on the nucleus, is known as an *alpha particle,* one of the most important tools of nuclear physics.

helix

The locus of a point which rotates about a fixed axis and moves along the same axis at the same time.

hemi-

A prefix, used in Greek compounds, meaning *half*. (Analogous with the prefix *semi-* used in compounds of Latin derivation.)

hemihedral crystal

A crystal in which are found only alternate planes of the fundamental crystal form. Such crystals thus lack symmetrically opposite faces and have different properties in different directions.

hemimorphic crystal

A crystal whose axes of symmetry differ at opposite faces.

hemisphere

One of the two parts of a sphere obtained by cutting the sphere by a plane passing through its center.

henry

The electromagnetic unit of inductance, defined as the inductance present which will cause 1 volt to be induced if the current changes at the rate of 1 ampere per second. 1 henry = one billion (10^9) abhenrys.

Henry's law

The weight of a slightly soluble gas that dissolves in a definite weight of a liquid at a given temperature is very nearly directly proportional to the partial pressure of that gas. This holds for gases which do not unite chemically with the solvent.

heptagon

A regular polygon bounded by six straight lines.

Hero's formula

A formula expressing the area of the triangle as :

$$\sqrt{s\ (s-a)\ (s-b)\ (s-c)},$$

where a, b and c are the sides and $s = \frac{a+b+c}{2}$.

hertz

A unit of electrical frequency, mostly used in Europe; 1 hertz = 1 cycle per second.

Hertzian waves

The electromagnetic radiation now generally referred to as *radio waves.*

Hess' law of constant heat summation

The amount of heat generated by a chemical reaction is the same whether reaction takes place in one step or in several steps, since all chemical reactions which start with the same original substances and end with the same final substances liberate the same amounts of heat, irrespective of the process by which the final state is reached.

heterocyclic compound

An organic compound the molecules of which have a ring structure with an atom or atoms of an element or elements other than carbon in the ring.

heterodyne

The interaction of two waves of different frequencies; they are alternately additive and subtractive, thus producing two *beat* frequencies which are respectively the sum of and the difference between the two original frequencies.

heterodyne frequency

A *beat* frequency which is the sum or the difference of two signals.

heterogeneous catalysis

A catalytic process in which the catalyst constitutes a phase separate from the reaction mixture.

hexa-
A prefix meaning *six.*

hexagon
A regular polygon bounded by six straight lines.

hexagonal system
The crystallographic system which comprises crystals having parallel 6-fold rotation axes of symmetry or a single set of parallel 3-fold rotation axes of symmetry with planes of symmetry perpendicular thereto, and with or without other symmetry elements. The typical hexagonal crystal has two equal axes, inclined at an angle of 120 degrees and a third axis, perpendicular to their plane.

hexahedron
A polyhedron having six faces. (The *regular hexahedron* is referred to as a *cube.*)

hexode
A six-electrode vacuum tube containing an anode, a cathode, a control electrode and three additional electrodes ordinarily in the nature of grids.

hexose
A simple sugar *(monosaccharide)* containing six carbon atoms in the molecule; the general formula of the hexoses is $C_6H_{12}O_6$.

Hf
The symbol for the element *hafnium.*

Hg
The symbol for the element *mercury.*

higher mathematics
Includes all mathematics beyond ordinary arithmetic, algebra, geometry and trigonometry.

higher plane curves
A collective term for *algebraic curves* of a degree greater than 2, and for all *transcendental curves.*

highest common factor (H.C.F.)
The largest expression common to two or more given expressions, i.e., the highest number or quantity that will divide exactly, without remainder, into two or more given numbers or quantities. E.g., 3 is the H.C.F. of 6, 9 and 15; ab is the H.C.F. of a^2b^2, a^4b and ab^3.

histogram
In the graphical representation of frequency distributions, a bar graph with the bars centered over the class marks and having heights equal to the class frequencies.

historical geology
That branch of geology which, aided by other branches, deals with the history of the earth and aims to give a chronological account of the events in that history.

H. M.
The symbol for *harmonic mean.*

Ho
The symbol for the element *holmium.*

holmium
A trivalent metallic element, one of the rare earth group of the periodic table, discovered by Cleve in 1879. Atomic number 67, atomic weight 164.94. Symbol: Ho.

holo-
A prefix meaning *whole.*

homocyclic compound
An organic compound the molecules of which have a ring structure consisting solely of carbon atoms.

homogeneous catalysis
A catalytic process in which the catalyst and the reaction mixture constitute a single, homogeneous phase.

homogeneous compound expression
A mathematical expression in which all terms are of the same dimension. Thus $a^7 - 5ab^6 + 3c^3d^4$ is a homogeneous expression of seven dimensions.

homogeneous equation
An equation in which all the terms are of the same degree in the unknowns.

homographic
Two ranges of points A,B,C,D,E........and A′,B′,C′,D′,E′........are said to be homographs when a cross-ratio of any four points of the one is equal to the corresponding cross-ratio of the four corresponding points of the other.

homographic pencil
If two systems of concurrent lines are so related that the cross-ratio of any four rays of one system is equal to the cross-ratio of the corresponding rays of the other, then the two systems are said to form homographic pencils.

homologous
Corresponding in position, proportion, value or structure.

homologous series
In chemistry, a series of compounds in which the composition of each member can be represented by a general formula, as,

e.g., the sugars with their general formula $C_nH_{2n}O_n$ or the methane series C_nH_{2n+2}.

homologue

A point or line in one system of points or lines corresponding to a point or line in another similar system.

homomorphous

A term applied to substances which crystallize in a similar form but are chemically distinct.

homopolar crystals

Crystals where neighboring atoms share electrons; for instance, diamond, where the four outer (valence) electrons of each carbon atom are shared with its four nearest neighbours, thus completing an outer shell of eight electrons per atom.

homothetic

In geometry, said of figures which are mutually similar and similarly placed, as any two parallel plane sections of a cone.

Hooke's law

The law which states that within the limits of elasticity, the strain is proportional to the stress which produces it. ("Stress" means here compression, elongation or shear, and "strain" is the applied force per unit area.) The value of the stress at which a material ceases to obey this law is called the *limit of proportionality*.

horizontal

Parallel to a straight line drawn as a tangent to the surface of the earth, i.e., parallel to the plane of the horizon; at right angles to the position of a plumb line.

horizontal line

In orthographic projection, a line that is parallel to the horizontal projection plane.

horizontal plane

A plane that is at right angles to any straight line that passes through the center of the earth, i.e., at right angles to the plumb line.

horizontal projection

The representation of a point or object in space in top *view*, projected to a horizontal projection plane, i.e., in a view in which the lines of sight of the observer are vertical.

hormone

A generic term for a group of substances produced by the internal secretion glands of the human and animal bodies, and which act as catalysts and regulators of physiological processes.

Horner's method
A method of approximating irrational roots of a polynomial, consisting of successive transformations each of which diminishes the roots by a certain amount.

horsepower (HP)
A unit of power, for measuring the rate of work; 1 HP = 23.56 centigrade heat units per minute = 42.42 British Thermal Units = 550 foot-pounds per second, or 33,000 foot-pounds per minute, or 746 watts. Also referred to as *watts horsepower,* to distinguish it from the *metric horsepower.*

horsepower-hour
An engineering unit of work, equal to the work done when one horsepower acts for one hour, i.e., 1,980,000 foot-pounds; it is the rate of doing work of 33,000 foot-pounds per minute, or 550 foot-pounds per second. It is a gravitational unit and varies according to locality.

hot
In nuclear engineering, a widely used colloquial term meaning "highly radioactive."

hot atom
An atom, the nucleus of which is radioactive, which circumstance may effect certain chemical reactions which would not ordinarily occur.

hot atom chemistry
The study and application of the chemical reactions and properties of atoms that are radioactive or in a high state of excitation or have acquired high kinetic energy as a result of nuclear processes (referred to as *hot atoms*).

hot junction
In a *thermocouple,* the joint at the heated end.

"hot" laboratory
A nuclear physical or chemical laboratory in which the β and γ radiation exceeds 1 curie.

hour circle
In astronomy, any of the great circles of the *celestial sphere* which pass through the celestial poles.

hue
A property of colors, determined by the assortment and relative intensities of the component wavelengths.

humidify
To increase, by any process, the density of water vapour within a given space.

humidity

Water vapour within a given space. (See *absolute humidity, relative humidity, specific humidity.*)

humidity ratio

See *mixing ratio* and *specific humidity.*

hundred

A number followed by two digits and denoting an integer multiple of the square of 10 (i.e., 100).

hundredth

A fraction the denominator of which is 100; in the decimal notation, the second digit after the decimal point.

Huyghens principle

Every point of an advancing wave front can be considered as the center of a wavelet, and the wavelets emitted by the point combine to form a new wave front which envelops all such wavelets.

hydracid

A *hydrogen acid,* i.e., an acid which contains no oxygen.

hydrate

Substance holding water in the molecular form (H_2O), or as water of crystallization. There are mono-, tri-, tetra-, penta-, hexa-, and hepta-hydrates, representing respectively 1 to 7 molecules of H_2O to 1.0 of the substance.

-hydrate

A suffix indicating the presence of hydrogen and oxygen in the same proportion in which they form water (i.e., two hydrogen atoms to one oxygen atom).

hydrated

Containing chemically combined water.

hydraulics

The study of fluids in motion and of their use to do work.

hydride

Any compound of hydrogen with a more positive element, or with a member of the phosphorus group, or with a radical.

hydro-

A prefix indicating "water" or hydrogen.

hydrocarbon

A compound which consists of carbon and hydrogen.

hydrochloric acid (HCl)

Aqueous solution of hydrogen chloride, a strong acid, fuming when concentrated; the concentrated acid of common use is

35% HCl and of specific gravity 1.18. (Also called *muriatic acid.*)

hydrodynamics

That branch of mechanics which treats of the motions of fluids.

hydrofluoric acid (HF)

A clear, colourless fuming corrosive liquid, which owing to its characteristic reaction with silica, is used as a chemical reagent. The acid is stored in wax containers, and produces severe sores if allowed to come into contact with the skin.

hydrogel

A gel obtained from a colloid with the inclusion of water.

hydrogen

A monovalent, gaseous element in Group I of the periodic table, the lightest of all known elements, discovered by Cavendish in 1776. Atomic number 1, atomic weight 1.008, melting-point −259.14°C., boiling-point −252.7°C., density 0.08988 gram per liter at normal temperature and pressure. Stable isotopes: H^1 (natural abundance 99.98%) and H^2, called *deuterium* (0.02%). Radioisotope: H^3 (called *tritium*), a highly unstable isotope with a weak β-radiation, which occurs with an abundance of less than 10^{-10}%. The nucleus of the H^1 atom is a single proton, one of the fundamental nuclear particles. Symbol: H. (See also *orthohydrogen.*)

hydrogen acid

An acid that contains no oxygen. Also called *hydracid.*

hydrogen bomb

A colloquial generic term for *nuclear fusion bombs,* i.e., for bombs which utilize the process of *nuclear fusion* (q.v.) to release vast quantities of energy. Also referred to as *H-bomb.*

hydrogen equivalent

The number of grams of a given metal which will combine with or replace one gram of hydrogen or its equivalent.

hydrogen ion

A hydrogen atom which has lost its electron and therefore is positively charged, i.e., a *proton.*

hydrogen ion concentration

A measure of the acidity of a solution, usually indicated as the *pH* of the solution. (See *pH.*)

hydrogenation

The addition of or combination with hydrogen.

hydrokinetics
That branch of physics which treats of the motion of fluids under the effects of forces.

hydrology
The science which treats of the properties and phenomena of water in all its states, as well as of the relation of water and its phenomena to human life and activities.

hydrolysis
The addition of water to a compound causing it to split into two parts, one part chemically reacting with the H of the water and the other part with OH.

hydrometallurgy
The extraction of metals from their ores by *leaching* with an aqueous solvent which dissolves the metal without attacking the *gangue*. (Also called *wet metallurgy*.)

hydrometer
An instrument for measuring the specific gravity (density) of liquids.

hydrometrograph
An instrument for determining and recording the quantity of water discharged from a pipe, orifice, etc., in a given time.

hydronium ion
A hydrated proton, i.e., a combination of a hydrogen ion and a water molecule (H_3O^+).

hydrophile
A substance which absorbs water. (Cf. *lyophilic.*)

hydrophobe
A substance which repels water. (Cf. *lyophobic.*)

hydrophore
An instrument for obtaining specimens of water from any desired depth, as in a river, lake or ocean.

hydroponics
The process or technique of growing plants without soil, in nutrient solutions of the mineral salts which the plants need and which they would obtain from the soil.

hydrosol
A colloidal solution in which water is the *continuous phase.*

hydrosphere
The liquid part of the surface of the earth.

hydrostatic pressure
The pressure exerted by water or another liquid at rest.

hydrostatic testing

The testing of a hollow part, e.g. a tube for tightness by means of water, oil, or other liquid under pressure.

hydrostatics

That branch of physics which applies the principles of *statics* to water and other liquids, and which treats chiefly of the equilibrium, pressure and other effects and phenomena of water and other liquids.

hydroxide

A compound formed by an element with the radical, -OH.

hydroxy-

A prefix indicating the presence of the monovalent -OH radical.

hydroxyl

Containing the monovalent -OH group.

hydroxyl radical

The monovalent -OH group.

hygrometer

An instrument for measuring the percentage of moisture in air or in other gases.

hygroscopic

Said of any substance which readily absorbs moisture or has a particular affinity for moisture.

hyper-

A prefix meaning *over* or *above,* and used in general to indicate an excess, as, e.g., *hypersalinity,* meaning an abnormally high degree of salinity.

hyperbola

A *conic section* formed when both *nappes* of a *right circular cone* are cut by a plane parallel to the axis. It is the locus of a point which moves so that the difference of its distances from two fixed points (the *foci*) remains constant, i.e., the ratio of the distance of which from a fixed point (one of the *foci*) and from a fixed line (one the *directrices*) is a constant (called the *eccentricity* of the hyperbola) is a constant greater than unity.

hyperbolic cylinder

A cylinder the right section of which is a hyperbola.

hyperbolic functions

Functions, analogous to the *circular functions,* but related to the *rectangular hyperbola*. They are the *hyperbolic sine (sinh), hyperbolic cosine (cosh), hyperbolic tangent (tanh), hyperbolic cotangent (coth), hyperbolic secant (sech), hyperbolic cosecant (csch).*

hyperbolic involution
See *involution.*

hyperbolic logarithms
See *natural logarithms.*

hyperbolic paraboloid
A *paraboloid* that can be held so that those of its sections lying parallel to one of the coordinate planes are hyperbolas, whereas those lying parallel to the other coordinate plane are parabolas.

hyperbolic system of logarithms
The *natural* or *Napierian* system of logarithms (q.v.).

hyperboloid
A quadric surface some plane sections of which are hyperbolas.

hyperboloid
A surface resulting from the revolution of a hyperbola about its axis (the *y axis*).

hyperboloid of one sheet
A hyperboloid every section of which that lies parallel to a coordinate plane is a hyperbola or an ellipse.

hyperboloid of revolution
A term applied to special cases of the *hyperboloids of one sheet* and of *two sheets,* viz., to those hyperboloids in which the sections which ordinarily are ellipses become circles.

hyperboloid of two sheets
A surface which is cut in hyperbolas by planes parallel to two of the three coordinate planes, and in ellipses by planes parallel to the third coordinate plane.

hyperons (Y-particles)
A generic term for subatomic particles the mass of which is greater than that of a *neutron,* but less than that of a *deuteron.*

hypersonic
Of or pertaining to the speed of an object moving at Mach 5 or greater relative to surrounding fluid.

hyperspace
A theoretical space containing more than three spatial dimensions.

hypertonic solution
A solution which has a greater *osmotic* pressure than another solution used as standard of comparison (usually blood).

hypo-

A prefix meaning *under-* or *below,* and used in general to indicate a deficiency, as, e.g., *hypoacidity,* meaning an abnormally low degree of acidity.

hypocycloid

A curve which is the locus of a point on a circle that rolls along the inside of a fixed circle. (Cf. *epicycloid.*) A closed hypocycloid which touches the inside of the stationary circle in four points is referred to as a *hypocycloid of four cusps* (also called an *astroid*).

hypotenuse

The side of a right triangle which is opposite the right angle.

hypothesis

An assumption, postulate or supposition; a statement based on a principle or rule that has not yet been proved.

hypotonic solution

A solution which has a lower *osmotic* pressure than another solution used for comparison (usually blood).

hypotrochoid

A curve which is the locus of a point on the radius, or extended radius, of a circle that rolls on the inside of a fixed circle.

hysteresis

The lagging of the magnetization of a magnetic substance behind the magnetizing force.

hysteresis

In metallurgy and foundry technique, a synonym for *cooling lag.*

hysteresis curve

A curve showing the relation between the *magnetizing force* and *flux density* in a sample of iron or steel, the curve being taken with ascending and descending values of magnetizing force. The resulting closed curve is called a magnetic *hysteresis loop.*

hysteresis loop

For a magnetic substance in a cyclically magnetized condition, a hysteresis loop is a curve (usually expressed in rectangular coordinates) showing, for each value of the magnetizing force, two values of the magnetic induction—one when the magnetizing force is increasing, and another when it is decreasing.

hysteresis loss

The power expended in a magnetic material as a result of magnetic hysteresis, when the magnetic induction is cyclic.

I

I

The symbol for the element *iodine.*

i

The imaginary unit, $\sqrt{-1}$.

-ic

In chemistry, (1) a suffix indicating a higher valence of an element or radical (e.g., ferric); (2) a suffix indicating acids in general.

ice

Water (H_2o) at or below the temperature of 0°C (32°F). Ice is less dense than water and therefore floats on water.

Iceland spar

A transparent calcite which has the property of producing double refraction and polarization of light. Its chemical formula is $CaCO_3$.

iconoscope

A camera tube in which a high-velocity electron beam scans a photoactive mosaic which has electrical storage capability. Used in television.

icosahedron

A polyhedron that has twenty faces. (In a *regular polyhedron* they are all congruent equilateral triangles.)

-id, -ide

In chemistry, a suffix indicating a compound derived from a negative or nonmetallic element, usually a binary compound (e.g., carbide, sulfide).

ideal gas

A gas which conforms to the characteristic equation of a gas (*"ideal gas law"*) : $pv = RT$, where p is the pressure and v the volume at a given temperature, T is the absolute temperature, and R is the *ideal gas constant,* $\frac{p_0 v_0}{273}$ (centigrade), where p_0 is the pressure and v_0 the volume at zero degrees C. (the numerical value of R is about 8.3144×10^7 ergs/mol°C.). For n moles of gas, the equation is PV = nRT; R is referred to as the *molar gas constant.*

ideal gas constant
See *ideal gas.*

ideal mechanical advantage
In mechanics, a measure of the utility of a machine; it is the ratio of the distance through which the input force acts to the distance through which the output force acts. (Abbreviation: IMA.) Often referred to also as *velocity ratio.* (Cf. *actual mechanical advantage.*)

ideal solution
A solution through which all the intermolecular forces are completely identical.

identical
Exactly alike in form, properties and/or value.

identical equation
An equation that reduces to the same value for both members regardless of the values assigned to the letters and symbols.

identity
A statement of equality which holds true for all values of the quantities involved in the expression. The sign denoting identity is $\equiv$. E.g., $x + 5 + 2x \equiv 3x + 5$.

idiomorphic crystal
A perfect crystal exhibiting geometric arrangement of plane faces, observed occasionally with certain constituents or intermetallic compounds in the microstructure of cast alloys.

illinium
Element, atomic number 61 (promethium), supposedly discovered by B. S. Hopkins in 1926.

illuminance
The density of *luminous flux* incident on a surface, equal to the quotient of the quantity of luminous flux by the surface when the latter is uniformly illuminated.

illumination
Luminous flux per unit area. (See *illuminance,* also *Lambert's law of illumination.*)

imaginary axis
In the *Argand diagram,* measured along the *y-axis.*

imaginary curves and surfaces
Curves and surfaces representing loci with imaginary coordinates, i.e., corresponding to imaginary values of the variables which satisfy the equations of the curves or surfaces in question.

imaginary number
A number that is the square root of a negative number or involves square roots of negative numbers.

imaginary unit
$\sqrt{-1}$, designated by *i*.

imbibition pressure
The pressure against which a colloid will imbibe a liquid, or conversely, that pressure which will be needed to force the dispersion medium out of a gel, or similar colloid.

immiscible
Incapable of being mixed to form a homogeneous substance; usually said of liquids (e.g., oil and water are immiscible).

impact
The action of two bodies in collision whereby the velocity of one or both bodies is changed. The sum of the moments of the two masses before and after impact is the same.

impact strength
The resistance a material is capable of developing against impact blows; usually expressed as the ft.-lb. of energy necessary to break a standard specimen.

impact test
A test to determine the resistance of a material to a suddenly applied stress, i.e., shock. The resistance is usually reported as the energy in ft.-lb. or kilogram-meters to fracture the test piece or as the energy per unit section behind the notch.

impedance
The total opposition to the flow of an alternating current, i.e. the combined effect of *resistance* and *reactance*. The magnitude of the impedance of an electric circuit to a completely specified periodic current and potential difference is the ratio of the effective value of the potential difference between the terminals to the effective value of the current, there being no source of power in the portion under consideration. It is expressed in ohms.

imperial standard pound
The mass of a platinum cylinder kept at the Bureau of Standards office. 1 lb. = 453.592 grams.

impregnation
The absorption of one substance by another.

improper fraction
A fraction the numerator of which is greater than its denominator; e.g., 7/3 or 131/13.

improper integral

An integral one or both of the limits of integration of which is not finite, or the integrand of which itself becomes infinite between or at the limits of integration.

impulse

In general, any force which acts over a relatively brief length of time. In mechanics, the average value of a force multiplied by its duration in time.

impurities

(1) Any extraneous undesirable material in a substance. (2) Those elements or compounds in metal or steel whose presence is accidental and which have not been added by design.

in parallel

Said of pieces of electrical apparatus when they are so connected that the current divides between them and re-unites after passing through them.

in phase

Said of periodic quantities which have the same *frequency* and *waveform* if they reach their maxima at the same time.

in series

Said of pieces of electrical apparatus when they are so connected that the same current flows in turn through each of them.

in vacuo

In a vacuum.

in vitro

Occurring outside a living organism; specifically, in a test tube.

in vivo

Occurring in a living organism.

incandescence

Emission of light owing to the high temperature of the emitting body.

incandescent

Made visible or luminous by heat.

incenter of a triangle or polygon

The center of a circle inscribed in a triangle or any polygon, the sides of which are all tangent to the circle.

inch-pound

A unit of energy equal to that required to lift 1 lb., 1 in. high.

inch-ton

A unit of energy equal to that required to lift 1 ton, 1 in. high.

incident radiation
The radiation which arrives at a given point or surface.

incident ray
With reference to refraction, diffraction, etc., a ray which arrives at a given point or surface from the source of the radiation, in contradistinction to a refracted ray, etc.

inclination of a line
In analytical geometry, the angle θ $(0 \leq \theta < 180°)$, measured counterclockwise from the positive X axis to a line, directed or not.

inclination of a straight line to a plane
The angle between the line (or the line produced) and its projection on the plane.

inclination of two planes to one another
The angle formed by the intersections of the two planes with a third plane which is drawn perpendicular to the line of intersection of the two planes.

inclined view
Any view, i.e., representation of a point or object in space on a surface, in which the lines of sight of the observer are neither vertical nor horizontal.

included angle
An angle confined between two straight lines.

inclusions
(1) In general, foreign particles in a substance.— (2) In metallurgy, nonmetallic particles found in steel or another alloy.

incommensurable
Said of two numbers the ratio of which cannot be expressed exactly by two whole numbers. Thus $\frac{1}{3} : \frac{7}{8}$ or 2.56 : 5.74 or $\sqrt{2:5}$ or $\frac{\sqrt{2}}{5}$. Having no common measure (integral or fractional).

incomplete diffusion
See *partial diffusion.*

incomplete quadratic
A quadratic where not all the terms of the type are present.

incompressibility modulus
A synonym for *modulus of elasticity* (in compression).

indefinite integral
Any function the derivative of which is the given function (called the *integrand*). The indefinite integral of the integrand $f(x)$ with respect to x is written by the notation $\int f(x)dx$.

independent variable

A quantity which may have any value we care to give it. The value of the corresponding dependent variable is fixed as soon as the value of the independent variable is known. Thus if in the equation $x \times 5 = y$ we give the x the value 7, then automatically y, the dependent variable, equals $7 + 5 = 12$. (Cf. *function.*)

indeterminate

Indefinite, not fixed, not precise, not determined. If the number of unknown quantities is greater than the number of independent equations, there will be an unlimited number of solutions, and the equations are said to be indeterminate.

index

(1) An indicator, usually expressed numerically, of the relation of one phenomenon to another.— (2) An indicating part of an instrument.

index of a logarithm

A synonym for *characteristic* of a logarithm (q.v.).

index of a radical

The small number or letter placed in the $\sqrt{}$ of the radical (or root) sign to indicate what root is sought. Thus in $\sqrt[3]{}$ the 3 is the exponent of the root and means third or cube root.

index of refraction

The ratio of the sine of the angle of incidence of light to the sine of the angle of refraction; it is a constant, independent of the magnitude of the angle of incidence.

index point

In *spectrographic analysis,* the index point is that concentration of the element being determined at which the intensities of the radiations producing the analysis and internal standard lines are equal.

indicator

A substance, which when added in small amounts to a chemical system, shows, by an easily discernible change of color or some other characteristic, a condition of the system or reaction; i.e., pH, etc.

indium

A trivalent metallic element in Group III of the periodic table, discovered by Reich and Richter in 1863. Atomic number 49, atomic weight 114.82, melting-point 155°C., boiling-point 1450°C., specific gravity 7.28. Symbol: In.

induced radioactivity
Artificial radioactivity, produced by bombardment by subatomic particles.

induced reaction
A chemical reaction which is speeded up by the addition of an inductor or a *promoter.*

inductance
(1) The resistance offered to a change in *e.m.f.* in a circuit, due to the self-induced e.m.f. caused by variation of the magnetic field, the direction of the latter e.m.f. being such as always to counteract changes in the current. – (2) An electrical unit in which this property is expressed.

induction
(1) The process which sets up electric currents in a conductor when it is placed into an electromagnetic field.– (2) See *electric induction* and *magnetic induction.*

induction period
In a chemical reaction that period before reaction commences.

inductive reasoning
A method of reasoning which starts out from specific cases, studies and analyzes them and reaches conclusions on the basis of the observations.

inductometer
An instrument or apparatus for ascertaining the force of electrical induction or for comparing the specific inductive capacities of various substances.

inductor
A substance the addition of which will accelerate a slow chemical reaction.

-ine
In organic chemistry, the usual ending of the names of *alkaloids,* indicating the presence of the *amine group.*

inequality
A statement of the fact that one quantity, a, is greater or less than another quantity, b (written as $a > b$, respectively $a < b$). If this statement holds true for all values of a and b, it is called an *absolute inequality;* if it is true only for certain values of a and b, it is referred to as a *conditional inequality.* (The idea of inequality applies only to *real* numbers.)

inert
Having no or almost no chemical action; very slow or totally unable to combine with any other substance.

inert gas
An alternative name for a *rare gas.*

inert oxide
An oxygen compound which forms neither an acid nor a basic compound.

inertia
The tendency of a body to remain in its condition of rest or motion. The force with which the body offers resistance to a change in its state of rest or motion is called *force of inertia.*

inertial force
The force produced by the reaction of a body to an accelerating force; it is equal in magnitude and opposite in direction to that accelerating force, and it ceases when the application of the accelerating force ceases.

inference
A conclusion arrived at or drawn from given premises or deduced from the facts; the act of reaching such conclusions.

infinite
Endless, boundless, unlimited; greater than any assignable quantity of the same kind.

infinite series
The indicated sum of an infinite number of terms.

infinitesimal
A quantity smaller than any assignable magnitude, but greater than zero.

infinitesimal calculus
See *calculus.*

infinity
In mathematics, that quantity which is greater than any assignable magnitude. (Symbol: ∞)

infinity
The extreme point of distance. In optics, the term is used to denote a distance sufficiently great so that light rays arriving from it are practically parallel to one another.

inflammable
Able to be set on fire, readily combustible. (In industrial terminology, the synonym *flammable* is more current.)

infrared radiation
Invisible rays beyond the red end of the solar spectrum, having wavelengths over 7000 Angstrom units, viz., between 7.7×10^{-5} and 0.4 cm.

infrasonic
Of a frequency below the range of audible sound.

ingredient
Any of the component parts of a mixture or compound.

inhibitor
In chemistry, any substance which stops or retards a chemical action.

initial line
In trigonometry, the stationary line forming one of the sides of an angle and coincident with the x axis of the rectangular coordinate system. (Cf. *standard position of an angle.*)

initial side of an angle
See *standard position of an angle.*

inorganic
(1) Not organic, i.e., not containing carbon or any carbon compound (other than carbonates or cyanides).— (2) In colloquial usage, the term is applied to substances which are not derived from or produced by living organisms.

inorganic chemistry
The chemistry describing the preparation, properties and reactions of all the elements, and of all their compounds, with the exception of the compounds of carbon. However, the metal carbides, together with oxides and sulphide of carbon are included in inorganic chemistry.

input
In general, energy fed into an apparatus, machine or electric circuit.

inradius
The radius of a circle inscribed in a triangle or any polygon in general all the sides of which are tangent to the circle.

inscribed
Said of a geometrical figure or solid placed into another geometrical figure or solid so that they have the greatest possible number of points in contact.

inscribed angle
(1) In a circle: an angle the sides of which are chords of the circle, and the vertex of which touches the circumferences of the latter.— (2) In a semicircle or segment: An angle the sides of which pass through the end points of the chord determining the arc.

inscribed circle
A circle drawn inside a triangle or other polygon every side of which is tangent to the circle.

inscribed sphere
A sphere so situated inside a polyhedron that every vertex of the polyhedron is tangent to its every vertex.

insolation
(1) Exposure to solar radiation.— (2) The rate at which radiant energy is incident directly from the sun per unit horizontal area at any place on or above the surface of the earth.

inspissate
To thicken by cooling or evaporation.

instrumental error
An error in the reading of an instrument caused by an imperfection or imperfections of the instrument itself. (When the imperfection is in the graduations on its scale, the error is called specifically a *scale error.*)

instrumentation
The art and technique of designing, installing and using precision instruments; also the precision instruments provided.

insulating refractories
Highly porous refractories, employed to reduce the absorption and radiation of heat in furnaces, kilns, etc.

insulation
The prevention of the flow of an electrical current or the retarding of the flow of heat. To this end the conductor is surrounded by a suitable non-conducting medium.

insulator
A *nonconductor* (q.v.).

integer
A whole number as opposed to a fraction. 2, 17, 1064 are integers. $3\frac{1}{3}$ or $\frac{7}{8}$ or 1.526 are not integers.

integral *(noun)*
See *definite integral, indefinite integral.*

integral *(adj.)*
Consisting of a whole number or an undivided quantity; not fractional.

integral calculus
The study of integration as such and its application to finding areas, volumes, centroids, equations of curves, solutions of differential equations, etc.

integral equation

An equation in which the unknown appears under an integral sign.

integrand

The function the integral of which is sought or calculated; in the expression $\int f(x)\,dx$ the term $f(x)$ is the integrand.

integrating meter

A generic term for measuring instruments which integrate the measured quantity with reference to time (e.g., the ampere-hour meter, watt-hour meter, etc.).

integration

The ascertaining of the integral (definite or indefinite) of a given function. The inverse process of *differentiation.*

intensifier

For X-rays, a substance placed in, on, or under a photographic emulsion or a fluorescent screen, or used in photographic development, to make the effects of X-rays more conspicuous.

intensity

In general, relative strength. In electromagnetic radiation or wave motion in general, the transfer of energy per second per unit area perpendicular to the direction of propagation; it is proportional to the square of the amplitude.

intensity level

In acoustics, the tenfold of the logarithm of the ratio of the intensity of a given sound to an intensity chosen as standard of reference. The intensity of a sound is expressed in units called *bel* and *decibel.*

intensity of an electric field

The force which would act on a unit positive charge placed at the given point of the electric field.

intensity of magnetization

Magnetic moment per unit volume (usually cubic centimeter), i.e., the magnetic moment of a magnet or magnetized body divided by its volume; its unit is the intensity of magnetization of a uniformly magnetic or magnetized body the magnetic moment of which is unity per cubic centimeter.

intensity of radiation

The flux of energy per unit time and per unit of cross-section perpendicular to the direction of propagation.

inter-

A prefix meaning *between* or *among.*

intercept *(verb)*

To cut off or delimit a portion of a line, surface or solid. E.g., a center angle of a circle is said to intercept any area of the circle cut off by its arm. The lengths of the x, y and z axes from the O point cut off by the plane passing through these axes are said to be intercepted by the plane.

intercept *(noun)*

In geometry, any point where a given line, plane or solid is cut off or bounded by another line, plane or solid.

interest period

In computing compound interest, the time elapsed between successive additions of interest to the principal.

interest rate

The percentage of the principal that is to be paid at specified intervals for the use of the money lent.

interface

The boundary surface between two media, phases or systems.

interference

The phenomenon produced when waves from two or more sources reach a point in space simultaneously, or when two more waves from the same source reach the same point through different paths.

interference fringes

The pattern of alternate dark and light bands produced by the *interference* of light owing to the fact that the interfering beams weaken each other or cancel each other out at some places, while reinforcing each other at others.

interferometer

An instrument for measuring wavelengths of light and for analyzing spectra. A beam of light from a large luminous area (as a sodium flame) is separated into two or more parts by partial reflections, the parts being subsequently re-united after traversing different optical paths.

intergalactic space

That part of the universe which lies beyond the limits of *interstellar space.*

interior and opposite angles

The two interior angles at any base of a triangle the sum of which is always equal to the exterior angle at the vertex to that base, are said to be interior and opposite to the exterior angle.

interior angle

The inside angles of any triangle or polygon. Also the inside angles formed by one straight line cutting two other straight lines.

internal friction

(1) The resistance of a metal to bending.—(2) The capacity of a solid to dissipate applied mechanical energy in the form of heat.

internal stress

(1) The resistance of a physical body to external forces. It is the unit internal resistance set up by external forces.—(2) In metallurgy, residual stress existing within a metal part due to the different effects of heating, cooling or working, or to the mechanical structure of the metal, or to constitutional changes in the metal itself.

international ampere

The unvarying electric current which deposits silver from a standard solution of silver nitrate at the rate of 0.00111800 gram per second. One international ampere equals 0.999835 absolute ampere.

international Angstrom (I.A.)

A unit very nearly equal to the *Angstrom* unit (10^{-8} cm.) but defined as being such that the red cadmium line at 15°C. and 760 mm. pressure would have a wave-length of 6438.4696 I.A. It is the unit adopted by the International Union of Solar Research for the measurement of the wave-lengths of light and X-rays.

international candle

A unit of luminous intensity, representing a specified fraction of the average horizontal candlepower of a group of 45 carbon-filament lamps kept at the Bureau of Standards when the lamps are operated at a specified voltage.

international coulomb

The quantity of electricity which passes any section of an electric circuit in one second, when the current in the circuit is one international ampere. One international coulomb equals 0.99985 absolute coulomb.

international farad

The capacitance of a capacitor of a charge of one international coulomb produces a potential difference between the terminals of one international volt. One international farad equals 0.99952 absolute farad.

International Geophysical Year (IGY)

A program of cooperative study of all earth sciences and all factors and elements that shape or in any manner pertain to the common physical environment of mankind. All scientific data acquired during this program (July 1957-January 1959) are to be shared by all participating nations which number more than fifty.

international henry

The inductance which produces an electromotive force of one international volt when the current is changing at the rate of one international ampere per second. One international henry equals 1.00048 absolute henrys.

international joule

The energy required to transfer one international coulomb between two points having a potential difference of one international volt. One international joule equals 1.00018 absolute joules.

international ohm

Defined as the resistance at zero degree centigrade of a column of mercury of uniform cross-section, having a length of 106.300 centimeters and a mass of 14.4521 grams. Experimental results show that one international ohm equals 1.00048 absolute ohms.

international radium standard

The 21.99 mg. of pure radium chloride prepared by Mme. Curie, and kept at Sèvres, France.

international standard atmosphere

See *standard atmosphere.*

international steam table calorie

A unit of heat, defined by international agreement as 4.11868 x 10^7 ergs.

international temperature scale

A scale of temperature fixed by international agreement; its basic fixed points are the boiling-point of oxygen (−182.97°C.), the freezing-point of water (0°C.), the boiling-point of water (100°C.), the boiling-point of sulfur (444.6°C.), the melting-point of gold (1063°C.).

international volt

The voltage that will produce a current of one international ampere through a resistance of one international ohm. One international volt equals 1.00033 absolute volts.

interplanetary space

That part of space which lies between the orbit of the moon and the outer limits of the solar system.

interpolation

The process of determining the value of a function between two known values, using some other principle than the law given by the function in question. The process or act of filling in the intermediate terms of a series of numbers or observations by numbers that follow the same law.

intersection

In geometry, the point or line which is common to two or more lines, planes or surfaces. The intersection of two lines, or of a line and a plane, is a *point,* the intersection of two planes is a straight *line,* the intersection of two surfaces is a straight *line* or a *curve.*

interstellar space

That part of space which extends from the outer limits of the solar system to the outer limits of our galaxy.

intra-

A prefix meaning *within.*

intrinsic energy

The energy assumed to be inherent in every substance and capable of being released in the form of heat in chemical reactions or in the form of other radiation in other reactions.

intrinsic radiance

Time rate of total emission of radiant energy per unit area of radiating surface.

invariant

In mathematics, a function of coefficients of a quantic that retains its form when the quantic is transformed linearly. An invariable quantity unaffected by the group of operations under consideration.

invariant

In chemistry, possessing no degree of freedom.

inverse

Inverted, reciprocal. Thus the inverse of $\frac{a}{b}$ is $\frac{b}{a}$; of $\frac{4}{5}$ is $\frac{5}{4}$.

inverse circular functions

A synonym for inverse *trigonometric function* (q.v.).

inverse cosecant

See *arc-cosecant.*

inverse cosine

See *arc-cosine.*

inverse cotangent

See *arc-cotangent.*

inverse function
The function resulting from expressing the independent variable in terms of the dependent variable, considering the latter as an independent variable.

inverse logarithm
A synonym for *antilogarithm* (q.v.)

inverse notation
If the sine of an angle B equals a number b (sin B = b), then the inverse notation is $\sin^{-1} b = B$ or $\sin^{-1}$ of b = the angle B. This applies to any other inverse trigonometrical function. Thus if tan c = d then $c = \tan^{-1} d$. Note: —1 is a symbol and not "minus one."

inverse of a number
The *reciprocal* of the number.

inverse piezoelectric effect
The contraction and expansion in quartz and certain other asymmetric crystals along an electric axis as a result of the mechanical strain set up in them as they are subjected to an electric field in that direction. (Also referred to as *indirect piezoelectric effect* and as *electrostriction*.)

inverse points
Two points, *a* and *b*, which lie on the same radius of a circle the center of which is *c*, so that $ac \times bc = c^2$.

inverse proportion
The mutual relationship of two variables the product of which is always a constant.

inverse ratio
The ratio of the reciprocals of the numbers or quantities involved. In the inverse ratio the antecedent becomes the consequent and the consequent becomes the antecedent. Thus for the ratio a : b the inverse ratio is b : a or $\frac{b}{a}$.

inverse secant
See *arc-secant*.

inverse sine
See *arc-sine*.

inverse square law of illumination
The illumination of a surface by a point source of light varies directly with the intensity of the source and inversely as the square of the perpendicular distance from the source of the surface.

inverse tangent
See *arc-tangent*.

inverse trigonometric equation
An equation in which an *inverse trigonometric function* of the unknown appears.

inverse trigonometric function
The angle the trigonometric function of which is the given argument. Thus, the inverse sine of x (written *arcsin x* or sin^{-1}) is the angle the sine of which is *x*, i.e., arcsin x = A and x = sin A. The inverse trigonometric functions, also called *antitrigonometric functions,* are the *arc-sine, arc-cosine, arc-tangent, arc-cotangent, arc-secant,* and *arc-cosecant* (q.v.). See also *principal value of an inverse trigonometric function.*

inversely proportional
Said of two variables the product of which is always a constant. Thus AB = m, which means that if A changes then B changes in such a manner that A × B always equals m.

inversion
A symmetry operation by which each point of a structure is replaced by a point the same distance from a fixed point (called the *inversion center* or *center of inversion*) but in the opposite direction therefrom.

inversion
In chemistry, (1) the change of a levorotatory compound into a dextrorotatory one, or vice versa; (2) the change of an isomeric compound to its opposite.

inversion center
See *inversion.*

invert sugar
A mixture of *glucose* and *fructose* obtained by acid hydrolysis of *sucrose,* so named because the optical rotation is inverted from *dextro* to *levo* after hydrolysis.

involute of a circle
A curve defined as the locus of a fixed point on a taut string as it is unwound from a stationary circular spool.

involute of a plane curve
A curve defined as the locus of a fixed point on a taut string which is unwound, while kept under tension, from the curve in question.

involution
The method of obtaining any power of any quantity. To obtain the power of a number, multiply the logarithm of the number by the index representing the power required; the product is the logarithm of the number required. If pairs of points A,A′; B,B′; C,C′ lie on a straight line so that their dis-

tances to a fixed point O are such that OA × OA′ = OB × OB′ = OC × OC′ = a constant k, then the pairs of points are said to be in *involution*. The pair of points is said to be *conjugate*. There will be two points F and F′, on either side of the fixed point O, the square of whose distance to O also equals the constant; these are called *double points*.

$$FO \times FO = FO^2 = k$$
$$F'O \times F'O = F'O^2 = k$$

When the constant, *k*, is negative $(OP \times OP' = -k^2)$, the involution is called *elliptic* or *overlapping;* when *k* is positive $(OP \times OP' = +k^2)$, the involution is said to be *hyperbolic* or *non-overlapping*.

Io

The symbol for *ionium*.

iodine

An element in Group VII of the periodic table, one of the group of halogens, discovered by Courtois in 1812. Atomic number 53, atomic weight 126.91, melting-point 113.5°C., boiling-point 184.35°C., specific gravity 4.94, valence 1, 3, 5, 7. Symbol: I.

iodine number

The number of grams of iodine absorbed by 100 grams of fat or oil. It is a measure of the amount of unsaturated acids present in fats and oils.

ion

An electrically charged atom or group of atoms, formed when an ordinarily neutral atom, or group of atoms, gains or loses an electron or electrons. The loss of an electron or electrons reduces the negative charge in the extranuclear atomic structure which normally offsets the positive charge of the protons in the nucleus, and therefore the atom which has lost an electron has a positive charge—a *positive ion*. Conversely, an atom which has gained an electron or electrons has a net negative charge and has become a *negative ion*.

ion antagonism

The mutual undoing of effects of various ions, e.g., calcium vs. sodium, magnesium vs. calcium, potassium vs. sodium.

ion concentration

The ion concentration of any species of ion is equal to the number of those ions, or of mols or equivalents of those ions contained in a unit volume of an electrolyte.

ion activity

The ion activity of any species of ion is equal to the "thermodynamic ion concentration," i.e., the ion concentration corrected for the deviations from the laws of ideal solutions.

ion migration

A movement of ions produced in an electrolyte by the application of an electric potential between electrodes.

ionic displacement

The exchange of charges by one element with another, through the process of oxidation or through that of reduction.

ionic reaction

A reaction between ions in solutions, usually taking place instantaneously.

ionium

A radioactive isotope of thorium ($_{90}Th^{230}$).

ionization

A process by which an atom (electrically neutral under normal circumstances) becomes electrically charged, due to losing or gaining one or more of its extranuclear electrons. An atom which has lost one or more extranuclear electrons is a *positive ion,* whereas an atom which has gained one or more extranuclear electrons is a *negative ion* (cf. *ion*).

ionization chamber

An apparatus for measuring the amount of ionization produced by radioactivity in air or some other gaseous medium. The apparatus consists essentially of an enclosure filled with air or some other gas, and containing two oppositely charged electrodes. When the air or gas is ionized by a radioactive emission or x-rays, the resulting ions are attracted to the electrodes, and the ionization current is measured and registered mechanically.

ionization coefficient

The number of ions or ion pairs produced in air or another gaseous medium by an ionizing particle or radiation per unit length (usually a centimeter) of its path.

ionization constant

See dissociation constant.

ionization energy

The energy necessary for removing an electron from an atom or molecule.

ionization potential

The minimum potential difference required for dislodging an electron from an atom or molecule so as to produce a positive ion.

ionize

To make an *atom* or *molecule* of a gaseous element, such as a *particle* of air, lose an electron, as by X-ray bombardment, and thus be converted into a positive ion, the freed electron then attracting itself to another molecule to form a negative ion.

ionogen

A substance which when dissolved in a specified medium (usually water) produces an electrolyte.

ionometer

An ionization chamber for measuring the intensity of ionizing radiations. The term is often used as a synonym for *roentgenometer.*

ionosphere

That region of the *atmosphere of the earth* which extends from the upper limits of the *stratosphere* to the *exosphere,* i.e., from an altitude of approximately 500 miles above ground. It is so named because it is a region of ionization of atoms, and its ionized layers reflect radio waves and short-wave communication waves, thus making radio and short-wave transmission possible. It is divided into layers, designated as *D layer, E layer* and *F layer.*

Ir

The symbol for the element *iridium.*

iridescence

The play of rainbow colors on the surface of a substance.

iridium

A metallic element in Group VIII of the periodic table, discovered by Tennant in 1803. Atomic number 77, atomic weight 193.1, melting-point about 2450°C., specific gravity 22.4, valence 2 or 4. Symbol: Ir.

iron

A metallic element in Group VIII of the periodic table. Atomic number 26, atomic weight 55.85, melting-point 1535°C., boiling-point 3000°C., specific gravity 7.87, valence 2 or 3. Symbol: Fe.

iron family

The elements manganese, iron, cobalt and nickel (atomic numbers 25-28). Also called *iron period.*

iron period

Found in the iron family in the periodic chart.

irradiation

(1) The process of exposing, or the exposure, to a radiation; (2) the amount of radiant energy incident on an irradiated object per unit area of surface per unit time.

irrational number

A number which cannot be expressed as an integer or a fraction representing a quotient of integers. E.g., $\sqrt{2}$, $\sqrt{3}$, $\sqrt[5]{6}$.

irregular polygon

A closed geometrical figure the sides and angles of which are not equal.

irregular polyhedron

A polyhedron the faces of which are not congruent.

irreversible process

A process which cannot be made to take place in the opposite sense to its course, and in which some property is always greater at or near the end of the process than at its beginning or early stages. An irreversible process continues until it is completed.

isentropic

Occurring without change in entropy.

iso-

A prefix indicating *equal,* e.g., isobar; the *o* is dropped and the prefix becomes *is-* before a vowel (e.g., isallobar).

isobar

(1) Any two or more atomic species having the same atomic weight. They may or may not have the same atomic number. (2) A line or diagram in a map showing equal barometric pressure.

isochore

A graph of the relationship between the pressure and the temperature of a fluid at constant volume.

isoelectric point

The point of zero potential; the pH value at which a substance is electrically neutral, i.e., has no electric charge.

isogonal

Having equal angles.

isogonal conjugates

See *isogonal lines.*

isogonal lines

Lines drawn through the vertex of an angle and forming equal angles with the bisector of the original angle. Also called *isogonal conjugates.*

isolated point
See *acnode.*

isomer
One of two or more compounds which have the same composition, but differ in properties because of different atomic arrangements. (See also *nuclear isomers.*)

isomerism
A change in the properties of a compound without change of state or of composition, due to a change in the number or arrangement of atoms in a molecule.

isomerization
A reaction or process which involves a rearrangement of a molecule, and in which a new substance is obtained with the same molecular formula, but with a different structural formula.

isometric
Of equal measure.

isometric crystal
A *cubic crystal.*

isometric drawing
This type of drawing takes as its points of departure three axes known as the isometric axes. The dimensions of the object to be drawn are marked off on these axes. Through the points marked off, lines are drawn parallel to the isometric axes. It is to be noted that angles in isometric drawings do not show their true size and therefore cannot be measured in degrees. Circles appear as ellipses in isometric drawings.

isometric system
See *cubic system.*

isomorph
One of a number of similarly shaped crystals.

isomorphism
The crystallization of two or more compounds in the same system of symmetry and very closely similar forms.

isomorphous phases
Phases with crystal structures of the same type.

isosceles trapezoid
A trapezoid which has two equal sides.

isosceles triangle
A triangle which has two equal sides, and therefore also two equal angles.

isostatic
Exposed to identical pressure from all directions.

isostere
One of two or more substances which have the same number and arrangement of electrons in their molecules (e.g., N_2O and CO_2), and consequently similar physical properties.

isosteric substances
Isosteres.

isosterism
The property or condition of being *isosteric.*

isothermal process
A process which takes place at a constant temperature.

isotonic solutions
(1) Solutions which have the same *osmotic pressure.* (2) Solutions which have the same osmotic pressure as normal blood serum.

isotope
Isotopes are elements or varieties of the same chemical element, the atoms of which have the same atomic number, i.e. the same number of nuclear protons, but different atomic weights or mass numbers, i.e. a different number of nuclear neutrons. In writing the same or symbol of an isotope, the rough atomic weight, or mass number, is written as a suffix to the same of the element or as a superscript to its symbol; thus, e.g., *uranium-235* is the name and U^{235} (or ${}_{92}U^{235}$) the symbol, of the uranium isotope of atomic number 92 and mass number 235.

isotropic body
A body in which the value of any given property is independent of the direction of measurement. A body may be isotropic with regard to one or more physical properties while not isotropic with respect to others.

-ite
(1) In chemistry, a suffix denoting a salt of an acid containing an element or radical in its negative lower valence (e.g., *sulfite* is a salt of sulfur*ous* acid).

-ite
In geology, a common ending of names of minerals.

J

joule

A unit of energy equal to 10^7 *ergs,* or a watt-second. Derived from the practical system of electrical units. It may be defined as the energy dissipated in one second by a current of one *ampere* flowing across a potential difference of one *volt.*

Joule effect

The heating effect caused by an electric current flowing through a resistance. (See *Joule's law, 3.*)

Joule's laws

(1) The intrinsic energy of a given mass of gas in a function of temperature alone; it is independent of the pressure and volume of the gas. (2) The molecular heat of a solid compound is equal to the sum of the atomic heats of its component elements in the solid state. (3) The heat produced by a current I passing through a conductor of resistance R for a time t is proportional to I^2Rt.

Joule-Thomson effect

The slight fall in temperature which occurs when a gas is allowed to expand without doing external work. The effect is due to energy absorbed in overcoming the cohesion of the molecules of the gas. (Also called *Joule-Kelvin effect.*)

K

K
The symbol for the element *potassium.*

K-capture
A radioactive transformation of an atomic nucleus, consisting of the capture by the nucleus of an extranuclear electron (usually from the K shell) which unites with a proton to form a new neutron; the result of the process is the reduction of the atomic number by one, and the release of a quantum of radiant energy, called *K x-ray.*

k-factor
See *multiplication factor.*

K-mesons
Subatomic particles, heavier than *L-mesons* and lighter than *protons.*

K x-ray
Radiant energy emitted by an atom as a result of a *K-capture.*

kalium
The Latin and German name of the element *potassium* (K).

katharometer
A device for detecting small quantities of a gas within a gaseous mixture, based on the great differences between the thermal conductivities of the different gases.

Kelvin
A name sometimes used for kilowatt-hours.

Kelvin effect
See *Thomson effect.*

Kelvin scale
See *absolute temperature.*

kenotron
A high-vacuum rectifier tube.

Kepler's laws
The three laws of planetary motion discovered by Kepler (1571-1630) that explain the movements of planets in terms of the Copernican concept. These laws are: 1. The orbit of every planet about the sun is an ellipse, the sun occupying

one focus. 2. A line from each planet to the sun sweeps over equal areas in equal times. 3. The squares of the lines required for the different planets to complete their orbits are proportional to the cubes of their mean distances from the sun.

kernel

A term introduced into chemistry by Lewis and Langmuir. According to Langmuir, "the kernel of an atom may be defined as that part of the atom which remains after the *sheath* is removed, e.g. the whole neon atom constitutes a kernel with 0 charge, since the neon atom has no sheath; the kernel of a fluorine atom consists of the nucleus and 2 electrons, the whole having 7 positive charges."

Kerr effect

(1) An electro-optical effect in which certain transparent isotropic substances become doubly refracting when subjected to the effect of an electric field at right angles to a beam of light. (2) The conversion of plane-polarized light into elliptically polarized light when it is reflected from the polished end of a magnet.

keto-

A prefix indicating the presence of the divalent =CO radical, also called the *keto* or *carbonyl group*.

ketol

See *ketone alcohol*.

ketone

An organic compound containing the bivalent *ketone group* =CO (which is also called *carbonyl group*).

ketone alcohol

A generic term for compounds which contain a keto (=CO) and a hydroxy (—OH) group.

kev

The symbol for 1000 electron volts.

kg

The abbreviation for *kilogram*.

killed steel

Steel that has been completely *deoxidized* by the addition of manganese, silicon, and sometimes aluminum, before casting so that there is practically no evolution of gas during solidification, and sound ingots are obtained.

killing

In metallurgy, especially in foundry terminology, a term for *deoxidation*.

kilo-
A prefix meaning 1000.

kilocalorie (kcal)
1000 calories, i.e. the amount of heat required for raising the temperature of 1 kg. of water from 14.5°C. to 15.5°C. (The terms "large calorie" and "great calorie" are obsolescent; the term "kilogramcalorie" is a misnomer.)

kilocycle
Measure of frequency of electric discharge or alternating current. One thousand cycles.

kilogram (kg)
A unit of weight in the metric system, equal to 1,000 grams or 2.2046 pounds avoirdupois. It is the weight of a platinum rod preserved in Paris as the standard unit of the metric system of weights.

kilo-line
A unit of *magnetic flux,* equal to 1000 maxwells.

kilometer (km)
1000 meters, equal to 1094 yards, or 0.6214 mile.

kilometric waves
Radio waves the wavelengths of which are between 1000 and 10,000 meters.

kilovolt (kV)
1000 *volts.*

kilovolt-ampere (kVA)
A measure, used instead of *kilowatts,* of power in alternating-current circuits; the product of number of kilovolts multiplied by the number of the amperes in the circuit.

kilowatt
A unit of measure of electrical power equal to 1,000 watts (approximately 1⅓ horsepower). See also *watt.*

kilowatt-hour (kWh, kwhr)
A unit of electric energy; 1 kWh = 1 kilowatt of power used for 1 hour (i.e. 1000 watt-hours), equivalent approximately to 1⅓ horsepower acting for 1 hour.

kilo-X-unit (kX, kXU)
1000 X Units, equal to 10^{-3} centimeter.

kindling temperature
The lowest temperature at which a given substance can catch fire and burn; it depends not only on the nature and constitution of the material, but on various other external factors as well.

kinematic viscosity

The ratio of viscosity to density. It is measured and expressed in units called *stoke.*

kinematics

That part of mechanics which treats of different kinds of motion and of the modes of strains in elastic bodies without reference to the forces involved.

kinescope

Cathode ray tube used as picture tube in a television receiver.

kinetic energy

The energy which a mechanical system possesses by virtue of its motion. The kinetic energy of a particle at any instant is $mv^2/2$, where m is the mass of the particle and v is its path velocity at that instant. The kinetic energy of a body at any instant is the sum of the kinetic energies of its several particles.

kinetic theory of gases

The mathematical explanation of the behavior of gases on the assumption that gases consist of *molecules* which are in ceaseless motion in space, the *kinetic energy* of the molecules depending upon the temperature of the gas. The molecules are considered to be perfectly elastic particles which collide with each other and with the walls of the containing vessel. The pressure exerted by a gas on the walls of the vessel is due to the collisions of the molecules with it.

kinetics

That part of mechanics which treats of the effect of forces in changing the motion of bodies.

Kirchhoff's laws of electric circuits

(1) The algebraic sum of the electric currents meeting at any point in a circuit is zero. (2) In a closed electrical circuit the algebraic sum of the products of the current and resistance in each part of the circuit equals the electromotive force present in the circuit.

Kirchhoff's radiation law

The ratio of the emissive power to the absorption coefficient is the same for all substances, and is affected only by the temperature and by the frequency of the radiation.

klystron

A velocity-modulated type of electron tube, i.e., a tube in which oscillations are generated by the bunching of electrons as the result of velocity modulation of the electron stream.

km
The abbreviation for *kilometer.*

Kohlrausch's law
If complete ionization has occurred in an electrolyte, the conductivity of the electrolyte equals the sum total of the values of conductivity of the individual component ions.

Kr
The symbol for the element *krypton.*

krypton
A zerovalent element in Group O of the periodic table, one of the rare gases, discovered by Ramsay and Travers in 1898. Krypton is a colorless, odorless, monatomic gas, occurring in the atmosphere as 1 part in about 10^6. Atomic number 36, atomic weight 83.80, melting-point −157°C., boiling-point 151.7°C., density 3.708 per liter at normal temperature and pressure. Symbol: Kr.

kVA
The abbreviation for *kilovolt-ampere.*

kw, kW
The abbreviations for *kilowatt.*

kWh, kwhr
The abbreviation for *kilowatt-hour.*

kX, kXU
The abbreviation for *kilo-X-unit.*

L

L

The symbol for *lambert.*

***l*-**

In chemistry, the abbreviation for *levo-*.

λ

The symbol for *wavelength.*

L-mesons (light mesons)

A collective designation for the *mu-mesons* and *pi-mesons.*

La

The symbol for the element *lanthanum.*

labeling

The admixture to a non-radioactive substance of a minute quantity of one of its radioisotopes; being an isotope of the same chemical element, this admixture is chemically indistinguishable from the rest of the substance, but can be detected and traced by radiation counters because of its radioactive property. (Cf. *tracer.*)

labile

Unstable.

laevorotatory

See *levorotatory.*

lag

(1) In general, a time interval or delay. (2) The phase angle by which the electric current in an induction circuit is behind the electromotive force.

lake

An insoluble coloring agent, a pigment made by absorbing animal, vegetable or coal-tar coloring matter from an aqueous solution by means of a metallic base.

lambert (L)

A unit of brightness equal to $1/\pi$ candle per square centimeter, and therefore is equal to the uniform brightness of a perfectly diffusing surface emitting or reflecting light at the rate of one lumen per square centimeter.

Lambert's law of absorption
As an electromagnetic radiation passes through a homogeneous absorption medium, each layer of an equal thickness absorbs an equal fraction of it.

Lambert's law of illumination
The *illumination* on a surface from a point source of light is proportional to the cosine of the angle formed by the light ray with the normal at the point of incidence, but inversely proportional to the square of the distance between the surface and the source when the light incides at right angles.

lamellar structure
One consisting of fine alternating parallel layers of two constituents, e.g. *pearlite.*

lamina
A thin sheet.

laminar flow
A particular type of *streamline* flow, in which the fluid moves in parallel layers, the velocities of the fluid particles within each layer *(lamina)* not being necessarily equal. The term is usually applied to the flow of a viscous liquid near solid boundaries, when the flow is not turbulent.

Langmuir adsorption isotherm
An equation expressing the relationship between the quantity of gas adsorbed by a given mass of adsorbent and the pressure.

lanthanides
See *rare earth elements.*

lanthanum
A trivalent metallic element, first member of the group of rare earths in the periodic table, discovered by Mosander in 1839. Atomic number 57, atomic weight 138.92, melting-point 826°C., specific gravity 6.15. Symbol: La.

large calorie
An obsolescent synonym for *kilocalorie.*

latent energy
(1) *Potential energy.* (2) *Latent heat.*

latent force
Potential energy.

latent heat
The heat absorbed by a substance, without change in temperature, while passing from the liquid to the vapor state, or from the solid to the liquid state, and released by the substance

while passing from the vapor to the liquid state or from the liquid to the solid state.

latera recta

The plural of *latus rectum.*

lateral area

The sum of the areas of all the lateral faces of a prism, pyramid or frustum of a pyramid.

lateral chromatic aberration

In optics, a lens defect which impairs the sharpness of images off the axis owing to the fact that light of different colors produce different magnifications. (Cf. *longitudinal chromatic aberration.*)

lateral edges

In a prism or pyramid, the intersection of any two lateral faces.

lateral face

In a prism or a pyramid, any of the faces which are not the bases.

latitude

The latitude of any place on the earth's surface is the angular distance of that place measured from zero degrees to ninety degrees north or south of the equator.

latitude of a place

The angular distance of the place north or south of the equator, measured from there from 0° to 180°. It is marked N. or S. according as the place is north or south of the equator.

lattice

See *space lattice.*

lattice constants

The lengths of the sides of the unit cell in a *space lattice,* given in *Ångström units.*

lattice energy

See *heat of solution.*

lattice parameter

The distance between successive points in a space lattice measured in a direction parallel to one of the three principal axes. A space lattice has three parameters of which all may be different, all may be equal, or two may be equal.

latus rectum of a hyperbola

The length of the focal chord perpendicular to the transverse axis, as extended.

latus rectum of an ellipse

The focal chord perpendicular to the major axis.

latus rectum of a parabola
The focal chord perpendicular to the axis.

Laue diffraction pattern
See *Laue pattern.*

Laue method
For crystal analysis, a method using heterochromatic X-rays, a small fixed crystal and a plane screen, usually a photographic plate; the *Laue spot* is the image (of a pinhole or of the small crystal itself) characteristic of the resulting diffraction pattern.

Laue pattern (Laue diffraction pattern)
The photographic record of diffraction beams produced when a single crystal, which diffracts rays in all directions, is hit by x-rays proceeding from a narrow slit or minute orifice.

Laue photograph
A photograph made by the *Laue method.*

Laue spot
See *Laue method.*

Lauritsen electroscope
An electroscope devised by C. C. Lauritsen, in which the gold leaf of the *gold-leaf electroscope* is replaced by a quartz fiber, using the elasticity of the fiber as a restoring force; the position of the fiber is read with the aid of a low-power microscope.

lava
(1) Fluid, molten rock material which has been forced or has poured out of the interior of the earth onto the surface, and often has lost a great deal of its original content of gas and water while solidifying. (2) The same substance in the fully solidified stage.

law of conservation of energy
Energy can neither be created nor destroyed, and the sum total of all energies in a closed system always remains constant, and merely a transformation of energy from one form to another can take place.

law of conservation of mass
The total quantity of mass and energy in a system always remains constant. Thus, in ordinary molecular reactions, the total mass of the reactants must always equal the mass of the products, and in nuclear reactions the total mass plus energy of the reactants must always equal the total mass of the products plus the energy liberated.

law of conservation of matter

The *law of conservation of mass* as it applies to ordinary molecular reactions in which the quantity of mass transformed into energy is negligible.

law of conservation of momentum

In any reaction involving collisions of particles, vector sum of the momentums of the colliding particles must be the same before and after the collision.

law of constant proportions

See *law of definite proportions.*

law of cooling

See *Newton's law of cooling.*

law of cosines

See *cosine law.*

law of definite proportions

In every sample of any one compound substance, the proportions by weight of the constituent elements are constant. (Also called *law of constant proportions.*)

law of diffusion

See *Graham's law.*

law of distribution

See *distribution law.*

law of Dulong and Petit

(1) The specific heats of the elements are inversely proportional to their atomic weights. (2) The atomic heat of all solid elements is identical, namely a constant 6.2, equal to 5.96 calories per gram-atom per degree C. Atomic heat = specific heat × atomic weight = 6.2.

law of electrostatic attraction

See *Coulomb's law.*

law of equivalent proportions

A law of chemistry, according to which the proportions in which two chemical elements separately combine with the same weight of a third chemical element equal the proportions in which those two elements combine with each other.

law of exponents

Used in algebraic division and multiplication. In division, the law states that the exponent of any factor in a quotient is equal to its exponent in the dividend minus its exponent in the divisor. In multiplication: to find the exponent of any number in a product, add the exponents of that number in the multiplicand and the multiplier.

law of illumination
See *Lambert's law of illumination.*

law of mass action
When a reversible reaction reaches a state of equilibrium, the quotient, found by dividing the product of the active masses on one side of the equation, by the product of the active masses on the other side, equals a constant. The quotient so found varies with temperature and pressure. In single-phase solution the active mass may be regarded as the *mole fraction* of a constituent in that phase.

law of multiple proportions
When two elements combine in more than one proportion, that is, to form more than one compound, the weights of the first element which combine with a fixed quantity of the other element show a ratio of small integers.

law of octaves
It was first noted by J. A. R. Newlands, that when the elements were arranged in the ascending order of their atomic weights each eighth element was a "kind of repetition of the first". He provisionally termed this relationship, the law of octaves. However, the inaccuracy of the atomic weights at that time, about 1866, masked the truth of this finding and little attention was paid to it until the *periodic table* arranged by Mendeleeff had been accepted.

law of partition
See *distribution law.*

law of proportionality
The relationship between stress and strain (called *Hooke's law*).

law of reciprocal proportions
The same as the *law of equivalent proportions.*

law of reflection
The angle of reflection equals the angle of incidence.

law of signs
A conventional rule in addition accepted by mathematicians for multiplying positive and negative numbers. It is stated as follows: To multiply one signed number by another, determine the product of their absolute values and make the product positive if the multiplicand and multiplier have like signs, and negative if they have unlike signs.

law of sines
The ratio of a side of a triangle to the sine of the angle opposite that side is the same for all three sides.

law of sines of spherical trigonometry

In a spherical triangle the sines of the angles are proportional to the sines of their opposite sides.

law of the equivalence of mass and energy

See *Einstein's law of the equivalence of mass and energy.*

law of the lever

If there is equilibrium for two weights or forces, the weights or forces are to each other inversely as their lever arms, or what is equivalent, the products of the weights by their lever arms are equal, or the sum of the moments of all the forces about the fulcrum is equal to zero.

law of the mean

In differential calculus, the theorem that if $f(x)$ has a unique derivative at all points in the interval (a,b), there exists at least one interior point such that

$$\frac{f(b) - f(a)}{b -} = f'(x_1), \text{ where } a < x_1 < b$$

Geometrically stated, if $f(x)$ is continuous and differential in the interval between two points of intersection of a curve with the x-axis, the curve has at least one point where the derivative is zero, i.e., where the tangent is parallel to the secant. (Also called *mean value theorem.*)

law of the photochemical equivalent

See *Einstein's law of the photochemical equivalent.*

law of the refraction of light

When light passes from a medium into a denser medium, its path is deviated *toward* the normal; when it passes from a denser medium into a less dense one, its path is deviated *away from* the normal. The ratio of the sine of the angle of incidence to the sine of the angle of refraction is constant; this constant is called the *index of refraction,* also referred to as *Snell's law.*

laws of thermodynamics

First law of thermodynamics: Energy can neither be created nor destroyed. *Second law of thermodynamics:* All systems tend to approach a state of equilibrium. (Hence, heat cannot pass spontaneously from one body to another, hotter body.) *Third law of thermodynamics:* The *entropy* of a perfect crystalline substance at $0°K$ is zero.

law of universal gravitation

Newton's law of gravitation, stating that every particle of mass attracts every other particle, and the force of the attraction, is directly proportional to the product of their masses and inversely proportional to the distance between them.

L.C.M.

Least common multiple (q.v.).

Le Chatelier's law

A system in equilibrium, if disturbed by external factors such as temperature and pressure, will adjust itself in such a way that the effect of the disturbing factors will be reduced to a minimum.

leaching

The process of washing or extraction of soluble constituents from insoluble material, as, e.g., the soluble metals from an ore by dissolving in a solvent which does not affect the *gangue,* followed by the precipitation of the metal from the solvent. (Also called *lixiviation.*)

lead

A metallic element in Group IV of the periodic table. Atomic number 82, atomic weight 207.21, melting-point 327.5°C., boiling-point 1620°C., specific gravity 11.34, valence 1 or 4; an efficient shield against hard radiations. Symbol: Pb. Stable isotopes: Pb^{204} (natural abundance 1.5%) Pb^{206} (23.6%), Pb^{207} (22.6%), Pb^{208} (52.3%); with the exception of Pb^{204} these lead isotopes are the stable end products of the three natural *radioactive series.*

least common denominator

The least common multiple of the denominators of two or more fractions. Also called the lowest common denominator. (Abbreviation L.C.D.)

least common multiple

The least common multiple of two or more numbers is the product of all the different prime factors of the numbers, each raised to the highest power in which it occurs in any one number. The least common multiple of

$12 = 2^2 \cdot 3$, $8 = 2^3$, 2, $9 = 3^2$ is $2^3 \cdot 3^2 = 72$.

Also called lowest common multiple. (Abbreviation: L.C.M.)

least-time principle

See *Fermat's principle.*

left-hand rule

A method of determining direction of magnetic lines of force. The first three fingers of the left hand are spread so as to form rectangular axes. Then pointing the center finger of the left hand in the direction of the current and the forefinger with the flux, the thumb will give the direction of thrust. Using the right hand in the same manner will give the direction of induced current.

left-handed elliptically polarized wave

An elliptically polarized wave in which the rotation of the direction of displacement is counterclockwise for an observer looking in the direction in which the wave is traveling.

left-handed quartz

Quartz which rotates the plane of polarization to the left, i.e., counterclockwise, as seen when looking back toward the source of light.

legal ohm

See *international ohm.*

legend

A description, explanation, table of symbols and other information appearing on or under a map, chart, graph or illustration, for its better understanding and interpretation.

Leisegang phenomena

If one compound, such as potassium chromate is dissolved in a gelatin gel and a solution of another substance, such as silver nitrate is allowed to diffuse into the gel, the precipitate which forms (silver chromate) forms a series of concentric rings, separated by more or less clear portions of the gel. These rings are called Leisegang rings. No completely satisfactory explanation has been offered.

lemniscate

A curve representing the plane locus of the vertex of a triangle as the product of the two adjacent sides remains equal to one-fourth of the square of the third side which is of an invariable length. Also called *lemniscate of Bernouilli.*

lens

A portion of a refracting medium consisting of any transparent substance bound by two spherical surfaces having a common axis.

lens distortion

In optics, an aberration that affects the position of images off the axis, produced by a lens defect as a result of which objects at different angular distances from the axis undergo different magnifications. Usually referred to simply as *distortion.*

lens system

Two or more lenses arranged to work in combination with one another, as though they formed one optical unit.

Lenz's law

The current induced in a circuit as a result of its motion in a magnetic field is in such a direction as to oppose the charge which induces it, i.e., in such a direction as to exert a mechanical force opposing the motion.

lepton

A collective term for the *electron, positron, neutrino, antineutrino* and *photon.*

lever

A rigid bar used to lift weights by placing the bar against a support called the fulcrum, and applying a force or weight. A lever is said to be of the first, second, or third type depending on whether the fulcrum is under the bar and between the weights, under the bar at one end, or above the bar at one end. (See *law of the lever.*)

lever arm

The distance of a weight or line of action of a force from the fulcrum of the lever.

levo-

A prefix, abbreviated as *l-,* indicating that the substance the name of which it precedes is *levorotatory.*

levo-compound

A *levorotatory compound.*

levorotatory

Turning to the left; producing a counterclockwise turn. (Abbreviation: *l-,*) (Also written *laevorotatory.*)

levorotatory compound

An *optically active* substance which turns the plane of polarization of light to the left. (Also written *laevorotatory.*)

Li

The symbol for the element *lithium.*

lifting condensation level

See *condensation level.*

light

Electromagnetic radiation capable of producing optical sensation, i.e. having a wavelength between 4×10^{-5} and 7.7×10^{-5} cm.; often designated more specifically as *visible light,* to distinguish it from infrared and ultraviolet radiation, which are frequently referred to also as infrared "light" and ultraviolet "light." The speed of light is, in round figures, 186,000 miles a second.

light alloy

Any alloy of a low specific gravity; e.g., magnesium and aluminum alloys.

light flux

See *luminous flux.*

light mesons

See *l-mesons.*

light metal

Any metal having a low specific gravity, as, e.g., beryllium, magnesium or aluminum.

light quantum

A photon.

light-sensitive

Exhibiting a photoelectric effect when irraditated by visible light. Also called *photosensitive.*

light vector

A purely diagrammatic line representing a light wave traveling through space at any given point and at any desired instant; the direction and length of this line represent the direction and the displacement of the electric field the vibration of which produces the wave of light.

light-year

An astronomical unit of distance, equal to the distance of the travel of light in 1 year (in vacuo), i.e., approximately 5.8783×10^{12} miles.

lignite

The softest sort of coal, commonly known as *brown coal,* somewhat similar to peat.

like terms

In mathematics, terms which differ only in the numerical coefficient. Thus 3b, 7b, 10b, or $5b^2c$, $8b^2c$, etc.

limaçon

A curve which is the locus of a point on a line remaining at a given distance from the point where that line intersects a stationary circle, as the line revolves about a point on that circle. Also referred to as *limaçon of Pascal.*

limit

That which terminates, circumscribes, restrains or confines. The circle is the limit of its inscribed and circumscribed polygons of an indefinitely increasing number of sides. 1 (one) is the limit of the sum of n terms of the series $\frac{1}{2} + \frac{1}{4} + \frac{1}{8} + \ldots$

limit of error

In decimals, if we know a quantity is correct to two decimal places, then there is a possible error in the third figure after the decimal point of half its value, or 0.005, which may have to be added or subtracted from the quantity. This is known as the limit of error and is written ± 0.005 (plus or minus 0.005).

limit of proportionality

The value of stress at which a material ceases to obey *Hooke's law,* i.e., which no longer produces a proportional strain in the material. Also called *proportional elastic limit stress* or simply *proportional limit.* It is usually assumed to coincide with the *elastic limit.*

limits of integration

In a definite integral, the values between which the function called the *integrand* is integrated. Thus, in the expression $\int_b^a f(x)dx$, we refer to *a* as the upper limit, and to *b* as the *lower limit of integration.*

line

A one-dimensional configuration which is the locus of a point; the path of a point in motion. (In common usage, the term *line,* unless specified otherwise, denotes a *straight line.*)

line of centers

A straight line that connects the centers of two circles.

line of force

(1) An imaginary line in an electric or magnetic field, coinciding in direction with the field intensity at each point. (2) A unit of magnetic flux, equal to a *maxwell.*

line of intersection

The line common to two or more planes, i.e., the line in which they meet, intercept or cut each other.

line of magnetic flux

A line drawn in a graphical representation of a magnetic field so that its direction at any point is the same as the direction of the field at that point. The line is also commonly used as a unit of magnetic flux, one line being equal to one *maxwell.* (Also referred to as *line of magnetic induction.*)

line of magnetic force

See *magnetic line of force.*

line of magnetic induction

See *line of magnetic flux.*

line of sight

The line connecting the eye of the observer with a point or object intended to be represented on a surface.

line pair

In *spectrographic analysis,* the term includes an analysis line and the internal standard line with which it is compared.

line segment
A straight line having a definite length.

line spectrum
A spectrum, of atomic origin, showing sharply defined lines.

linear accelerator
A high-voltage apparatus for accelerating ions and subatomic particles.

linear equation
A *linear function* or *expression* equated to zero. Its graph is a straight line.

linear expansion
The increase of the size of a body, solid or liquid, restricted to taking place in one dimension only.

linear expression
In mathematics, an expression which contains no higher power of the unknown or variable quantity than the first power.

linear function
The simplest polynomial in two variables, other than a constant, of the first degree.

linear interpolation
The procedure of *interpolation* in which the three points the ordinates of which are the two known values and, respectively, the unknown value lie on one straight line.

liquid contraction
The shrinkage of molten metal as it cools.

liquid crystals
A melt or solution of a crystalline substance which shows orientation, as by *double refraction,* indicating a shaping or grouping of molecules along different axes which do not show the same optical magnetic or electrical properties in all directions.

liquid shrinkage
Contraction in volume of a mass of molten metal or alloy as it cools from a higher temperature to that of solidification.

liquid solution
A gas or gases, liquid or liquids, solid or solids dissolved in a liquid solvent.

liquidoid
A term advocated to indicate the beginning of the formation of a new phase from a solid solution, suggestive of *liquidus* which refers to the beginning of the formation of a solid phase from a liquid solution. The term *solidoid* is advocated to indi-

cate the end of the phenomenon, suggestive of *solidus* which marks the end of solidification.

liquidus

A line in a binary *phase diagram* or a surface on a ternary phase diagram, representing the temperatures at which freezing begins during cooling, or melting is completed on heating, under equilibrium conditions, i.e. the line or surface above which all alloys in the system are completely molten.

liquostriction

The change in dimensions of a solid body on immersion in a liquid.

liter

(1) A unit of volume in the C-G-S system; 1 liter = 1 kilogram of pure water at 4°C. (i.e. its maximum density), equal to 1000.028 cubic centimeters.

literal coefficient

A letter which has a fixed value (usually the letters at the beginning of the alphabet are used for this purpose). For instance, to save the time in writing down 7,842,653 each time in a sum, or if its value although fixed is unknown, then this number may be called a or b or c, etc.

lithium

A monovalent metallic element in Group I of the periodic table, one of the alkali metals, and the lightest of all known metallic elements, discovered by Arfvedson in 1817. Atomic number 3, atomic weight 6.940, melting-point 186°C., boiling-point above 1200°C., specific gravity 0.534 at 20°C. Symbol: Li.

lithosphere

The solid crust of the earth.

litmus

A coloring matter used as a reagent to test the acidic or alkaline nature of liquids and solutions.

lixiviation

A synonym for *leaching*.

local action

The voltaic effect produced in one of the plates of a primary cell owing to the impurities in the plate; in other words, it is the loss of otherwise usable chemical energy by currents which flow within the cell regardless of its connection to an external electric circuit.

loci
The plural of *locus.*

locked-up stress
See *residual stress.*

locus (*plural:* loci)
Any system consisting of points or straight or curved lines which satisfies a given condition or given conditions.

locus of a point
In elementary geometry, the geometrical figure every point of which satisfies a given geometric condition or conditions, and which can thus be regarded as the path of a point that moves in accordance with that condition or those conditions.

logarithm
The logarithm of a number is the power to which it is necessary to raise a given number, called the *base,* to produce the number. The base of a given system of logarithms is the same for all numbers; 2 is the logarithm of 100 to the base 10, written $\log_{10}100 = 2$; and -2 is the logarithm of .01 to base 10.

logarithmic curve
A curve which represents the locus of equation $y = \log a^x$, where a is greater than unity.

logarithmic equation
An equation in which the logarithm involves the unknown quantity (e.g., $\log x + 5 = 10$).

logarithmic scale
A scale of measurement in which an increase of one unit represents a tenfold increase in the quantity measured.

logarithmic spiral
A curve the vectorial angle of which is proportional to the logarithm of the radius vector. Also called *equiangular* and *logistic spiral.*

logistic spiral
See *logarithmic spiral.*

logometer
A scale for measuring chemical equivalents.

longitude of a place
The arc of the equator intercepted between the *prime meridian* and the meridian of the place, measured from the prime meridian toward the east or west from 0° to 180°.

longitudinal chromatic aberration
In optics, a lens defect which impairs the sharpness of the image owing to the fact that light of different colors is brought to a focus in different planes, or different distances from the

lens. Usually referred to simply as *chromatic aberration* unless it is intended to distinguish it specifically from the *lateral chromatic aberration* (q.v.).

longitudinal wave

Wave in which the particles of the medium in which it propagates vibrate in parallel to the direction of the propagation of the wave.

loops

Points, lines or surfaces in a standing wave system where some characteristic of the wave has maximum amplitude.

Lorenz's law

The ratio of the thermal and electrical conductivities of all metals is proportional to the *absolute temperature.*

Loschmidt number

The number of molecules contained in one cubic centimeter of an ideal gas at the temperature of 0°C. and the pressure of 760 millimeters of mercury, viz.; 2.687×10^{19}. (In European literature, the term "Loschmidt number" is frequently used to designate *Avogadro's number.*)

loudness

A property of intensity of a sound or auditory sensation which depends on the amplitude of the acoustic disturbance of the medium.

lower limit of integration

In a definite integral it repersents the lower limit between which the integration is performed.

lowest common multiple

See *least common multiple.*

lox

Liquid oxygen used in mining and space flight devices.

Lu

The symbol for the element *lutetium.*

lumen (lm)

The unit of luminous flux. It is equal to the flux through a unit solid angle (steradian) from a uniform point source of one candle, or to the flux on a unit surface all points of which are at unit distance from a uniform point source of one candle. For some purposes the kilolumen, equal to 1,000 lumens, is a convenient unit.

lumen-hour (lm-hr)

The unit of quantity of light. It is the quantity of light delivered in one hour by a flux of one lumen.

lumen-second

A unit of light measurement, used for measuring brief flashes of light; one lumen-second equals the quantity of light delivered in one second by a flux of one lumen.

luminance

Luminous intensity per unit area emitted by an extended source of light. It is measured and expressed in candles per square foot or square meter.

luminescence

The emission of visible light at a temperature below that of *incandescence,* i.e., induced by reasons other than high temperature.

luminous density

The luminous energy present in a unit volume of space.

luminous flux

The time rate of the flow of visible light, i.e., the quantity of visible radiation per unit time. Its unit is the *lumen.*

luminous intensity

Luminous intensity, of a source of light, in a given direction, is the solid-angular flux density in the direction in question. Hence, it is the luminous flux on a small surface normal to that direction, divided by the solid angle (in steradians) which the surface subtends at the source of light. Mathematically a solid angle must have a point as its apex; the definition of luminous intensity therefore applies strictly only to a point source. In practice, however, light emanating from a source whose dimensions are negligible in comparison with the distance from which it is observed may be considered as coming from a point.

lune

A portion of a sphere which is bounded by the halves of two great circles of the sphere. (The angle at which these two bounding semi-circles meet is called the *angle of the lune.*)

lutetium

A trivalent metallic element, one of the group of rare earth elements in the periodic system, discovered by Urbain and Welsbach in 1907. Atomic number 71, atomic weight 174.99. One of the very few elements below atomic number 82 which have naturally radioactive isotopes. Symbol: Lu.

lux (lx)

The practical unit of illumination in the metric system; equivalent to the "meter-candle." It is the illumination on a surface one square meter in area on which there is a uniformly distributed flux of one lumen, or the illumination produced at a surface all points of which are at a distance of one meter from a uniform point source of one candle.

lye

(1) Originally the name given to an alkaline solution obtained by leaching wood ashes with water. (2) A solution of sodium or potassium hydroxides.

lyophilic colloidal system

A *colloidal system* in which there is a strong attraction between the continuous phase and the dispersed phase. (Also referred to as a *lyophile.*) If the continuous phase is water, the system is said to be *hydrophilic.*

lyophobic colloidal system

A *colloidal system* in which there is very little attraction between the continuous phase and the dispersed phase. (Also referred to as a *lyophobe.*) If the continuous phase is water, the system is said to *hydrophobic.*

M

M

A chemical symbol indicating any metal in general, usually a monovalent one.

m

(1) The abbreviation for *meter.* (2) See *slope.*

μ

(1) The symbol for *micron;* (2) a prefix indicating 10^{-6} *(micro-)*.

M.A.

Abbreviation of *mechanical advantage.*

μ meson

See *meson.*

m-curie

See *millicurie.*

macerate

To soak or steep a solid in a liquid.

mach

A unit of speed measurement for a moving object equal to the speed of sound in the medium in which the object moves. (Named for the Austrian physicist, Ernst Mach.) Thus, *mach .5* is a speed equal to one-half the speed of sound in the medium, *mach 1* is a speed equal to that of sound in the medium. (Mach speed has no meaning in an element that does not conduct sound.)

mach wave

A shock wave set up by an object traveling with a mach number above unity.

macro-

A prefix meaning *large* or *big.*

macro axis

The long axis in a crystal in which there are three unequal axes.

macrograph

A graphic reproduction of any object which has not been magnified more than ten diameters. (Cf. *photomacrograph.*)

macrography

(1) The preparation, study and interpretation of *macrographs.* (2) The study of *macrostructures* in general.

macroscopic

(1) Visible either to the naked eye or under low magnifications (up to about 10 diameters). (2) Observed or performed by the unaided eye. (Synonymous with, although less preferable than, *megascopic* in this sense.)

macrostructure

A structure or structural feature that can be seen or distinguished by the unaided eye or under magnifications of less than ten diameters.

magma

Mobile rock material occurring naturally within the earth, consisting to a considerable extent of a liquid silicate-melt phase of high temperature and a number of solid phases. The igneous rocks are considered to have solidified out of this substance.

magnesia

Magnesium oxide, MgO.

magnesium

A divalent metallic element in Group II of the periodic table. Atomic number 12, atomic weight 24.32, melting-point 651°C., boiling-point about 1110°C., specific gravity 1.75. Symbol: Mg.

magnet

A body which produces a magnetic field external to itself, and which, when appropriately suspended in a magnetic field, assumes a uniform, determined direction, and is capable of attracting or repelling the same or another metal. (See *electromagnet, permanent magnet, temporary magnet.*)

magnetic alloys

Alloys of iron with silicon, nickel or other elements, which possess high permeability and low *hysteresis* losses.

magnetic circuit

A closed path of magnetic flux, the path having the direction of the magnetic induction at every point.

magnetic compass

A compass that depends entirely on the earth's *magnetic field* for its directive force, without the aid of any elaborate mechanism. Its indicator needle always points in the direction of the earth's magnetic lines of force, and therefore to the magnetic north pole.

magnetic crochet

An abrupt change in the magnetic field of the earth, produced on the daylight side by currents generated in the upper layers of the atmosphere.

magnetic dip

The angle between the direction of the earth's magnetic field and the horizontal. (Also called *angle of dip.*)

magnetic equator

A great circle on the earth, every point of which is 90° from the magnetic poles; its plane is at right angles to the magnetic axis of the earth.

magnetic field

A region in the neighborhood of an electric current or of a permanent magnet, in which forces due to the current or magnet can be detected. It is a vector field of magnetizing force. At every point the divergence of the magnetic intensity of a current is zero, so that the magnetic field of a current is a tubular field. As generally used a magnetic field indicates the region throughout which the magnetizing force values are of significant magnitude with respect to the conditions under consideration.

magnetic field strength

The force which produces the *magnetic flux density* at any point of the *magnetic field;* it is the force per unit *N pole.* Synonymous with *magnetic force* or *magnetizing force.*

magnetic flux (Φ)

A condition produced in the medium in the neighbourhood of electric currents or magnets. The amount of magnetic flux through any area is measured by the quantity of electricity caused to flow in an electric circuit of given resistance bounding the area when the circuit is removed from the *magnetic field.* It is the total number of magnetic lines in a magnetic field, and it equals the ratio of the *magnetomotive force* to the *reluctance.* Unit magnetic flux in the electromagnetic system is that flux the removal of which from a circuit of unit resistance causes one electromagnetic unit of electricity to flow in the circuit. The CGS unit of magnetic flux is called *maxwell.*

magnetic flux density

The normal *magnetic flux* induced by a *magnetizing force.* It is the product of the field strength and the permeability, expressed in terms of x lines of magnetic flux per unit area;

the unit used is the *gauss* (1 gauss = 1 line of flux per square centimeter).

magnetic force

The force which produces the *magnetic field density* at any point of a magnetic field. Synonymous with *magnetic field strength.* See also *magnetizing force.*

magnetic hysteresis

The lagging of the *magnetic flux density* behind the *magnetizing force* during reverse magnetization.

magnetic hysteresis

In steelmaking, this term designates the property of a magnetic material by virtue of which the magnetic induction for a given magnetizing force depends upon the previous condition of magnetization.

magnetic hysteresis loop

A closed figure formed by plotting the *magnetizing force* against the *magnetic flux density* when the former is taken through a complete cycle after initial magnetization, the magnetic hysteresis loss being proportional to the area of this loop.

magnetic induction

(1) The process which magnetizes a body capable of magnetization, as a result of its being merely placed in a magnetic field. (2) See *magnetic flux density.*

magnetic intensity

The magnetic intensity produced by an electric current at any point in the neighborhood of a circuit in which an electric current is flowing can be computed on the assumption that every infinitesimal length of circuit produces at the point an infinitesimal magnetic intensity, and the resulting magnetic intensity at the point is the vector sum of the contributions of all the elements of the circuit.

magnetic line of force

An imaginary line in a magnetic field which at every point has the direction of the *magnetic flux* at that point.

magnetic meridian

At any place, the line in which the earth's magnetism exerts its force.

magnetic moment

The product of *magnetic intensity* and the distance. The unit magnetic moment is that of a magnet the two poles of which are of unit pole strength, of opposite signs, one centimeter distant from each other.

magnetic permeability

A measure of the ease with which the magnetic lines of force can pass through a substance for a given magnetizing force. (Analogous to *conductivity* in electricity.) Quantitatively, it is expressed as the ratio between the *magnetic flux density* produced and the *magnetizing force* producing it; the symbol used to designate this ratio is the Greek letter μ (mu).

magnetic polarization

The magnetic moment per unit volume.

magnetic pole

The magnetic poles of a magnet are those portions of a magnet toward which the external magnetizing force tends to converge or diverge. The size and position of the poles of a magnet depend upon its shape and characteristics. For a long, thin magnet, the poles occupy small regions near the ends, so that for many purposes the poles of such a magnet may be considered as concentrated at points.

magnetic potential difference

The line integral of a magnetizing force between two points in a *magnetic field.*

magnetic reluctance

See *reluctance.*

magnetic retentivity

The maximum value of the *residual induction.*

magnetic rotation *(of polarized light)*

When a plane-polarized beam of light passes through certain transparent substances along the lines of a strong magnetic field, the plane of polarization of the emergent light is different from that of the incident light. On looking from north to south along a line of magnetic intensity, the rotation is clockwise. Also called *Faraday effect.*

magnetic saturation

The condition of a magnetic substance the magnetization of which has reached its highest possible value.

magnetic storm

A sudden violent disturbance of the earth's magnetic field, shown by erratic variations of the compass needle, and sometimes by serious interference with radio, telegraph and telephone communication. Such storms occur about a day after a large sun-spot has crossed near the center of the sun. They are believed to be caused by streams of electrified particles shot out from the vortex of the sunspot.

magnetic susceptibility

The ratio of the intensity of magnetization to the magnetic field strength present in a magnetic circuit; it is a measure of the attraction which a magnet exerts on a given substance. Also called coefficient of *magnetization.*

magnetic viscosity

The time lag between the increase of a *magnetic field* and the resultant magnetization. It occurs in most ferromagnetic substances and is accounted for, almost completely, by *eddy-current* formation.

magnetics

That branch of science which deals with the laws of magnetic phenomena.

magnetism

The science of magnets and their fields of force; the property of certain bodies to attract or repel other bodies, the force being distinct from gravitational and electrostatic forces, exhibited by lodestone, liquid oxygen and iron in a *magnetic field.*

magnetization curve

A curve showing the variation in *magnetic flux density* from zero to saturation as the *magnetizing force* is increased.

magnetizing field

The region in the neighborhood of a permanent magnet or a current-carrying conductor in which magnetic forces can be detected.

magnetizing force

The force which produces a *magnetic flux density* (B) expressed as a number of *lines of force* per sq. cm.

magnetochemistry

A branch of physical chemistry dealing specifically with the magnetic properties of chemical substances.

magnetocrystalline energy

See *anisotropy energy.*

magnetoelastic energy

The part of the energy of a crystal which arises from the interaction between the magnetization and the mechanical strain of the lattice. The magnetoelastic energy is defined to be zero for an unstrained lattice.

magnetographic inspection
A term covering various methods for revealing defects in ferromagnetic materials in which the principles of magnetism are applied.

magnetometer
An instrument for measuring the magnitude and sometimes also the direction of a magnetic force.

magnetomotive force (mmf)
The force which tends to produce a magnetic field. It is an analogous force to the electromotive force. The magnetomotive force acting in any closed path in a magnetic field is the line integral of the magnetizing force around the path. In any closed path, the magnetomotive force resulting from a current is directly proportional to the current which links the path. It is defined as work per unit pole, and it is measured in *gilberts.*

magneton
A unit of magnetic moment in nuclear physics.

magneto-ohmmeter
An ohmmeter embodying a magneto-generator.

magnetostatics
That branch of physics which deals with magnetic fields which are not subject to variation with time.

magnetostriction
The change in dimensions occurring in magnetic materials brought about by magnetization or conversely, the changes in magnetic properties produced by mechanical stresses.

magnetostrictive
Changing in dimensions when placed in a magnetic field.

magnification
The ratio of the size of an image to that of the object. Magnification is usually expressed in diameters.

magnitude
Quantity, size, amount, greatness. The comparison of two or more quantities of the same class, with respect to their relative values.

magnitude of a dihedral angle
The magnitude of the plane angle formed by the intersection of the faces of the dihedral angle with a plane perpendicular to the edge of the dihedral angle.

magnitude of angles
See *angular magnitude.*

magnitude of stars
The astronomical measurement of the relative brightness of

stars, expressed on a *logarithmic scale,* a star of any given magnitude being approximately 2.5 times brighter than one of the next magnitude, and 2.5 x 2.5 brighter than a star two magnitudes dimmer, etc.

major arc

An *arc* that is greater than a semicircle.

major axis of an ellipse

The longer of the two axes, i.e., the one which passes through the two foci.

major segment of a circle

A segment of a circle the area of which is greater than half the area of the entire circle.

malleability

A measure of that property of a substance which permits it to be rolled or hammered into a desired form.

manganese

A metallic element, in Group VII of the periodic table, discovered in 1774. Atomic number 25, atomic weight 54.91, melting-point 1260°C., boiling-point 1900°C., specific gravity 7.2, valence 2, 3, 4, 6, 7. Symbol: Mn.

manometer

An instrument for measuring the pressure of gases.

mantissa

The number placed after the decimal point in a logarithm or the fractional part of the logarithm expression. The whole number in front of the decimal point is called the *characteristic.* The mantissa in the logarithm tables is always a positive number.

mantle

The first inner layer of the earth made up of solid rock. This goes down to about 1800 miles below MOHO (*Mohorovicic discontinuity*).

martensite

A microconstituent of steel characterized by its acicular structure. Essentially, a non-equilibrium condition of *alpha iron* formed directly from undercooled *austenite* when steel is cooled very rapidly from the hardening temperature.

maser

The designation coined from the initials of the phrase "Microwave Amplification by Stimulated Emission of Radiation," denoting a device introduced in 1957 which, among several other applications in physics, chemistry, radio and television communication, is expected to increase the utility of radio telescopes a hundredfold.

mass

The mass of a body or particle is usually defined as the quantity of matter contained in the body, or as a property measurable in terms of the inertia of the body, i.e., of the resistance which the body offers to acceleration. According to Einstein's Special Theory of Relativity, the mass of a body or particle varies with its velocity, according to the equation

$$m = \frac{m_0}{\sqrt{1 - \frac{v^2}{c^2}}}$$

where m is the mass of the body or particle when it is moving with the velocity v, m_0 is its mass when at rest (called *rest mass*), and c is the velocity of light.

mass-absorption coefficient

With reference to x-rays in a substance, the rate of decrease per unit mass traversed of the natural logarithm of the intensity of a parallel beam; usually written μ/ρ. It is often more convenient than the absorption coefficient because its use does not require knowledge of the density ρ.

mass action

See *law of mass action.*

mass defect

The divergence of the mass number of an isotope from an integer, i.e. is the difference between the sum total of the masses, or atomic weights, of the sub-atomic particles composing an atom and the atomic weight of the atom as a whole. The mass defect is the *binding energy* expressed in mass units, on the basis of 1 a.m.u. = 931 Mev. The mass defect divided by the mass number equals the *packing fraction* of the atom.

mass deficit

An alternative name for *mass defect* (q.v.).

mass-energy equivalence principle

See *Einstein's law of the equivalence of mass and energy.*

mass number

The mass number of an isotope is the same as its *rough atomic weight,* i.e. the integer closest to its atomic weight; it indicates the total number of nucleons in the nucleus of an atom of the isotope, i.e. the combined number of its protons and neutrons.

mass-scattering coefficient

With reference to x-rays in a substance, the rate of decrease, per unit mass traversed, of the natural logarithm of the intensity of a parallel beam, caused by scattering alone; usually written σ/ρ; the excess of the *mass-absorption coefficient,* μ/ρ over σ/ρ is called the true mass-absorption coefficient, μ'/ρ.

mass spectrograph

An apparatus for determining the relative masses of isotopes of an element, and for sorting electrically charged particles in general. A small amount of the substance to be studied is gasified, and the gas is admitted into a vacuum tube where its molecules are ionized by electrons emitted by a thermionic cathode and accelerated; the various types of ions so formed are sent through a combination of electric and magnetic fields which deflects them to different degrees according to their masses, and those of different mass are brought to focus at different places. (Also referred to as *mass spectrometer.*)

mass spectrometer

See *mass spectrograph.*

mass spectrum

The spectrum of a substance obtained by using an electric or magnetic field to deflect a beam of *positive rays* emerging from a tube containing a small amount of the gas or ionized metallic atoms intended to be examined. The amount of deflection depends on the ratio of the mass of a particle to its atomic charge. Every element has characteristic mass-spectrum lines.

mass susceptibility

In reference to magnetism, the magnetic susceptibility (i.e., the response of a substance to magnetic attraction) per unit mass, expressed as the ratio of the magnetic susceptibility to the density. Also called *specific susceptibility.*

mathematical astronomy

The study of astronomy from a mathematical point of view with major interest on the gravitational forces involved.

mathematical signs and symbols

$+$	plus, the sign of addition : also of positive (Elec. and Mag.), and compression (Eng.).
$-$	minus, the sign of subtraction : also of negative (Elec. and Mag.), and tension (Eng.).
$\times$	the sign of multiplication.
$\div$	the sign of division.
$:$	is to
$::$	as — the signs of proportion. Thus a : b :: c : d equal $\frac{a}{b} = \frac{c}{d}$
$:$	is to
$\because$	because.
$\therefore$	therefore.
$=$	equals ; the sign of equality.
$\propto$	varies as.
∞	infinity.
$\sqrt{}$	square root.
$\sqrt[3]{}$	cube root.
$\sqrt[4]{}$	fourth root, etc.
$\sqrt[n]{}$	nth root.
$\approx$	equals approximately.
$\equiv$	is equal to in all circumstances. Is congruent.
$>$	is greater than.
$\ngtr$	is not greater than.
$<$	is less than.
$\nless$	is not less than.
$\parallel$	is parallel to
$\perp$	is perpendicular to
$\angle$s	angles.
$\angle$	angle.
R$\angle$ or L	right angle.
$\triangle$	triangle.
$\square$	square.
$\boxed{}$	rectangle or parallelogram
$\odot$	circle.
$\bigcirc$	circumference.
◠ or $\frac{1}{2}$ $\odot$	semicircle.
◗	quadrant.
$\frown$	arc.
$\sim$	difference.
0	the cipher, zero.

$^{\circ}$	degrees, ′minutes, ″seconds, ‴thirds.
′	feet, ″inches.
R′	radius in degrees of minutes $= 3437'.75$.
R°	radius in degrees of arc $= 57^{\circ}.2958$ or one radian.
c	constant.
d	differential (in calculus)
$\int$	integration (in calculus).
E	modulus of elasticity.
F or f	functions.
g	gravity.
k	coefficient.
M or μ	modulus.
n	any number
e	base of hyperbolic logarithms.
π	ratio of circumference to diameter $= 3.14159$.
R, or r	radius.
Σ	sum of finite quantities
ω	angle of velocity
H.C.F.	Highest common factor.
L.C.M.	Lowest common multiple.
R.H.S.	Right hand side.
L.H.S.	Left hand side.

mathematics

The science which includes *arithmetic, algebra, trigonometry* and *geometry*. It treats of the exact relations existing between numbers, quantities, magnitudes or operations, and of the methods by which, in accordance with these relations, quantities sought are found from others known or supposed. It is subdivided into various fields, such as *abstract mathematics, applied mathematics, mixed mathematics, pure mathematics, mathematics of finance,* etc.

mathematics of finance

The study of mathematical problems and operations in the field of banking, insurance, etc.

matrix (*plural:* matrices)

A rectangular array of terms called *elements,* written with either a parenthesis or double vertical lines on each side of the array, and used to facilitate the study of problems in which the relation between these elements is fundamental.

matrix mechanics

The quantum mechanics of Heisenberg, expressed in the mathematical notation of matrices (see *matrix*).

matter

Any physical entity which possesses mass and occupies space.

maximum

The greatest quantity attainable in any given case or calculation. On a curve it is the point where the increasing value of the ordinate starts to diminish. On a wavy curve there are as many maxima as there are waves even if the waves are uneven. It is the highest point of each wave.

maximum stress

Calculated by dividing the load in pounds required to pull the material apart by its cross-sectional area expressed in square inches.

maxwell

The CGS electromagnetic unit of magnetic flux, obtained from the law connecting magnetic flux and magnetic induction; 1 maxwell = 1 line of force = 1 gauss × square centimeter $= 10^{-8}$ weber.

Me

The chemical symbol for *methyl.*

mean

See *arithmetic mean, geometric mean, harmonic mean.*

mean

In meteorology, the sum of a set of individual values of a quantity, divided by the number of the values included in the set.

mean calorie

One-hundredth of the amount of heat required to raise the temperature of 1 gram of water from 0°C. to 100°C.

mean free path

The average distance traveled by a molecule between successive collisions with other molecules, or in general by a particle between successive collisions with another particle, in the system; it is inversely proportional to the molecular diameter or particle cross-section.

mean proportional

If three quantities, a, b, c are in continued proportion, then a:b::b:c; or $\frac{a}{b} = \frac{b}{c}$ or $ac = b^2$; in this case b is said to be the mean proportional between a and c.

mean square velocity

The average value of the squares of all the velocities in a system of particles.

mean stress

The midpoint of the *range of stress.*

mean value theorem

See *law of the mean.*

mean velocity

The average value of the velocities of a system of particles.

means

(1) The terms of a progression between any two terms are designated as the means between those two given terms. The means are designated, analogous to the progression, as *arithmetic means, geometric means,* etc.– (2) The second and third terms of a *proportion.*

measurand transmitter

In a *telemetering system,* the apparatus which receives the electrical impulses from the telemetering pickups and processes them for transmission through the *telemetering link.*

measurement of angles

See *angular magnitude.*

mechanical advantage

The ratio of the output to the input of a machine, i.e., of the force exerted by the machine to the force applied to it.

mechanical equivalent of heat
Heat units converted into mechanical units: one calorie equals 41.855,000 ergs, or one British thermal unit (B.t.u.) equals 778 foot-pounds.

mechanical equivalent of light
Radiant flux expressed in mechanical units: one lumen at the wavelength of 555 m equals 0.0015 watt. (The reciprocal, about 660 lumens per watt, is also referred to by the same term.)

mechanical impedance
In a mechanical system, the complex quotient of the alternating forces applied to the system by the resulting alternating linear velocity in the force at its point of application; measured in *mechanical ohms.*

mechanical metallurgy
The science concerned with the relationships among the measured mechanical properties of metals and their mechanical behaviour in service, and with the way these properties and service characteristics vary with chemical composition, structure and temperature. The field also includes mechanical processing such as *rolling* and *drawing.*

mechanical mixture
A mixture of two or more substances such that they retain their individualities; e.g., a mixture of sand and sugar.

mechanical ohm
The unit of measurement of *mechanical impedance, reactance* and *resistance;* one mechanical ohm equals one dyne per centimeter per second.

mechanical properties
Those properties which reveal the reaction of a material to an applied force and which are determined by mechanical means involving deformation and/or destruction. They include *tensile-, compression-, torsion-, bend-, hardness-, cupping-, impact-* and *fatigue-tests.* The properties vary considerably, not only with the composition of the material, but with the method of manufacture, mechanical working and heat-treatment.

mechanical reactance
In a mechanical system, the imaginary component of the *mechanical impedance;* measured in *mechanical ohms.*

mechanical resistance
The opposition of a body to forces which tend to produce

motion. In a mechanical system, the real component of the *mechanical impedance;* measured in mechanical ohms.

mechanics

The science of the effects of forces on bodies at rest or in motion. More customarily called *dynamics,* and usually subdivided into *statics, kinematics* and *kinetics.*

median

The value of the middle term of a series; in frequency distribution, that value of the variable which has a relative cumulative frequency of 0.50 or a cumulative frequency of $N/2$.

median of a trapezoid

The straight line joining the midpoint of a side and the opposite vertex.

median of a triangle

The straight line joining any vertex or corner of a triangle to the middle point of the side opposite that vertex or corner. The three medians of a triangle meet in one point called the *centroid* of the triangle.

medium (*plural:* media)

In physics, any substance or space through which a force or radiation is transmitted.

mega-

(1) A prefix meaning *large.*— (2) A prefix to names of units, meaning *one million.* (Abbreviated to *meg-* when prefixed to word beginning with a vowel.)

megabarye

A unit of pressure equal to one million dynes per square centimeter.

megacycle

The measure of frequency of high-frequency electric alternating current or oscillatory discharge. One million cycles.

megaline

Ten million maxwells.

megascope

A modification of the magic lantern, used especially for throwing a magnified image of an opaque object on a screen, solar or artificial light being used.

megascopic

(1) Visible to the unaided eye.— (2) Observed or performed by the unaided eye.— (Synonymous with *macroscopic,* in both senses.)

megavolt

One million volts.

megawatt
A unit of electrical power equal to one million *watts.*

megohm
The electrical unit of resistance equal to one million *ohms* and, therefore, used only for high resistance values.

melt
To fuse or liquefy.

melting
A synonym for *fusion;* applied in particular to the change of ice to water.

melting point
The point at which a solid changes into a liquid state. The table of chemical elements indicates the melting points of the various elements. Sometimes called *fusion point.*

members
The sides of an equation, inequality or identity.

mendelevium
A transuranic element, not found in nature. Atomic number 101, mass number of the most stable known isotope 256. First produced by Ghiorso, Harvey, Choppin, Thompson and Seaborg in 1955, by high-energy (41 Mev) cyclotronic alpha-particle bombardment of einsteinium ($_{99}E^{253}$). Symbol: Mv.

Mendeleyeff's law
The "periodic law" based on successive atomic weights as related to the properties of the elements, discovered by the Russian chemist Mendeleyeff in 1869. The "periodic law" states that the chemical properties of the elements are periodic functions of their atomic weights. This law has been superseded by the "new periodic law" which is based on *atomic number* (Moseley number) rather than atomic weight.

meniscus
The curved surface of a liquid in a vessel. A liquid which wets the surface of the vessel, e.g. water, has a depressed meniscus; other liquids, e.g. mercury, have a raised or convex meniscus.

mensuration
The art of measuring or of finding the length of lines, areas of surfaces, and volumes of solids, from certain simple data of lines and angles.

Mercator map
A map of the earth or other spherical object produced by the *Mercator projection.*

Mercator projection
A method of depicting a sphere on a plane, in which the me-

ridians are represented as being equidistant, parallel to each other and perpendicular to the parallels of latitude. Each parallel of latitude is expanded about in proportion to the secant of its latitude, and each small arc of the meridian is expanded in the vicinity of each parallel in the same ratio. All rhumb lines (loxodromic curves) appear as straight lines. (The Mercator projection is a so-called *cylindrical projection.*)

mercurial barometer

An instrument for measuring atmospheric pressure. It consists basically of a glass tube, one end of which is sealed, the other one open. It contains a liquid, usually mercury, the displacement of which with the changes in atmospheric pressure can be read on a scale or dial.

mercury

A metallic element, in Group II of the periodic table. Atomic number 80, atomic weight 200.61, melting-point −38.87°C., boiling-point 356.8°C., specific gravity 13.546 (at 20°C.), valence 1 or 2. Symbol: Hg.

mercury vapor lamp

An evacuated gas tube which contains mercury which is vaporized by heat; the mercury vapor gives an incandescent arc between the electrodes, serving as a source of ultraviolet radiation as well as of an intense visible light.

meridian

A great circle on the surface of the earth or another heavenly body, which passes through the north and south poles.

mero-

A prefix meaning *part, partial* or *fractional.*

meso-

A prefix meaning *between* or *middle.*

meso-compounds

These chemical compounds have an even number of asymmetric carbons, the members of any pair having a rotation tendency equal to each other, but opposite in sign, so that the total effect is of no optical activity.

mesomerism

The phenomenon which occurs when an atom takes up a spatial position intermediate between two other theoretically possible positions.

meson

A generic name for a number of sub-atomic particles of mass ranging from about 200 to 1,000 times that of electrons. Most of them carry a negative or positive charge equal to that of an electron. The best known are the *pi-meson* and the *mu-meson,* collectively called *L-mesons*. The former is of mass 273 and decays in 2.6 × 10^{-8} seconds into a mu-meson and a neutrino. The mu-meson (or *muon*) has a mass of 207 and an average life of 2.2 × 10^{-6} seconds, decaying into an electron plus 2 neutrinos. The *pi-meson* (or *pion*) is thought to be the bond holding neutrons and protons together in the nucleus of an atom. Other types of mesons are designated collectively as *K-mesons* (as the *tau-meson,* the *theta-meson,* etc.).

mesosphere
A layer of the atmosphere, defined by Wares as the stratum extending from an altitude of approximately 250 miles to that of approximately 600 miles, between the *ionosphere* and the *exosphere,* whereas Chapman applies this term to designate the *chemosphere.*

mesothorium 1 ($MsTh_1$)
A radioactive isotope of radium ($_{88}Ra^{228}$).

mesothorium 2 ($MsTh_2$)
A radioactive isotope of actinium ($_{89}Ac^{228}$).

metabolism
The sum total of the chemical processes taking place in a living organism.

metal
An element that, as a rule, when untarnished, has lustre, is malleable, conducts heat and electricity. It can replace the hydrogen of an acid and so form a *base.* Except in very thinnest layers, it is opaque. Most metals are solid at ordinary temperatures, *mercury* being the exception, but at high temperatures most metals are fusible. Metals are electropositive and their atoms readily lose *electrons.* Commercially, *brass, Britannia metal,* etc., are called metals, but in the scientific sense they are alloys.

metal leaf
An extremely thin sheet of a metal.

metal physics
A term used to describe the whole range of subjects associated with the structure, physical properties and theories of metals and alloys.

metallic
Pertaining to, consisting of, or resembling a metal or metals.

metallic conduction
The flow of electrons in a metal, without producing any chemical change in the conductor.

metallography
The branch of science which deals with the study of the structure and constitution of solid metals and alloys, and the relation of this to properties on the one hand and manufacture and treatment on the other.

metalloid
Having the nature or properties of a metal.

metalloid elements
Certain elements which share the properties of metals and non-metals; specifically, they have the physical properties of metals. E.g., antimony, bismuth, etc.

metallurgy
Art and science applied to metals. The term covers extraction from ores, refining, alloying, shaping, treating, and the study of structure, constitution and properties.

metamorphic rock
A general geological term for rocks formed in the solid state from previously existing rocks through chemical, physical or mineralogical changes.

metamorphism
Chemical and physical changes occurring in igneous and sedimentary rocks produced by interaction together with heat and pressure.

metarals
A term suggested by Howe for true phases, i.e. definite chemical compounds or solid solutions. He suggested dividing all microscopical constituents of iron and steel into metarals and *aggregates.*

metastable
Energized beyond the most stable state, but not yet become unstable. The metastable state is a condition of pseudo-equilibrium, which is apt to change for a more stable or a less stable one, under certain conditions.

metastable atom
An atom which has been *excited* but is unable to return to the normal state by releasing its excitation energy in the form of radiation until it is further stimulated by a collision or other external influence.

metatectic
An occasionally used synonym for *peritectoid.*

metathesis
A chemical reaction in which an element or radical changes places with an element or radical in another compound without any change in valence. (See *double replacement.*)

meteoric iron
The only form in which iron is found native. It varies in composition, but is usually alloyed with an appreciable amount of nickel, ranging up to 20%.

meteorograph
(1) A self-registering apparatus which records simultaneously the values of two or more meteorological elements, such as air pressure, temperature, humidity, precipitation, etc.—(2) A *radiosonde* (q.v.).

meteorology
That branch of physics which treats of weather and atmospheric phenomena related to the weather.

meter (m)
A unit of length in the C-G-S system; 1 meter = 100 centimeters = 39.3701 inches. (Also written as *metre.*)

meter
A measuring instrument; the word is usually used as a suffix, as in *voltmeter, ammeter,* etc.

meter-candle
A unit of illumination, also called *candle-meter,* equal to one *lux.*

meter-kilogram (mkg, m-kg)
The unit of work in the C-G-S system; 1 mkg = the work required to raise 1 kilogram through 1 meter against the force of gravity.

meter-kilogram-second system
See *MKS system of units.*

method of averages
A method of *curve fitting* based on the computation of pairs of *average points* for each of two sets into which the available data are divided.

method of least squares
A method of *curve fitting* based on the principle of holding the sum of the squares of deviations or errors to a minimum, since the value for which this sum is the lowest is the most

accurate one that can be obtained from the data under consideration.

methyl

The monovalent CH_3 radical; symbol: Me.

metric calorie

The amount of heat required for raising the temperature of one gram of water through one degree C, either from 3.5°C to 4.5° (called the *standard* or *4° calorie*) or from 14.5°C to 15.5°C (called the *normal* or *15° calorie*). Also called *gram calorie* or *small calorie.*

metric carat

See *carat.*

metric horsepower

A unit of power, used in Europe, except for Great Britain; it equals 75 meter-kilograms per second (about 0.9863 U.S. horsepower), or approximately 736 watts. (See *Watts horsepower.*)

metric system

A system of weights and measures depending upon the *meter,* the unit of length from which the original factors are derived. It is based on the decimal system, in that each ascending value is the tenfold of the value immediately preceding it, and the values are identified by the following prefixes:

Prefix	*In abbreviations*	*Value*
tera-	T	10^{12} × the unit
giga-	G	10^{9} × " "
mega-	M	10^{6} × " "
kilo-	k	10^{3} × " "
hecto-	h	10^{2} × " "
deka-	dk	10^{1} × " "
deci-	d	10^{-1} × " "
centi-	c	10^{-2} × " "
milli-	m	10^{-3} × " "
micro-	μ	10^{-6} × " "
millimicro-	mμ	10^{-9} × " "

metric ton

1000 kilograms, equal to 0.9842 long ton = 2204.6 lb.

metric waves

Electromagnetic radiation having wavelengths between 1 meter and 10 meters.

metrology
The theory and practice of precision measurement in the field of mechanical engineering.

mev, m.e.v.
Abbreviations for one million electron volts. I mev = 1.6020 $\times 10^{-6}$ erg.

Mg
The symbol for the element *magnesium.*

mg
The abbreviation for *milligram.*

mho
A unit of measurement of transconductance which is measured in reciprocal ohms. "Mho" is "ohm" spelled backwards. An mho is equal to one billion (10^9) abmhos. The unit is frequently encountered in discussions of the transconductance of radio tubes.

mica
The designation of a group of mineral silicates with varying compositions, crystalline in structures, and readily separable into thin sheets, which are rather elastic, by splitting along the natural cleavage lines by hand; they are transparent, and efficient heat insulators. Used for electric insulation and as a substitute for glass in doors and windows of various heating equipment, in goggles, etc.

micelle
A particle of the dispersed phase in a colloidal system.

Michelson interferometer
A device for calibrating finely divided scales, strain gauges, precision dial indicators, and other length-measuring equipment, using the wave-length of light as the unit of measurement.

micro-
(1) A prefix meaning *small.*— (2) A prefix to names of units, meaning *one one-millionth.* (Symbol: μ.)

microammeter
An *ammeter* calibrated in microamperes.

microanalysis
The determination of chemical composition where the quantities involved are of the order of not more than one-tenth of those measured by the usual analytical methods.

microanalytical reagent
A standard of purity which indicates that a reagent is suitable for *microanalysis.*

microbalance
An extremely sensitive balance for very accurate work and highly precise measurement.

microbar
The pressure of 1 milligram on 1 cm^2 (equal approximately to a pressure of 10^{-6} atmosphere, or 1 dyne of force). Some writers use the term *microbar* for *bar.*

microburette
A burette, designed to deliver drops of approximately 0.01 milliliter.

microcalorie
See *small calorie.*

microcalorimeter
An instrument used to measure very small quantities of heat.

microchemistry
A chemical technique involving extremely sensitive reactions and measuring devices as in *microanalysis.*

microchronograph
A device for recording and measuring very brief intervals of time.

microcleanness
The extent or quantity of non-metallic inclusions observed by examination under a microscope.

microconstituents
The components of a *microstructure* observed under a microscope.

microcosmic salt
Sodium ammonium hydrogen phosphate.

microcurie (μ curie)
10^{-6} curie, i.e. 3.7×10^4 disintegrations per second.

microdurometer
A machine designed for measuring the hardness of very thin specimens of metal.

microetching
The *etching of* samples for examination under the microscope.

microfarad
A practical unit of capacitance, equal to one one-millionth of a farad.

microfiltration
See *ultrafiltration.*

microgram (μg)
One thousandth of a milligram, i.e., one one-millionth of a gram. Also called *gamma* and written as γ.

micrograph
A graphic reproduction of any object magnified more than 10 diameters (usually 200 or more).

micrography
(1) The preparation, study and interpretation of *micrographs*. – (2) The study of *microstructures* in general.

microhardness
Hardness determined by the use of very low loads and diamond pyramid indenters. It is claimed that by means of specially developed instruments it is possible to measure impressions with width of the order of the *micron*, e.g. it enables the hardness of microconstituents to be determined.

microhm
An electrical unit of resistance, equal to one one-millionth of an ohm.

micro-inch (μ inch)
One one-millionth of an inch.

micromanometer
A sensitive device for measuring gas pressures in the 0 to 150 *micron* range. It measures the absolute pressures of condensible as well as non-condensible gases independently of their composition.

micromaster
A micrometer capable of measuring accurately 0.001 millimeter over a range from zero to 25 mm.

micrometer
A precision measuring instrument equipped with a screw of fine graduations showing very minute movements of the screw, used for measuring very small angles, dimensions or displacements.

micrometer scale
An auxiliary scale used to read small angles or magnitudes.

micromicro-
A prefix meaning one one-millionth of one one-millionth (10^{-12}), written for short as $\mu\mu$.

micromicron ($\mu\mu$)
One one-millionth of a *micron*, i.e., 10^{-12} millimeter.

micromillimeter
See *millimicron*.

micron
A unit of length equal to a millionth of a meter (0.001 mm.) = 0.000039 in. It is used for expressing minute distances or wavelengths and is denoted by the Greek letter μ.

microphone

A device which converts sound energy into electrical energy.

microradiography

The process of passing X-rays through a thin section of an alloy in contact with photographic emulsion and then magnifying the *radiograph* 50 to 100 times to observe the distribution of alloying constituents.

microreaction

A chemical reaction between minute amounts of reagents, performed under the microscope.

microscope

An optical instrument used for obtaining highly magnified images of small objects. A *simple microscope* has one lens or lens system, a *compound microscope* has two.

microscopic

(1) Invisible to the unaided eye.– (2) Pertaining to or performed by a microscope.

microscopy

The study and examination of matter by means of a microscope.

microspectroscopy

A method of identifying metallic constituents; it consists of drilling out the minute portion to be analysed, flowing collodion over the resulting chips, and transferring the collodion together with the chips to a pure-carbon electrode for analysis in a standard spectrographic arc.

microstructural change

The alteration in solid metals or alloys involving modification of the microstructure. It is usually associated with constitutional changes, but may occur independently of these, as for example, when crystals are deformed by working of the metal or when deformed crystals recrystallize.

microstructure

A structure or structural feature that cannot be distinguished by the unaided eye.

microtome

An instrument for cutting very thin sections of specimens.

microwave

A general term for radio waves having wavelengths shorter than approximately 30 centimeters and ranging down to 1 millimeter, and of frequencies ranging from approximately 300 to 300,000 megacycles.

mid-point
The point at which a line segment is bisected.

migration
The displacement of particles, usually in a continuous process, as the migration of ions in an electrolyte when current is passed through it.

mil
(1) A unit of angular measurement, defined as an angle equal to 1/6400 of a full revolution. Thus 6400 mils = 360°.—(2) A linear unit, equal to 1/1000 inch, used for measuring the diameter of wires.

milli-
A prefix meaning *one-thousandth.*

milliammeter
An *ammeter* calibrated in milliamperes.

millibar
A unit of pressure, equal to 1/1000 bar, i.e., to 1000 dynes per square centimeter. In meteorology, the values of atmospheric pressure are generally expressed in millibars, one millibar being equal to 0.0295299 inch (0.750076 millimeter) of mercury; thus, 1.013 represents standard atmospheric pressure.

millicurie (m-curie)
One-thousandth of a curie, i.e., that quantity of any radioactive isotope which undergoes 3.7×10^7 disintegrations per second.

milligram (mg)
One thousandth of a gram, equal to .0154 grain.

millileter (ml)
One thousandth of a *liter.*

millimeter (mm)
One-thousandth of a meter, equal to .03937 inch.

millimicron ($\mu\mu$ or mμ)
One-thousandth of a micron, i.e., one-millionth of a millimeter. (Also called *micromillimeter.*)

millions
Figures followed by six digits, and representing multiples of the 6th power of 10.

mineral
A body produced by the processes of inorganic nature, i.e., an inorganic substance or fossilized organic matter, having a definite chemical composition and, if formed under certain favorable conditions, a certain characteristic molecular structure as exhibited in its crystalline form and other physical properties.

A mineral must be a homogeneous substance, even when minutely examined under the microscope; furthermore, it must have a definite chemical composition capable of being expressed by a chemical formula.

mineral oils

Oils, consisting of hydrocarbons, obtained from petroleum and its derivatives.

mineralogy

The science of the nature, occurrence, formation and uses of minerals.

minimum

The least quantity obtainable in a given case or calculation. It is also the point on a curve where the diminishing value of the ordinate (y) starts to increase. On a wavy curve there are as many minima as there are waves even if the waves are uneven. It is the lowest point of each wave.

minimum time path

The *brachistochronic path.*

Minkowski world

The totality of the four-dimensional continuum known as space-time and measured off in space-time coordinates.

minor arc

An *arc* that is less than a semicircle.

minor axis of an ellipse

The shorter of the two axes, i.e., the one perpendicular to the major axis.

minor elements

A synonym for *trace elements.*

minor segment of a circle

Any segment of a circle the area of which is less than half the area of the whole circle.

minus

The word denoting the negative sign (–). It means that the quantity behind this sign must be subtracted or taken away from the preceding quantity.

minute

(1) A unit of time, equal to 60 seconds.– (2) A unit of angular measure, equal to 1/60 degree.– (3) A unit of centesimal angular measure, called *centesimal minute,* equal to 0.01 grade.

mirage

The appearance of distant objects, seen as distorted, inverted or suspended in the sky above the apparent horizon. It is caused by reflection by layers of air of different density, especially when the surface of the land or the sea, and the air in contact with it, is considerably warmer than the air a few yards above, making the density of the air increase upward.

mirror
Any smooth, highly polished surface for reflecting light.

miscellaneous
Mixed or composed of various kinds.

miscibility
The mutual solubility of two or more substances.

miscible
Capable of mixing or dissolving.

mixed acids
A general term for mixtures of two or more acids.

mixed crystals
An obsolete term synonymous with *solid solutions.* It refers to a homogeneous solution of two or more crystallized bodies in the solid state and does not include mechanical mixtures of different substances.

mixed mathematics
That branch of mathematics which treats of the relations of quantities as applied to astronomy, mechanics, optics, etc.

mixed number
The sum of an integer and a proper fraction, e.g., $3\frac{3}{4}$.

mixed quantity
A whole number together with a fraction, thus $2\frac{3}{4}$ or $a - x + \frac{x^2}{a + x}$

mixing ratio
In meteorology, the mass of water vapor per unit mass of absolutely dry air, expressed in grams per gram or kilogram of absolutely dry air. This value is not identical with the *specific humidity,* although numerically it is very nearly equivalent to it.

mixture
A substance consisting of two or more ingredients which retain their own individual properties.

MKS (mks) system of units
An absolute system of units, proposed by Giorgi, which is based on the meter, kilogram and second, and which is extended to the electrical units by the measurement of current by its magnetic effects and of potential difference by the power

unit current. The mechanical units of the system are developed by means of the recognized equations of mechanics using unity as the proportionality factor in each equation, and inserting a new unit of force, named *newton*. In this system, the *joule* is the unit of energy, and the *watt* the unit of power.

MKSA (mksa) system of units

The system of practical and absolute electrical and magnetic units, based on the *meter, kilogram, second* and *ampere* as fundamental units.

ml

The abbreviation for *milliliter*.

mm

The abbreviation for *millimeter*.

mμ**,** $\mu\mu$

Abbreviation for *millimicron*.

$\mu\mu$

(1) The symbol for *micromicron*. (2) A prefix meaning *micromicro-*.

mmf

The abbreviation for *magnetomotive force*.

Mn

The symbol for the element *manganese*.

Mo

The symbol for the element *molybdenum*.

Möbius strip

A "one-sided" surface, formed by giving a strip of paper, cloth, etc. a half-twist and pasting its ends together.

mode

In frequency distributions, the class mark having the greatest frequency.

moderated reactor

A nuclear reactor in which a *moderator* substance is used. (Also referred to by the obsolescent term, *moderated pile*.)

moderation of neutrons

In nuclear physics and reactor technique, the reduction of the speed of neutrons by producing collisions between them and the nuclei of the atoms of material used as *moderator*.

moderator

A material used to "slow down" neutrons in a nuclear reactor, i.e. to reduce their speed to *thermal velocities* by causing them to collide repeatedly with the nuclei of the atoms of the moderator without being captured by the latter. Approximately 20 such collisions with nuclear protons will reduce the speed of a fast neutron to a thermal velocity. Moderators are an important part of every uranium reactor, primary as well as secondary, since slow ("thermal") neutrons are less prone to be captured by U^{238} nuclei, and therefore in a moderated reactor there remain more free neutrons to sustain and propagate the fision chain reaction of U^{235}. The most efficient moderators are logically the light elements, in other words one of the elements at the very beginning of the periodic table, the atomic weight of which is approximately the same as the atomic weight of a neutron, i.e., hydrogen, deuterium, helium, lithium, beryllium, boron and carbon.Of these, carbon is the most suitable from all theoretical as well as practical points of view, and it is therefore the moderator most extensively used, in the form of graphite prepared in compact blocks, machined to close tolerances and baked at high temperatures to help eliminate impurities which might act as neutron absorbers.

modulated quantity

In electronics, a combination of two or more oscillating quantities which result in the production of new frequency components not present in the original oscillating quantities.

modulated wave

A wave of which either the amplitude, frequency or phase is varied in accordance with another wave. A combination of two or more waves which results in the production of frequencies not present in the original waves, these new frequencies being ordinarily made up of sums and differences of integral multiples of the frequencies present in the original waves.

modulating wave

See *modulation.*

modulation

The process, or the result of the process, by which the amplitude or another characteristic of wave is varied as a function of the instantaneous value of another wave. (The first wave is called the *carrier wave,* the second one is referred to as the *modulating wave.*)

modulation factor

In an amplitude-modulated wave, the ratio of half the difference between the maximum and minimum amplitudes to the average amplitude.

modulus

A number or a quantity that is the measure of a force or of a function. A constant factor in a function of a variable quantity.

modulus of common logarithms
The number by which natural (Napierian) logarithms are multiplied to change them into common (Briggs') logarithms, viz.: 0.4342944819. . . . . .

modulus of a complex number
The absolute value of the complex number, i.e., the numerical value of the length of the vector which represents the complex number.

modulus of cubic compressibility
See *volumetric modulus of elasticity.*

modulus of elasticity
See *elastic modulus;* cf. also *volumetric modulus of elasticity* and *Young's modulus.*

modulus of resilience
The amount of energy absorbed when one cubic inch of material is stressed to its elastic limit.

modulus of rigidity
In a torsion test, the ratio of the unit shear stress to the displacement caused by it per unit length in the elastic range.

modulus of rupture
The ultimate strength or the breaking load per unit area of a specimen tested in torsion or in bending (flexure). In tension, it is the *tensile strength.*

modulus of volume elasticity
See *volumetric modulus of elasticity.*

MOHO
Popular term used by geologists to describe the *Mohorovicic discontinuity.*

Mohorovicic discontinuity
The boundary layer of the earth above which is probably basaltic rock and below which is the *mantle.*

Mohs' scale
A method, published in 1822 and still in use, for determining comparative *hardness,* by testing against ten standard minerals: (1) talc, (2) gypsum, (2) calcite, (4) fluorite, (5) apatite, (6) orthoclase, (7) quartz, (8) topaz, (9) corundum, (10) diamond. Thus, a mineral with hardness 5 will scratch or abrade fluorite, but will be scratched by orthoclase.

moisture
Wetness of a substance, measured and expressed as the percentage of water contained.

mol
A synonym for *mole* (See *gram-molecule.*)

molal solution
In chemistry, a solution that contains one mole (gram molecule) of solute per 1000 grams of solvent.

molal volume
The volume occupied by one mole of a substance at the temperature of 0°C and a pressure of 760 millimeters of mercury. For an ideal gas, this volume is 22.4 liters. (Also called *gram-molecular volume.*)

molar
Relating to and expressed in *gram-molecules.*

molar gas constant
The constant R in the *ideal-gas law* expressed for n moles of gas as $PV = nRT$, where P is the pressure, V is the volume, and T is the temperature in degrees of the absolute scale; R is equal in this case to 0.08204 liter-atmospheres.

molar heat
The amount of heat required to raise the temperature of one mole of a given substance by one degree; it equals the *specific heat* of the substance multiplied by its *molecular weight.*

molar refraction
The product of the *specific refraction* of the substance times its molecular weight, i.e., $\frac{n^2 - 1}{n^2 + 2} \times \frac{M}{d}$, where n is the *index of refraction* of the substance (at the given wavelength), M is its molecular weight, and d is its density.

molar solution
In chemistry, a solution that contains one mole of solute per liter of solution.

molar surface
The surface of a sphere the mass of which equals one mole.

molar weight
Gram-molecular weight.

mole
The molecular weight of a substance expressed in grams, i.e., that quantity of the substance which weighs a grams, where a equals the *molecular weight* (in units of atomic weight) of the substance. Also called *gram-molecule* and *mol.*

mole fraction
The proportion of the number of molecules that a constituent bears to the total number of molecules in the same phase. Where compositions are stated in weight per cent, the mole

fractions are calculated by first dividing the weight per cent of each constituent by its molecular weight and then dividing each result so obtained by the sum of all the results.

mole volume

See *molecular volume.*

molecular

Relating to or having the property of a molecule or molecules.

molecular adsorption

The adsorption of a gas by a solid where the atoms of the gas are assumed to have condensed as liquid in the pores of the solid and can be removed by pumping.

molecular compound

A chemical compound formed by the chemical combination of two or more whole molecules.

molecular concentration

The concentration of a solution expressed in terms of gram-molecules or *moles* in a given volume.

molecular conductivity

The electrical conductivity of a solution divided by the number of moles contained in a unit volume of the solution.

molecular formula

A chemical formula indicating the elements as well as the number of atoms of each element composing a given molecule.

molecular formula weight

See *formula weight.*

molecular heat

Specific heat multiplied by molecular weight.

molecular percent

See *molecular ratios.*

molecular proportion

See *molecular ratios.*

molecular ratios

The ratios or proportions in which the various molecules occur in a substance. It is obtained by dividing the weight percent of each molecular "species" (oxides, etc.) by its molecular weight. Recalculation to a basis of 100 molecules yields the *molecular percent.*

molecular reaction

Any reaction between molecules, which is generally slower than the reactions between ions.

molecular reflectivity

The product of the *specific reflectivity* of the given substance by its molecular weight.

molecular refractivity

The product of the *specific refractivity* of the given substance by its *molecular weight.*

molecular spectrum

A spectrum of a substance in the molecular state, i.e., a spectrum produced by molecular vibrations.

molecular structure

The internal structure of a molecule, indicating the atoms which compose it, and the *valence bonds* which link them together.

molecular volume

The volume occupied by one mole, i.e., one gram-molecule of a substance, equal to its molecular weight divided by its density. In the case of a solid or liquid, this ratio varies according to the atomic volume of the substance, but it is constant (22.414 liters) for all gases at normal temperature and pressure. (Also called *mole volume.*)

molecular weight

The weight of a molecule of a substance relative to the weight of a hydrogen atom, i.e., expressed in terms of multiples of 1/16 of the atomic weight of oxygen. The molecular weight of any substance can be computed by adding the atomic weights of all the atoms constituting a molecule of it.

molecule

The smallest part of a substance that can exist separately and still retain its chemical properties and characteristic composition identical with those of the substance as a whole; the smallest combination of atoms that will form a given chemical compound. Atoms are linked into molecules by their outer orbital electrons.

molybdenum

A metallic element in Group VI of the periodic system, discovered by Scheele in 1782 (although its discovery is accredited also to Hjelm). Atomic number 42, atomic weight 95.95, melting-point 2550°C., boiling-point about 3200°C., specific gravity 10.2 (at 20°C.), valence 2 or 6. Symbol: Mo.

moment

A product of force and distance. The product obtained by multiplying a force by the perpendicular distance from a point to the line of action of the force, is called the moment of the force about the point. If the action of the force is such as to tend to produce motion about the point in the same direction as the hands of a clock move (clockwise rotation) as viewed by the observer, the moment is said to be negative. If the tendency to rotate is counterclockwise, the moment is said to be positive. For balance, the sum of all moments acting on a point must be zero.

moment arm of a force

The perpendicular distance from the axis to the line of action of the force.

moment of inertia

The resistance of a body or particle to angular acceleration; its magnitude depends on the axis of rotation in question, being the sum of the products of the mass and the square of the radius for each and every particle of the body, i.e., it equals $\Sigma\, mr^2$, where m is the mass of the body or particle, and r is the perpendicular distance from the axis.

moment of momentum

A synonym for *angular momentum.*

moment of torsion

The ratio of the magnitude of a torque to the angle of twist which that torque produces.

momentum

The products of mass and velocity. Unless otherwise specified, the term *momentum* signifies *linear momentum,* which is the product of the mass and linear velocity of a body or particle.

monatomic

Consisting of one single atom.

Monel metal

An alloy of high resistance to rust, consisting of about 67–68% nickel, 28% copper, 2–2.5% iron and 1–2% manganese.

mono-

A prefix meaning *one, single.* (Before a vowel it is frequently shortened to *mon-*.)

monobasic

Said of an acid in which the hydrogen atom is replaceable by only one basic atom or radical.

monochromatic

Having only one color, i.e., more precisely, only one wavelength or frequency or radiation.

monoclinic

Crystals having only parallel 2-fold axes of symmetry, or parallel planes of symmetry, or both of these with the axes perpendicular to the planes. The typical monoclinic crystal has three unequal axes, only one of which is perpendicular to two others.

monohydrate
A solid midway between dry and crystallized, containing one molecule of water of hydration; that is, it contains less water of crystallization than the crystallized form but has more water than when in the dry form.

monoisotopic
Said of a chemical element which occurs in nature in only one stable isotopic form, i.e., in which the natural abundance of one isotope is 100%. (E.g., cobalt.)

monomer
(1) One single molecule.—(2) A substance which consists of one single molecule.

monomial
An algebraic expression of one single term, or a part of an algebraic expression connected or separated by mathematical signs only. E.g., 5x.

monomorphous
Existing in only one crystalline form.

monomolecular
Pertaining to, consisting of or involving only one molecule. Also called *unimolecular.*

monomolecular reaction
A reaction in which only one molecule undergoes a change. Also called *unimolecular reaction* and *reaction of the first order.*

monosaccharides
A collective designation of any *simple sugar,* i.e., any *pentose* or *hexose.*

monotropic
Existing in only one stable physical form, any other form obtainable being unstable under all conditions.

monovalent
Having a *valency* of one; said of an element whose atom is capable of combining only with a single atom of another element. Also called *univalent.*

mordant
(1) A metallic hydroxide or oxide which is deposited on the fibers of a textile from the solution of its salts, and which is capable of forming an insoluble colored compound, with the dyestuff. (2) In metallurgy, an acid by which the biting is done in *etching.*

mosaic structure

A structure of units presumed to exist in metals and alloys, composed of unit blocks about $10^{-3} - 10^{-4}$ centimeter square. In metals it is characterized by a regularity of arrangement of the atoms on a *space lattice* of individual crystalline grains.

Moseley number

A synonym for *atomic number.*

Moseley's law

The square root of the frequency of a given line in the characteristic x-ray spectra of elements increases by a constant amount from one element to the other, when the elements are arranged in the order of ascending atomic numbers. (Cf. *periodic table.*)

mother liquor

The solution that remains after the removal of the substances crystallizing readily and in a regular form.

mother metal

The molten alloy just before final solidification and freezing out of the solid.

motor

Short for *electric motor* (q.v.) . Incorrect usage applies the term also to the power plant of a motor vehicle, the proper term for which is *engine.*)

m. p.

The abbreviation for *melting point.*

$MsTh_1$, $MsTh_2$

The symbols for *mesothorium 1* and *mesothorium 2,* respectively.

mucilage

A generic term for salts, either with metals or organic bases, of polysaccharide sulfuric acid esters, which gel with water.

multi-

A prefix meaning *many.*

multilateral

Having several or many sides.

multilineal

Having several or many lines.

multinomial

An expression of more than one term, as $4a + 5b + 7cd + 8abc$.

multiphase alloys

Alloys possessing heterogeneous structures.

multiple
Any number which contains another given number as an exact divisor. Example: 6, 9, 12, and 30 are multiples of 3.

multiple ionization
Ionization in which the atom loses or gains more than one extranuclear electron.

multiple reflection
The reflection of light by two or more reflectors or mirrors several times in succession.

multiplicand
The number or quantity to be multiplied by another, the multiplier.

multiplication
The process of repeating or adding to itself any given number, or quantity a desired number of times. Thus three times five or 3×5 is adding 3 to itself five times or 3 plus 3 plus 3 plus 3 plus 3, or $3 + 3 + 3 + 3 + 3 = 15$.

multiplication factor (k)
The ratio of the number of neutrons produced by nuclear fission to the number of neutrons originally present in a fission reactor—in other words, the number of neutrons produced in a chain reactor for every neutron absorbed or otherwise lost. If the value of this factor drops below 1, the nuclear fission ceases to be a self-sustained chain reaction and will eventually die away; on the other hand, if it exceeds 1, the chain reaction will eventually become uncontrollable. In order to obtain a useful controlled chain reaction, the multiplication factor must be kept at a value of k equal to or slightly greater than 1. (Referred to also as *k-factor* and as *reproduction factor*.)

multiplication sign
Any symbol used to indicate that numbers or quantities are to be multiplied. The symbol $\times$, the period, and the mere juxtaposition of two letters, or a number and a letter, can all be used as signs of multiplication.

multiplier
The number or quantity by which another (the *multiplicand*) is multiplied.

multiply
To add any given number to itself a given number of times. Thus multiply 3 by 5 means add 3 to itself 5 times; $3 \times 5 = 3 + 3 + 3 + 3 + 3 = 15$.

mu-meson (μ meson)
See *meson.*

muon
A *mu-meson.*

muriatic acid
See *hydrochloric acid.*

muscovite
Common *mica,* potassium aluminum silicate.

mutarotation
The change in the *optical rotation* of a substance, usually observed in freshly prepared solutions of certain sugars.

mutual inductance
The common property of two associated electric circuits which causes a change of flux in one to result in an induced electromotive force in the other, and which determines, for a given change of current in one circuit, the electromotive force induced in the other.

mutual induction
The creation of the existence of *mutual inductance* when the relative position of two inductors causes the magnetic lines of force from one to link with the turns of the other.

mutual-inductive coupling
Coupling of two circuits by means of their *mutual inductance.*

myriad
(1) The numeral 10,000. (2) In general, an expression denoting any immense or incalculably great number.

Mv
The symbol for the element *mendelevium.*

N

N

The symbol for the element *nitrogen.*

n

(1) The symbol for the frequency of a wave motion. (2) A symbol for the *neutron.* (3) The symbol for *n unit.*

$_0n^0$

A symbol for the *neutrino.*

$_0n^1$

A symbol for the *neutron.*

***n*-leafed rose**

A combination of loops having the shape of rose petals which touch each other in the 0 points of the coordinate system. The graph of the equation $r = a \sin n\theta$ or $r = a \cos n\theta$, where n is a positive integer. When n is odd, the loops number n, when it is even, there are $2n$ loops.

N pole

In a magnet, the *north-seeking pole,* i.e., the pole which always points toward the magnetic north.

n unit (n)

A unit for measuring neutron-induced ionization; 1 n is that amount of neutrons which produces the same amount of ionization in a Victoreen 100 r thimble chamber as would be produced by 1 r (roentgen unit) of gamma-rays.

Na

The symbol for the element *sodium.*

nadir

That point on the celestial sphere directly below the observer.

napier

See *neper.*

Napierian (*also* Naperian) logarithms

See *natural logarithms.*

nappe

One of the two parts of a conical or pyramidal surface which are separated by the *vertex.*

narcotic

(1) Possessing the property of inducing a stupor, insensibility to pain, or sleep. (2) Any substance possessing such property.

narrow-angle diffusion

Diffusion in which light is scattered in all directions from the diffusing medium, while the intensity is notably greater over a narrow angle in the general direction which the light would take by regular reflection or transmission.

nascent

Said of a chemical substance at the instance of its formation.

nat

In the designations of the *natural trigonometric ratios* (q.v.) the abbreviation for *natural;* e.g., nat sin = natural sine.

natrium

The Latin and German name of the element *sodium* (Na).

natural cosecant, natural cosine, natural cotangent, natural secant, natural sine, natural tangent

See natural *trigonometric ratios.*

natural frequency

(1) A constant frequency of a vibrating system in the state of *natural oscillation.* (2) In an electrical circuit, a frequency that equals the reciprocal of $2\pi\sqrt{LC}$, where L is the inductance in henrys, and C is the capacity in farads. (3) In an antenna, the lowest resonant frequency without added inductance or capacitance.

natural gas

Gas present in underground pockets and emerging from volcanoes, encountered in caves, as well as released frequently in the course of operations of drilling oil wells. It is used for illuminating, for heating, and for industrial purposes. Its basic constituents are hydrogen, methane (marsh gas), carbon dioxide, nitrogen, and certain hydrocarbons.

natural logarithms

The logarithms the base of which is $e = (2.7182818284\ldots\ldots)$. Also called *Napierian* or *hyperbolic logarithms.*

natural number

Any positive integer.

natural oscillation

An oscillation of a vibrating system which may take place in the absence of any external force.

natural period

The period of frequency of the free oscillations of a body or system.

natural trigonometric ratios

The trigonometric ratios taken in a right triangle with equal legs and having a hypotenuse the length of which is unity (1), so that the ratio or proportion of one side to the other is always an exact or natural fraction. If the two legs are designated *a* and *b*, the principal natural trigonometric ratios are: *natural sine* $= \frac{a}{1} = 1$; *natural cosine* $= \frac{b}{1} = 1$; *natural tangent* $= \frac{a}{b}$; *natural cotangent* $= \frac{b}{a}$; *natural secant* $= \frac{1}{b}$; *natural cosecant* $= \frac{1}{a}$.

Nb

The symbol for the element *niobium.*

Nd

The symbol for the element *neodymium.*

Ne

The symbol for the element *neon.*

needles

(1) Elongated crystals, tapering at each end to a fine point, as those typical of *martensite.* (2) Metal powder particles of elongated form, resembling needles.

negative

Minus, to be subtracted. Any number or symbol preceded by the minus sign (—) is said to be a negative number or quantity.

negative acceleration

A synonym occasionally used for *deceleration,* i.e., retardation or loss of velocity, either spontaneous or deliberately induced.

negative angle

In trigonometry, an angle formed by a clockwise revolution of the *terminal line* about the vertex. (Cf. *standard position of an angle.*)

negative catalyst

A *catalyst* which retards a chemical reaction.

negative charge

The type of electric charge which is present in a substance the atoms of which have more than their normal number of electrons.

negative electricity

The kind of electricity which predominates in a body composed of resin after it has undergone electrification by rubbing with wool.

negative eyepiece

An *eyepiece* consisting of two lenses, both of which are flat on one side and convex on the other.

negative glow

A luminous zone which appears in a *discharge tube,* at slight pressure, between the *cathode dark space* and the *Faraday dark space.*

negative ion

An atom which has more than its normal number of electrons.

negative lens

See *divergent lens.*

negative proton

See *antiproton.*

negative quantity

In mathematics, any quantity preceded by a minus sign.

negative specific heat

Some authors use this term to designate the quantity of heat which must be extracted from certain substances under certain conditions in order to raise their temperature (e.g., the specific heat of saturated water vapor is a negative quantity, viz., —1.07).

negative trigonometric angles

We divide the circle into four quadrants by a horizontal and vertical diameter. All sine angles above the horizontal diameter (1st and 2nd quadrants) are positive and below that diameter (3rd and 4th quadrants) are negative. All cosine angles right of the vertical diameter (1st and 4th quadrants) are positive and left of that diameter (2nd and 3rd quadrants) are negative. This information gives the following table for all the trigonometrical ratios in the 4 quadrants.

	1st Quadrant	2nd Quadrant	3rd Quadrant	4th Quadrant
Sine	+	+	—	—
Cosine	+	—	—	+
Tangent	+	—	+	—
Cotangent	+	—	+	—
Secant	+	—	—	+
Cosecant	+	+	—	—

negative valence

The valence of an atom which gains an electron or electrons. Its numerical value equals the number of the electrons which the atom has gained as a result of a chemical action.

neodymium

A metallic element, a member of the rare earth group of the periodic system, discovered by Welsbach in 1885. Atomic number 60, atomic weight 144.27, melting-point 840°C., specific gravity 6.96, valence 3 or 4. Symbol: Nd.

neon

A zerovalent gaseous element in Group O of the periodic table, one of the rare gases, discovered by Ramsay and Travers in 1898. Atomic number 10, atomic weight 20.183, melting-point –248.67°C., boiling-point –245.9°C., density 9.0035 grams per liter at normal temperature and pressure. Neon is a colorless, odorless, monatomic gas, occurring in ordinary air as approximately 18 parts per 1,000,000 by volume. Symbol: Ne.

neper

A unit used to express the relationship or ratio between two amounts of power (acoustic, electric, etc.) as an interval on a logarithmic scale. In acoustics, the neper is equal to 0.8686 bel or 8.686 decibels. (The synonym *napier* is not regarded as good usage.)

nephelometer

An apparatus for measuring the amount of incident light scattered by suspended particles in a solution, which when compared with the scattering in a standard solution, permits the determination of the quantity of the matter present in suspension.

nephelometry

A method of quantitative analysis, viz., of the determination of the concentration of dispersed particles or of matter suspended in a solution by measuring the amount or degree of scattering of light produced by them, based on the measurement of the intensity of the *Tyndall cone.*

neptunium

A highly unstable and radioactive element, not found to occur in nature, produced artificially; the first of the transuranic elements, and the first transuranic element to be discovered, in 1940, by Segré, Seaborg, Kennedy and Wahl, etc. McMillan and Abelson were the first ones to show that the neutron-irradiation of uranium produces the isotope $_{93}Np^{239}$. The atomic number of neptunium is 93, the mass number of its most stable isotope is 237; it has mostly the valence 4, at times also 2, 3, 5 or 6. Symbol: Np.

neptunium series

A *radioactive series* which starts with the artificial plutonium isotope $_{94}Pu^{241}$ and ends with the stable bismuth isotope $_{83}Bi^{209}$. It has been named *neptunium series* after its longest-lived member, $_{93}Np^{237}$, or $(4n + 1)$ series, since the mass numbers of its members are always $(4 \times n) + 1$, where n is an integer. Also referred to as *neptunium family*, and as *(4n + 1) series* or *family*.

Nernst effect

The development of an electromotive force in a strip of metal in magnetic field when heat passes through it across the magnetic lines of force; the electromotive force is usually at right angles to the flow of heat as well as to the magnetic lines of force.

Nernst law, Nernst theorem

See *distribution law*. See *law of distribution*.

Nernst theory

An explanation of the development of electrode potentials, based on the supposition that an equilibrium is established between the tendency of the electrode material to pass into solution and that of the ions to be deposited on the electrode.

Nessler tubes

Matched glass cylinders with strain-free, clear glass bottoms for comparing color density; used in colorimetric analysis.

Nessler's reagent

A strongly alkaline solution of potassium mercuric iodide. It is used as a test for minute quantities of ammonia, which impart a yellow color to it.

network

In topology, a set of points, called *vertices*, with a set of *paths*, also called *arcs*, connecting them. (Also referred to as a *finite graph*.)

neutral

Neither acidic nor basic; if referring to an aqueous solution, having a *p*H value of 7, i.e. that of pure water.

neutral atmosphere

An atmosphere which is neither oxidizing, reducing, nor carburizing, i.e. in which neither hydrogen nor oxygen is in excess. The term is generally used in connection with furnace conditions.

neutral flame

A welding term designating a gas flame in which the portion used is neither oxidizing nor reducing.

neutral point

In chemical analysis, that point at which acidity and alkalinity are completely balanced.

neutral solution

In chemistry, a solution in which the number or concentration of hydrogen ions equals that of the hydroxyl ions, and which therefore is neither acid nor basic. A solution with pH = 7.0—7.07, at ordinary temperatures.

neutralization

The chemical reaction between an acid and a base in such precise proportions that the characteristics of each disappear. The solution so produced contains a salt.

neutrino ($_0n^0$)

A subatomic particle, first postulated by Fermi (in 1934) in order to explain certain apparent contradictions to the law of the conservation of energy in β-emission (the continuous background of the γ-ray spectrum).

neutron ($_0n^1$, n)

A sub-atomic particle, recognized as one of the fundamental nuclear particles, discovered by Chadwick in 1932. The mass of the neutron is 1.00893 a.m.u. (corresponding to 939.4 Mev), or 1.6848×10^{-24} gram, i.e. very slightly more than the mass of a proton, but it has no electric charge; neutrons are important and very practical agents for the production of nuclear reactions, fission and fission chain reaction, for, being electrically neutral, they are not repulsed by the positively charged protons in the atomic nuclei, and therefore are much better suitable for use as "projectiles" for nuclear bombardment than the positively charged α-particle, proton and deuteron. Free neutrons may be captured by atomic nuclei, and the resulting nuclear reaction usually produces a nucleus of a different element or isotope.

neutron cross-section

In nuclear physics, a measure of the degree to which a substance will absorb neutrons.

new candle

See *candle.*

newton

The unit of force in the MKS system of units; 1 newton is defined as the force which produces an acceleration of 1 meter per second per second in a mass of 1 kilogram.

newton-meter

A unit of work in the MKS system of units; one newton-meter

is the work done by a force measuring 1 newton when exerted through a distance of 1 meter.

Newtonian mechanics

The classical mechanics, based on Newton's laws of motion.

Newton's law of cooling

The rate at which a hot body cools is proportional to the difference in temperature between the body and its surroundings.

Newton's law of gravitation

Every particle of mass attracts every other particle, and the force of the attraction is directly proportional to the product of their masses and inversely proportional to the distance between them.

Newton's laws of motion

(1) Every body continues in its state of rest, or of motion with a constant velocity in a straight line, unless acted upon by some external force. (2) Change of momentum is proportional to the force and to the time during which it acts, and is in the same direction as the force. (3) To every action there is an equal and opposite reaction.

Newton's rings

Interference fringes formed in the air layers between optical surfaces.

nickel

A metallic element in Group VIII of the periodic table, discovered by Cronstedt in 1751. Atomic number 28, atomic weight 58.69, melting-point 1452°C., boiling-point 2900°C., specific gravity 8.85, valence, 2 or 3. Symbol: Ni.

nicol

A *Nicol prism.*

Nicol prism

Two prism-shaped segments of Iceland spar cemented together with Canada balsam; when ordinary light is passed through it, it causes the emitted light to vibrate in only one plane, i.e., to be plane polarized.

nine-point circle

The circle that bisects the sides of a triangle and bisects the lines between the orthocenter (point of intersections of the altitudes) and the vertices and passes through the feet of the altitudes. The center of the circle is half-way between the orthocenter and the point of intersection of the vertical lines from the midpoints of the sides of the triangle.

niobium

A metallic element in Group V of the periodic table, discovered by Hatchett in 1801. Atomic number 41, atomic weight 92.91, specific gravity 8.5, melting-point 1950°C., valence 5. The natural abundance of Nb^{93} is .00001% Radioisotopes: Nb^{91}, Nb^{92}, Nb^{94}, Nb^{95} and Nb^{97}. Formerly also called *columbium* (*Cb*). Symbol: Nb.

nital

A solution of nitric acid in methyl or ethyl alcohol, 1.5% by volume, used as an *etching* agent in ferrous metallography.

niton

Element of atomic number 86; now known as *radon.*

nitration

The process or operation of introducing the *nitro group* into an organic compound, usually with nitric acid or with some mixture containing nitric acid.

nitric acid (HNO_3)

A colorless, water-soluble liquid. In concentrated form it is called *aqua fortis;* in less concentrated form, it is used for dissolving metals, and in dilute solutions as an etching reagent.

nitro group

The monovalent $—NO_2$ radical.

nitrogen

A gaseous element in Group V of the periodic table, discovered by Rutherford in 1772. Atomic number 7, atomic weight 14.008, melting-point −209.86°C., boiling-point −195.8°C., density 1.25055 grams per liter at normal temperature and pressure, valence 3 or 5. Nitrogen is the major constituent of the atmosphere, occurring in ordinary air to the extent of 78.03% by volume and 75.5% by weight. Symbol: N.

nitrogen cycle

The passage of atmospheric nitrogen to the soil and from one compound to another in the course of its utilization by plant and animal life, and its return to the soil.

nitrogen fixation

In general, any process through which atmospheric nitrogen is transformed into useful compounds.

nobelium

A transuranic element, not found in nature. Atomic number 102. First produced by a team of Swedish, American and British scientists at the Nobel Institute of Physics in Stockholm by high-energy (110-120 Mev) cyclotronic bombardment of curium ($_{96}Cm^{244}$) by carbon ($_{6}C^{13}$) nuclei. Symbol: No.

noble gases

See *inert gases.*

noble metals

Metals, e.g., the members of the *gold* and *platinum* families, which do not tarnish or corrode in air or water and which are resistant to corrosion by all but the most powerful acids, such as *aqua regia.* It is owing to their characteristic corrosion-resistance that the noble metals occur *native.*

noble potential

A potential substantially cathodic to the standard hydrogen potential.

nodal point

In physics, any of the points, or areas, where two waves of the same wavelength and amplitude and traveling through a medium at identical speeds in directly opposing directions reinforce each other. These points (also called *nodes*) are half a wavelength distant from each other, and are situated halfway between the *antinodal points* (or *antinodes*) . Also referred to as *nodal region.*

nodal region

See *nodal point.*

node of a curve

The point where the curve crosses itself or becomes tangent to itself; thus it may be a *crunode* or an *acnode.*

node of a wave

See *nodal point.*

noise

(1) In general, any unwanted or undesirable sound. (2) Any unmodulated radio signal received, as in radio astronomy.

noise level

The strength of the acoustic noise perceived at a given location, or of the radio noise received at a given point in the circuit; usually measured and expressed in decibels.

nomogram

A combination of graded scales, arranged customarily parallel or perpendicular to each other, capable of being combined so as to solve an algebraic equation or several algebraic equations. Also called *alignment chart.*

nonagon

A polygon bounded by nine straight lines.

noncentral quadrics
A subdivision of the *nondegenerate quadric surfaces,* including the *elliptic paraboloid* and the *hyperbolic paraboloid.*

nonconductor
A substance which will block the propagation of heat, electricity, or another physical process or phenomenon.

noncoplanar forces
Forces (or *vectors*) the lines of action of which do not lie in the same plane.

nondegenerate quadric surfaces
The *central quadrics* (real ellipsoid and imaginary ellipsoid, hyperboloid of one sheet, hyperboloid of two sheets), and the *non-central quadrics* (elliptic paraboloid and hyperbolic paraboloid).

nondestructive tests
Methods of examining the soundness of materials which do not involve the destruction of the specimen under test, as in mechanical testing.

nonelectrolyte
A substance which does not conduct electricity when in solution; a substance not yielding *ions* in solution.

nonequilibrium
A synonym for *disequilibrium.*

nonferrous
Not containing any iron.

nonferrous metal
A generic term for all metals except for iron, steel or another alloy containing iron.

normal
Perpendicular or at right angles; vertical; regular, uniform or standard; average or mean.

normal
(1) A measure of concentration of a solution, equal to one gram equivalent of solute per liter of solution. (2) In organic chemistry the designation applied to straight carbon chain aliphatic compounds.

normal calorie
The amount of heat required for raising the temperature of one gram of water from 14.5°C. to 15.5°C. Also called *15° calorie.*

normal gravity
The value of gravity at sea level, based on a formula in which the earth is regarded as a *spheroid.*

normal pressure

Standard pressure, usually taken to be equal to that of a column of mercury 760 millimeters (about 30 inches) in height and weighing approximately 14.7 pounds per square inch.

normal salt

A salt in which there are no replaceable hydrogen atoms or hydroxide radicals (i.e., in which all hydrogen atoms of the acid have been replaced by a metal, or all hydroxide radicals of the base have been replaced by an acid radical).

normal solution

A solution of a reagent used in volumetric analysis, such that one litre of the solution contains one gram equivalent of the reagent. The normal solution of acid contains one gram of replaceable hydrogen ions per liter; that of a soluble metallic hydroxide contains 17 grams of hydroxide radical per liter.

normal temperature and pressure

The temperature of 0°C. and the pressure of 760 mm. of mercury. Abbreviation: *n.t.p.*

normality

In chemistry, a term used for expressing the concentration of solutions. (See *normal solution.*)

north-seeking pole

That pole of a magnet which always points toward the magnetic north. Referred to as *N pole.*

notation

The expressing of numbers by means of characters, letters or figures.

nought

A figure, character or symbol (written 0) and has the same meaning as *zero* or *cipher*. It means the absence of all magnitude or quantity, however small, except when in combination with other digits or numbers.

Np

The symbol for the element *neptunium*.

***n*th power**

The number or quantity obtained by multiplying the given number by itself *n* times.

***n*th root**

That quantity the *n*th power of which (i.e., the quantity multiplied by itself *n* times) equals the given expression or quantity.

n. t. p.

The abbreviation for *normal temperature and pressure.*

nuclear

Relating or pertaining to an atomic nucleus or to atomic nuclei. In popular parlance the adjective "atomic" is very frequently used instead of "nuclear" (e.g. "atomic" energy, "atomic" physics, etc.), but in scientific terminology a careful distinction must be made between these two concepts.

nuclear bombardment

The striking of the nucleus of an atom by *subatomic particles.*

nuclear charge

The number of electronic charges on the nucleus of the atom; the nuclear charge is always equal to the number of protons in the nucleus, i.e. to the atomic number (Z) of the isotope.

nuclear chemistry

The study of reactions which produce a transmutation of one isotope into another. In these reactions the atomic nuclei are combined, whereas the reactions and processes of ordinary chemistry concern the extranuclear electrons only.

nuclear cross section

The effective area of the target nucleus exposed to bombardment by nuclear particles (neutrons, alpha particles, etc), used to express the probability of a particular nuclear process or reaction. The nuclear cross section is expressed by $\frac{a}{bc}$, where a is the number of processes occurring, b the number of incident particles, and c the number of target nuclei per cm^2.

nuclear energy

The energy liberated by *nuclear fission.*

nuclear energy plant

Any installation designed to produce and ultilize *nuclear energy* for practical, industrial, etc., purposes.

nuclear fission

The splitting of an atomic nucleus into two fragments of more or less equal mass, as a result of the capture of a bombarding neutron. The only elements known to be capable of undergoing nuclear fission are uranium, plutonium, protactinium and thorium (*fissile materials*).

nuclear forces

The forces assumed to be operating in the atomic nuclei. They are pictured as very short-range forces, being effective at a range not exceeding 10^{-13} cm., within which they counterbalance the repulsive electrostatic ("coulomb") forces of the nucleons. The nuclear forces seem to be independent of the charges of the nucleons concerned, i.e. to be acting in the same way between proton and proton, neutron and neutron, and proton and neutron.

nuclear fusion

A nuclear reaction in which light nuclei combine to form a nucleus of a higher mass number. This process can be said to represent the opposite to *nuclear fission*. The fusion of nuclei of low atomic number releases large amounts of energy.

nuclear isobars

Atoms of isotopes of different chemical elements which have the same mass number but different atomic numbers, i.e. the same total number of nucleons but differently divided between protons and neutrons; e.g. the hydrogen isotope $_1H^3$ (*tritium*), the nucleus of which contains 1 proton and 2 neutrons, and the helium isotope $_2He^3$, which has nuclei composed of 2 protons and 1 neutron. Since nuclear isobars are isotopes of different chemical elements, their chemical properties are different, and therefore they can be separated chemically.

nuclear isomers

Isotopes of the same element, having the same mass number and the same atomic number, but differing in radioactive properties, such as different *half-lives* (e.g. $_{47}Ag^{106}$ with half-lives of 24.5 minutes and 8.2 days). Nuclear isomers are rare among the naturally radioactive elements, but occur frequently among artificial radioisotopes.

nuclear mass

The mass of the nucleus equals approximately, but never exactly, the sum of the masses of the protons and neutrons composing it.

nuclear particles

See *subatomic particles.*

nuclear physics

The science of the atomic nuclei, sub-atomic particles and nuclear reactions. The boundary line between nuclear physics and *nuclear chemistry* is necessarily rather vague and arbitrary.

nuclear reaction

Any change in the composition and physical characteristics of an atomic nucleus produced, directly or indirectly, by its irradiation or bombardment by high-energy α-particles, protons, deuterons, slow or fast neutrons, or high-energy radiation (γ-rays).

nuclear structure

The number, nature and arrangement of nucleons in the nucleus of an atom.

nucleon

A collective term for the fundamental building blocks of atomic nuclei, i.e., the *proton* and the *neutron,* the combined total number of which in an atomic nucleus always equals the mass number of the isotope. The term is usually extended to include the *antiproton* and *antineutron* as well.

nucleonics

The study and science of the behavior of nucleons.

nucleus

The central core of the atom, about which the orbital extranuclear electrons revolve. The nucleus is composed of protons and neutrons, collectively called *nucleons.* The number of the protons in the nucleus equals the atomic number, the number of neutrons equals the difference between the mass number and the atomic number, and the combined total of all the nucleons equals the mass number of the isotope.

nucleus of a crystal

In metallurgy and crystallography, any of the points at which crystals begin to grow during the solidification of a substance.

nuclide

Any species of atom that exists for measurable length of time and has a nuclear structure distinct from that of any other species of atom. Thus, isotopes of the same element are distinct nuclides.

number

A symbol denoting how many times a unit is to be taken or considered. Commonly, a synonym for *numeral.*

numeral

A symbol representing a number.

numerate

To number; to state or indicate or write numbers in the order of size or magnitude.

numerator

The name given to the number or quantity above the line in a fraction. It is also called the *dividend.*

numerical coefficient

An arithmetical number placed before a variable as a multiplier; e.g., in the expression 7x, 7 is the numerical coefficient of x.

O

O
The symbol for the element *oxygen.*

ob-
A prefix indicating *inversely, in opposite sense.*

objective
The lens, or set of lenses, opposite to the eyepiece in a microscope which forms an image of the specimen.

oblate spheroid
A spheroid that is flattened at the poles.

oblique
Said of a line which is neither perpendicular nor horizontal. (Synonym: *slanting.*)

oblique angle
Any acute, obtuse or reflex angle, i.e., any angle that is not an even multiple of 90°.

oblique prism
Any *prism* that is not a *right prism,* i.e., any prism the base of which is not perpendicular to any edge.

oblique spherical aberration
See *coma.*

oblique triangle
A triangle no angle of which is a right angle.

obtuse angle
Any angle greater than a right angle (90°) but smaller than a straight angle (180°).

obtuse triangle
A triangle one of the angles of which is an obtuse angle.

occlusion
The retention or absorption of fluids by solids, i.e., the adhesion of a gas or liquid on a solid or their entrapment within a solid.

octades
Periods of eight figures each of which was a scale of notation introduced by Archimedes.

octagon
A polygon bounded by eight straight lines.

octahedron

A polyhedron that has eight faces. (In a *regular octahedron* they are all congruent equilateral triangles.)

octane

An oily hydrocarbon with eight carbon atoms (C_8H_{18}) present in petroleum.

octane number

The "antiknock" rating given a gasoline, obtained by comparison with a standard or reference fuel. (Also called *octane rating.*)

octaves

See *law of octaves.*

odometer

A revolution counter which, attached to or connected with the wheel of a vehicle, can be used to measure the approximate distance traveled.

oersted

The CGS electromagnetic unit of magnetic field strength or intensity. At any point in a vacuum the value of the magnetic intensity in oersteds is equal to the force in dynes exerted on a unit magnetic pole placed at that particular point.

offset

In surveying, the perpendicular distance of a point from a chosen *base line.*

ohm

The practical unit of electric resistance; 1 ohm is defined as the resistance which allows 1 ampere of current to flow through a conductor at 1 volt potential; 1 ohm = 10^9 abohms.

ohmic resistance

In electricity, resistance to the flow of direct current.

ohmmeter

An instrument for measuring the electrical resistance of conductors and insulating materials (insulants) and provided with a scale graduated in *ohms* or *megohms.*

Ohm's law

The law stating that the current in an electric circuit is directly proportional to the electromotive force in the circuit. Ohm's law does not apply to all circuits. It is applicable to all metallic circuits and to many circuits containing an electrolytic resistance. Ohm's law was first enunciated for a circuit in which there is a constant electromotive force and an unvarying current. It is applicable to varying currents if account is taken of the induced electromotive force resulting from the self inductance of the circuit and of the distribution of current in the cross-section of the circuit. ($E = IR$ or $R = \frac{E}{I}$).

-oid

A suffix meaning "like," "resembling," or "having the form of." E.g., *spheroid* means sphere-like; *conoid* means "in the form of a cone," etc.

oil

Chemically, a liquid of considerable viscosity, composed to a major extent or wholly of carbon and hydrogen, immiscible in water, soluble in ether and most organic solvents, usually combustible.

oil geology

That branch of geology which treats of the geologic factors involved in the origin, accumulation and exploitation of oil deposits. (Also called *petroleum geology*.)

oil of vitriol

See *vitriol.*

oleic acid

An unsaturated liquid fatty acid, found in most fats and oils.

one-dimensional configuration

A line, since it has only length, but no width or height.

one-minute wire

Galvanized wire which will withstand immersion in a standard neutral copper sulphate solution for one minute without penetration of the zinc coating.

one-sided surface

See *Möbius strip.*

opacity

The measurable ability of a substance to obstruct by absorption the transmission of radiant energy such as light. Opacity thus is the degree of nontransparency.

open circuit

An electric circuit which does not provide a complete path for the flow of current.

open circuit potential

The measured potential of a cell from which no significant current flows in the external circuit.

open-hearth furnace

A regenerative, reverberatory furnace for the production of "open-hearth steel." The furnace most commonly used today.

operation

The execution of an act pursuant to certain rules or principles.

operational calculus

A mathematical procedure for solving ordinary linear equations with constant coefficients and some of the partial differential equations of theoretical physics.

optical pyrometer

See *vertical angles.*

optic axis

In crystallography, a direction through a doubly refracting crystal in which light traversing the crystal undergoes no *double refraction.* (Cf. *optical axis.*)

optical

Pertaining to phenomena produced by or involving visible light, to vision, or to optics in general.

optical activity

The ability possessed by certain chemicals to rotate the plane of polarized light when the latter is passed through the substance.

optical axis

In a lens, the straight line passing through the centers of curvature of the surfaces; light rays passing along this direction are neither refracted nor reflected. (Cf. *optic axis.*)

optical density

For a transparent or translucent medium, the logarithm of the ratio of the intensity of an incident ray to the intensity of the transmitted ray. Thus of two mediums the one having a greater *index of refraction* for a wavelength has a greater optical density for that wavelength.

optical dispersion

The variation of the *index of refraction* with the wavelength of light.

optical glass

Glass carefuly manufactured to precise specifications to obtain purity, transparency, homogeneity and exactly correct optical properties.

optical illusion

An error to the average eye in the perception of certain lines, angles and spaces.

optical isomerism

A type of *isomerism,* observed in certain substances having asymmetric molecules, in which the substances (the *optical isomers*) have identical structural formulas and general properties, but affect polarized light differently.

optical path

The distance traveled by light multiplied by the index of refraction of the medium.

optical pyrometer

An instrument for the measurement of temperature through which the observer sights the heated object and compares its incandescence with that of an electrically heated filament whose brightness can be regulated. Optical pyrometry is based on the knowledge of the emissivity of the hot body.

optical rotation

The number of degrees of rotation that an optically active substance produces in the plane of polarized light for a given length of passage.

optical surface

A reflecting or refracting system.

optical system

A number of lenses, or of lenses and prisms, so arranged as to refract, or to refract and reflect light, in order to perform some definite optical function.

optical window

The bands of electromagnetic radiation to which the atmosphere is not opaque.

optically active

Said of any substance which possesses the property of producing a rotation of the plane of polarization while it transmits polarized light.

optics

The science that treats of light and vision.

optimatic pyrometer

An instrument for measuring the temperature of hot moving bodies, using two vacuum photocells illuminated by the surface of the body under investigation and by the filament of a comparison lamp; the current in the filament when a perfect match in brightness is obtained is a measure of the temperature of the body.

orbit

The path traveled by a particle under the influence of gravitational attraction or another force.

orbital electron

One of the extranuclear electrons circling in an orbit about the atomic nucleus. Also called *planetary electron.*

order
The degree of an expression, equation, operation etc. as expressed by an ordinal number (first, second, third, etc.).

order of a reaction
See *reactions of the first, second, third and fourth order.*

order of a vertex
In topology, the number of distinct *arcs* have the vertex in question as an end point.

order of magnitude
The value of a number of physical quantity indicated approximately, usually within a power of 10. (E.g., 5×10^6 and 9×10^6 are the same order of magnitude, and both are 3 orders of magnitude greater than 7×10^3).

ordinal number
A number that denotes sequence or the relative order of something in a set or series as, *first, second, third,* etc. Ordinal numbers are used also to express fractions (e.g., *one-third, two-fifths,* etc.) except for "second" which is replaced in this sense by "half."

ordinary ray
In *double refraction,* that one of the refracted rays which has the greater *index of refraction* and remains fixed as the birefringent crystal revolves.

ordinate
The perpendicular from a given point on a graph or map grid to the horizontal center line or abscissa. A perpendicular from a given point to the x axis.

ore
A natural metalliferous mineral or mixture of minerals from which valuable or useful substances can be extracted.

organic
(1) Containing carbon or some carbon compound (except for carbonates and cyanides, which are regarded as inorganic). (2) In colloquial usage, pertaining to, derived from or produced by living organisms.

organic acid
Any compound containing the carboxyl radical (—COOH); derived from a hydrocarbon by replacing three hydrogen atoms by an oxygen atom and a hydroxyl radical (—OH).

organic chemistry
That branch of chemistry dealing with organic compounds, i.e., with the carbon compounds which form the basis of all living matter.

organism
A living creature; any human, animal or vegetable body or structure.

origin
A fixed point from which measurement or motion commences.

origin of a system of coordinates
The point of intersection of the axes of the system.

ortho-
A prefix meaning *straight* or *at right angles.*

orthocenter of a triangle
The point of intersection of the three altitudes.

orthogon
A geometric figure all angles of which are right angles—i.e., a rectangle.

orthogonal
Right-angled; relating to or determined by right angles.

orthogonal circles
Two intersecting circles, the tangents of which at the points of intersection are at right angles.

orthogonal projection
Projection using lines of projection (*projectors*) perpendicular to the plane on which the projection is shown.

orthographic map
A map produced by *orthographic projection,* in which the equator and the meridian that is 90 degrees from the meridian and determines the plane of projection are diameters, whereas all the other meridians appear as ellipses, and all parallels of latitude as parallel straight lines.

orthographic projection
The process of projecting accurate outlines of views of an object and properly arranging them. This method may best be explained through a process of tracing views of an object on the sides of a transparent box, and then manipulating the sides of the box (and consequently the traced views) in a mechanical method simulating the theory controlling the location and arrangement of the views necessary to present a picture of the object. (See *orthogonal projection.*)

orthohydrogen
The *ortho* form constitutes about 75% and the *para* form constitutes about 25% of ordinary *hydrogen.* The two nuclei are spinning in antiparallel directions in the para form, and in parallel directions in the ortho form.

orthorhombic
A form having three axes intersecting each other at a right angle but all of unequal length. The rectangular and rhombic prisms, and the rhombic octahedron belong to this system.

oscillate
To move or swing back and forth with a pendulum-like motion.

oscillating current
See *oscillatory current.*

oscillation
(1) The state of a physical quantity when, in the time interval under consideration, the value of the quantity is continually changing in such a manner that it passes through maxima and minima. (2) The process or state of oscillating (see *oscillate*).

oscillatory circuit
A circuit containing inductance and/or capacitance and resistance, so arranged or connected that a voltage impulse will produce a current which periodically reverses.

oscillatory current
(1) A current which alternately increases and decreases in magnitude with respect to time according to some definite law. (2) A current flowing in an *oscillatory circuit.*

oscillogram
See *oscillography.*

oscillograph
An apparatus for producing a continuous curve representing the instantaneous values of a rapidly varying electric quantity as a function of the time or another electrical quantity.

oscillography
The observation, recording and measurement of the changing instantaneous values of the current or voltage in an electric circuit when these quantities are either alternating or transitory. The instrument used for the purpose is an *oscillograph* and the graphical record obtained is an *oscillogram.*

oscilloscope
An instrument for making visible the presence and/or the nature and form of oscillations or irregularities of an electric current. (Several different types of instrument serving this general purpose have been designated by this name.)

osculating circle of a curve
The circle of closest possible contact with the curve at a point.

osmium

A metallic element in Group VIII of the periodic system, discovered by Tennant in 1803. Atomic number 76, atomic weight 190.2, melting-point 2700°C., specific gravity 22.48, valence 2 or 8. Symbol: Os.

osmosis

The diffusion of a fluid, i.e., the slow migration of its molecules, through a semipermeable membrane or a porous partition into another fluid of a different density; namely, the passage of a solvent from a dilute solution into more concentrated one. (See *endosmosis and exosmosis.*)

osmotic pressure

The force exerted by a solution on a semi-permeable membrane which separates it from a pure solvent and permits the molecules of the solvent to pass through it while it stops the passage of the molecules of the solution. It is the excess of the pressure on the solution side of the semipermeable membrane over the pressure on the solvent side which prevents the solution from passing through the membrane and into the solvent.

Ostwald's law

As the concentration of a weak electrolyte, such as acetic acid, decreases, the percentile ratio of ionization increases. (Also referred to as the *dilution law.*)

-ous

In chemistry, a suffix indicating the lower valance of an element or radical (e.g., ferrous).

out of phase

Said of periodic quantities which have the same *frequency* and *waveform* if they do not reach their maxima and minima at the same time.

outer space

(1) In contexts of currently developing practical aero-space activities, the space above the earth's atmosphere, or above its effective atmosphere. (2) Space beyond the limits of the solar system.

output

The energy delivered by an apparatus, machine or electric circuit.

overlapping

A synonym for *elliptic involution* (cf. *involution*).

overtone

In music, generally any of the *partials* except for the *fundamentals,* in a compound note. Specifically, a note of higher frequency (*pitch*) but lower intensity than the fundamental note, imposed on the fundamental to enhance the latter's tone quality.

overvoltage

The difference between the potential at which a gas, especially hydrogen is evolved from a solution and the potential of a reversible electrode of the same gas in the same solution.

Ox

The chemical symbol for *oxalic.*

oxalic

Containing the radical $(COOH)_2$; symbol: Ox.

oxidation

In the original and literal sense, the chemical combination of an element with oxygen; in an extended meaning, also its combination with a halogen or another electronegative element. In the modern, broadest sense, any increase in *valence* owing to the loss of one or more electrons (negative charges).

oxidation-reduction buffer

A substance the addition of which to a solution, compound or system gives the latter a resistance to any change in *oxidation-reduction potential.*

oxidation-reduction potential

The e.m.f. developable by a system capable of mutual oxidation and reduction. It is due to the flow of electrons from the oxidant to the reductant. Usually the potential is referred to an external standard electrode, as a standard hydrogen electrode. Example: ferrous sulfate—ferric sulfate.

oxide

The binary compound of oxygen with another element or a radical. Oxides are divided into four broad groups: *acid(ic) oxides, basic oxides, amphoteric oxides,* and *peroxides.*

oxidize

To combine with oxygen, to form an *oxide;* to increase the valence.

oxidizing flame

A welding term designating a flame in which the used portion has an oxidizing effect owing to an oxygen content in excess to that required to produce a *neutral flame.*

oxy-

A prefix indicating the presence of oxygen in a molecule.

oxy-acetylene

In welding, a term indicating the use of a mixture of acetylene with oxygen.

oxygen

A gaseous element in Group VI of the periodic table, discovered by Priestley in 1774. Atomic number 8, atomic weight 16.0000, melting-point −218.8°C., boiling-point −183°C., density 1.42904 grams per liter, specific density 1.056 with air taken as 1.00, valence 2. Oxygen is a colorless, odorless gas; it is the most abundant element, forming 21% by volume (23.15% by weight) of the atmosphere, about 89% by weight of water, and 46.6% by weight of the rocks of the earth's crust. Symbol: O.

ozone

A form of molecular oxygen (O_3), each molecule of which consists of three atoms; it is colorless, but has a characteristic odor. This allotropic form of oxygen can be produced by passing oxygen through a region of an electric discharge.

ozonide

The product formed by the addition of *ozone* to a double bond. It has antiseptic, bactericidal and oxidizing properties.

ozonosphere

A stratum of the upper atmosphere, at an altitude of approximately 40 miles, characterized by a comparatively high concentration of *ozone,* and which acts as an absorber of the ultraviolet radiation of the sun.

P

P

The symbol for the element *phosphorus.*

π

The Greek letter *pi* (pronounced *pie*). (See *pi.*)

π meson

See *meson.*

pacemaker

The substance whose rate of reaction in a series of linked reactions sets the pace for the series; lactic acid may be considered as the pacemaker in fermentation.

packing fraction

The ratio of the *mass defect* of an isotope to its mass number, i.e. $\frac{M - A}{A}$, where M is the atomic mass and A is the mass number. Since the mass number equals the total number of nucleons in an atom of the isotope, the packing fraction expresses the mass defect of the atom per nucleon. It is customary to indicate the packing fraction multiplied by 10,000, in order to obtain figures convenient to use in calculations.

packing loss

An alternative term for *mass defect.*

paleontology

The science dealing with fossil remains and life in remote geological ages.

palladium

A metallic element in Group VIII of the periodic table, discovered by Wollaston in 1804. Atomic number 46, atomic weight 106.7, melting-point 1554°C., boiling-point about 2500° C., specific gravity 12.17, valence 2 or 4. Symbol: Pd.

pantograph

A device by means of which a point is made to copy, on any desired scale, the pattern traced by another point.

pantograph

In electrical equipment, a sliding type of current conductor used for maintaining contact between electric street cars, etc. and an overhead contact wire.

para-

A prefix indicating *beyond* or *opposite.*

parabola

A *conic section* formed when the cutting is parallel to one, and only one, element of the cone. It is the locus of points the distance of which from a fixed point (the *focus*) and fixed line (the *directrix*) is always equal, i.e., the ratio of these two distances is always unity.

parabolic cylinder

A cylinder the right section of which is a parabola.

parabolic mirror

A concave mirror which has the form of a *paraboloid of revolution.* All light rays emanating from a source located at the focal point are reflected as parallel rays.

parabolic reflector

In optics, *a reflector* in which all sections parallel to a certain line are parabolic curves, and which brings distant parallel rays to a focus at a point, or produces a parallel beam when a light source is placed at its focus.

parabolic velocity

A velocity great enough to cause a revolving body to change from an elliptical to a parabolic orbit.

paraboloid

A surface formed by causing a parabola to revolve about its axis. (See *elliptic paraboloid, hyperbolic paraboloid, paraboloid of revolution.*)

paraboloid of revolution

A surface produced by causing a parabola to rotate about its axis, and the cross sections of which are circles.

parachor

A constant relationship for a given substance, between surface tension and density. This relationship is expressed as:

$$[P] = \frac{M\sqrt[4]{\gamma}}{\rho_l - \rho_g}$$

P is the parachor; M is the molecular weight; γ is the surface tension of the substance, in liquid form, at a given temperature; and ρ_l and ρ_g are the densities of the liquid and vapor at the same temperature. The parachor of a substance is therefore proportional to the molecular volume and has been used in determining the structure of molecules and polymers.

paraffin

(1) A white, waxy, tasteless, translucent solid, consisting of a mixture of solid hydrocarbons, insoluble in water, acids or

alkali, but soluble in alcohol, ether and chloroform; used as an ingredient of adhesives, as an absorbent, as general coating agent, as a general insulating agent, in explosives, candles, and ink. (2) See *paraffins.*

paraffins

The saturated aliphatic hydrocarbons of the methane series, of the general formula C_nH_{2n+2}. (The simplest member of the series is *methane,* CH_4.)

parahydrogen

See *orthohydrogen.*

parallax

In astronomy, the apparent displacement of an object, or the apparent difference in its direction of motion, if viewed from two different points.

parallel

Equidistant at all points; said of two or more lines or planes which can never meet or intersect no matter how far they are extended.

parallel beam of light

A theoretical concept of a beam of light coming from an infinite or at least astronomical distance (e.g., sunlight), so that its component pencils or rays neither converge nor diverge, but may be regarded as parallel to each other.

parallel connection

See *in parallel.*

parallel of latitude

A small circle of the earth parallel to the Equator.

parallelepiped

A prism the bases of which are parallelograms; a polyhedron bound by six parallelograms, the opposite pairs being equal and parallel.

parallelogram

A plane figure bounded by four straight lines. The opposite sides are parallel and equal. The opposite angles are equal. The diagonals bisect one another but are not equal.

parallelogram law

The rule for the addition (*"composition"*) of two vectors, consisting in that the vectors are regarded as adjacent sides of a parallelogram, the diagonal of which is their sum (*"resultant"*).

parallelogram of forces

If two forces act on a point but their lines of action are not identical the following procedure must be applied: Each force is represented by a line, the direction of the line being the line along which the force acts. The length of the line represents the magnitude of the force on some scale previously decided on. An arrowhead is always placed on the line to indicate the direction in which the force acts. The diagram is always constructed so that each of the two forces is acting away from the point representing the point on which the forces are acting. Through the opposite end of each line which represents a force, a line is drawn parallel to the other force. A parallelogram is thus formed. A diagonal line is then drawn from the point to the opposite corner. This diagonal line represents the resultant force.

paramagnetic substance

Any substance the *magnetic permeability* of which is slightly greater than unity, i.e., than that of air, and approximately independent of the *magnetizing force*. Such substance is therefore attracted to a magnet, that is, tends to move from the weaker to the stronger portion of magnetic field, and to take up a position with its greatest length parallel to the direction of the field.

paramagnetism

The property, exhibited by a paramagnetic substance, of having magnetic properties greater than unity (that of air) and being attracted to a magnet.

parameter

(1) A line or quantity serving to determine one point, line, figure or quantity out of a class of such things. (2) An independent variable in terms of which the coordinates of a variable point may be expressed. (3) An arbitrary constant characterizing by each of its particular values some particular member of a system of expressions, curves, surfaces, functions, etc. (4) A variable through functions of which may be expresssed other variables, as the coordinates of a locus. (5) In conics the 3rd proportional of any diameter and its conjugate, or in the parabola, to any abscissa and the corresponding ordinate. (Cf. *parameter of a conic.*)

parameter

The term as applied to crystals refers to the characteristic magnitude defining the unit cell or the position of a point within it. The edges of the unit cell, a, b, c and the angles, α (between b and c), β (between c and a), γ (between a and b), are its parameters. The edge of the unit cube, a_0, is the single parameter of a cubic lattice. When the fractional co-ordinates of a point within the unit cell may vary, the values of these fractions (x, y, z) are parameters of the structure.

parameter of a conic

A perpendicular line to the major axis drawn through the focus. The ends of this line terminate at the curve. Also called *latus rectum*. The word is used sometimes to denote the co-tangent of the angle which the tangent to a point on a parabola makes with the x-axis.

parametric equation of a curve

An equation expressing the coordinates of the curve as functions of a *parameter*.

parasite

See *parasitic current*.

parasitic current

In electricity, a current in a circuit, due to some unintentional cause, such as inequalities of temperature or of composition. (Also called *parasite*.)

parasitic oscillations

In electricity and electronics, unintended self-sustaining oscillations.

paraxial

Very close to the axis and forming a very small angle with it.

paraxial ray

In optics, a ray the path of which lies very near the axis of a lens and which intersects the lens surface at a point very close to its vertex and at nearly normal incidence.

parent element

Generally, an element which produces another element or isotope, whether by spontaneous radioactive decay or as target element in a nuclear bombardment, is called the parent of the newly produced element or isotope (which is called its *daughter* element). Specifically, with reference to a *radioactive series* or *family* of elements, the element or isotope starting the series is called its parent.

parent metal

In welding, the metal to be welded or cut. Also called *base metal*.

parentheses

Parentheses in algebraic equations indicate that the quantities inside them are to be combined before undergoing any other arithmetical operations.

parity

A symmetry property of a wave function. All nondegenerate states have a definite parity.

parsec

A unit of measure for interstellar space; the distance at which a star would have a *parallax* of one second of arc. One parsec equals 3.258 light years, or 206,265 astronomical units, or 19,-150,000,000,000 miles.

partial

In music, any part of a compound tone (chord or discord). The lowest partial is called the *fundamental,* and all others are the upper partials, regardless of the range of single tones embraced by the compound. When discriminated separately, they are sensed as pure tones.

partial diffusion

Diffusion of light in which the diffusing medium scatters part of the light incident upon it, but allows part of it to be regularly reflected or transmitted so that objects from which the light originally comes (e.g., the filament of a lamp) can be seen sharply defined by the reflected or transmitted light. (Also referred to as *incomplete diffusion.*)

partial fractions

A fraction which has in its denominator the product of various terms each containing the unknown quantity, x, may require to be split up into the sum of simpler fractions, each having only one of the terms in the denominator. As $\frac{x + 2}{(x - 4)\ (x - 2)} = \frac{A}{x - 4} + \frac{B}{x - 2}$. The simpler fractions are called partial fractions. Thus: $\frac{x + 5}{(x - 1)\ (x - 2)\ (x - 3)} = \frac{3}{(x - 1)} = \frac{7}{(x - 2)} + \frac{4}{(x - 3)}$

partial product

The product of one term of a compound expression with a term or number multiplying that expression. Thus in $(a + b - c)d$ the terms $ad + bd - cd$ are each partial products.

particle

In chemistry and in mechanics, a unit of matter which has the properties of mass and a determinable position, but not that of geometrical extension.

particle

In nuclear physics, one of the elementary building blocks of matter. (See *subatomic particles.*)

particle adaptation

The term applied in powder metallurgy when powder particles change their shape and position, but do not shrink.

particle density

In *powder metallurgy,* the mass of the actual volume of an individual particle.

particle size

In *powder metallurgy,* the controlling linear dimension of an individual particle as determined by analysis with screens or other suitable instruments.

particle size distribution

The percentages expressed in terms of weight of the various sizes of particles in a powder sample when classified in terms of size ranges and measured in terms of screen mesh or microns.

particle transformation

In powder metallurgy, the shrinkage of individual powder particles.

particle velocity

In a sound wave, the instantaneous velocity of a given infinitesimal part of the medium, with reference to the medium as a whole, due to the passage of the sound wave.

partition coefficient

See *distribution coefficient.*

partition law

See *distribution law.*

partition of numbers

Making up a number by adding together two or more numbers whose values are below that of the original number. Thus to form the partition of 5 we have $4 + 1$; $3 + 2$; $3 + 1 + 1$; $2 + 2 + 1$; $2 + 1 + 1 + 1$; $1 + 1 + 1 + 1 + 1$.

Pascal's law

If external pressure is applied to a confined fluid, the pressure increases at every point in the fluid by an amount that equals the external pressure applied.

passivating

In metallurgy, the oxidizing treatment applied to stainless steels and irons in order to increase their *passivity*. The best known treatment of this kind consists of treating stainless steel in nitric acid to remove surface impurities which might act as starting points of corrosive attack by electrochemical action.

passivator

In metallurgy, an *inhibitor* which changes appreciably the potential of a metal to a more cathodic value.

passivity

A property shown by substances under certain circumstances, consisting in an abnormal inactivity toward certain chemical reagents. Thus, iron immersed in a copper sulfate solution will normally enter solution and precipitate copper on its surface, but in the passive state this tendency is less marked, and its resistance to acid attack is very greatly increased.

Pauli exclusion principle

No two electrons can occupy the same position in the electron structure of an atom, i.e., have the same set of quantum numbers. This principle signifies that no electron orbit in the *electron structure* of the atom can be occupied by more than one electron, and in atoms having several electrons in the various electron *shells,* every electron must be in a different orbit.

Pb

The symbol for the element *lead.*

Pd

The symbol for the element *palladium.*

peak value

The maximum positive or negative value of an alternating or periodic quantity.

peat

A dark residual product of the partial decomposition of vegetable matter. (See *peat coal.*)

peat coal

(1) A decomposition product of vegetable matter, intermediate between *peat* and *lignite.*—(2) The product of the artificial carbonization of *peat.*

pedal curve

The locus of the foot of the perpendicular drawn from a fixed point to a variable tangent to a curve.

pedal line

The straight line which is the locus of the feet of the perpendiculars from any point on the circumference of the circumcircle of a triangle on the sides of the triangle.

pedal triangle

A triangle formed within a given triangle by joining the feet of the perpendiculars drawn from each corner of the given triangle to the opposite side.

Peltier effect

When a current flows across the junction of two dissimilar metals, it causes either an absorption or liberation of heat, depending on the direction of the current, at a rate proportional to the first power of the current.

pencil

A system of lines intersecting at one point. E.g., the envelope of a cone, or the spokes of a wheel.

pencil of involution

A pencil consisting of pairs of rays, such that any transversal cuts them in pairs of points forming an *involution.*

pencil of light (rays)

A group of light rays coming from a point source or converging toward a point. (A pinhole opening produces a pencil of light rays.)

pendulum

(1) A body so suspended as to swing freely under the action of gravity and of momentum (also referred to as *physical pendulum* or *simple pendulum*). (2) A body so supported by a spring below it as to vibrate under the action of gravity and the restoring forces of the spring.

Penetron

A gamma-ray instrument used to facilitate the detection of internal corrosion by measuring secondary radiation from gamma rays directed into the metal being tested.

penta-

A prefix indicating *five.*

pentagon

A polygon bounded by five straight lines.

pentahedron

A polyhedron with five faces.

pentane (C_5H_{12})

A liquid hydrocarbon, one of the constituents of petroleum.

pentode

A five-electrode vacuum containing an anode, a cathode, a control electrode, and two additional electrodes ordinarily in the nature of grids.

pentose

A simple sugar (*monosaccharide*) containing five carbon

atoms in the molecule; the general formula of the pentoses is $C_5H_{10}O_5$.

per-

A prefix indicating (1) *very, completely, thoroughly,* (2) *above* or *beyond,* (3) in chemistry, an indefinitely or relatively large amount or the largest possible proportion of the substance to the name of which it is prefixed.

percentage

A certain rate per hundred; the number of one-hundredths to be considered in a given case.

percentage composition

(1) A measure of concentration of a solution; it is the weight of solute in 100 grams of solution. (2) Percentage of each element in a given compound.

perfect crystal

A single crystal throughout which the arrangement of the atoms is completely uniform. Such crystal is capable of x-ray reflection.

perfect diffusion

Uniform scattering of the radiation in all directions by the diffusing medium.

perfect fluid

A hypothetical ideal substance the molecules of which offer no mechanical resistance, affording flow without any viscosity whatsoever.

perfect gas

See *ideal gas.*

peri-

A prefix meaning *about* or *around.*

perigon

An angle of 360°.

perimeter

The length of a closed curve or the sum of the lengths of the sides of a polygon.

period

(1) The interval of time required by a periodic motion to complete a cycle.— (2) See *periodic table.*

period of a function

Cf. *periodic function.*

period of the trigonometric functions

The values of the trigonometric ratios repeat again and again as the angle increases each time it has completed four right angles (respectively two right angles for tan and cotan) and for this reason are called *periodic functions.* Their period is 2 radians for the sine, cosine, secant and cosecant, and radians for the tangent and cotangent.

periodic function

A function the value of which is not changed by increasing or decreasing its arrangement by any multiple of a constant called the *period.* E.g., $(x \pm n.\ 2\ \pi) = x$.

periodic motion

Any motion that is repeated at equal intervals of time.

periodic spiral

A graphic arrangement of the elements in the sequence of their atomic numbers on a spiral which illustrates the similarities and interrelationships of their properties.

periodic table

A table of the chemical elements arranged according to a regular periodic recurrence of chemical and physical properties. D. Mendeléeff and L. Meyer (working independently) announced the *periodic law* for the first time in 1869, stating that the properties of the various chemicals as well as the forms and properties of their compounds depend on the atomic weight of the element and recur in regular cycles. Mendeléeff's *periodic system* of the elements was summed in a table in which the elements were arranged in the order of increasing atomic weights. In 1914, *Moseley* announced his law and proved that these periodic recurrences of similar properties of elements do not depend on the atomic weight, but on the atomic number. The modern version of the periodic table, based on Moseley's law, Bohr's atomic theory and Pauli's exclusion principle, lists the elements in the order of increasing atomic numbers. Since the atomic number of an element is always equal to the number of extranuclear electrons contained in its *electron structure,* the order in which the elements appear in the modern versions of the periodic table represents their arrangement in the order of increasing number of extranuclear, orbital electrons. This happens to be also the order of increasing atomic weights, with four exceptions (in the pairs aluminum and potassium, cobalt and nickel, tellurium and iodine, thorium and protactinium, the element having the higher atomic number has a lower atomic weight). See pages 544-5 for the latest version of the periodic table which includes also the artificially produced transuranic elements. It is divided into *periods* (horizontal rows) and *groups* (vertical columns); the groups, with the exception of the first and last one, are divided into *sub-groups* (*a* and *b*). Each group contains elements having similar physical and chemical properties; this is especially true of the elements in the same sub-group.

peripheral electron
An orbital electron in the outermost electron shell of the atom. Peripheral electrons are the source of the spectral lines of visible light and thermal radiation, and as *valence electrons* determine the chemical combination characteristics of the atom.

periphery
The *circumference* or *perimeter* of a plane figure.

peritectic reaction
An isothermal reversible reaction in binary alloy systems, in which a solid and a liquid phase react during cooling, to form a second solid phase. (Also referred to as *peritectic transformation.*)

peritectic transformation
See *peritectic reaction.*

peritectoid
The new solid phase formed by two solid phases in a binary alloy system in a *peritectoid reaction.* The peritectoid (occasionally referred to as a *metatectic,* is the inverse of a *eutectoid.*)

peritectoid reaction
An isothermal reversible reaction in binary alloy systems, in which two solid phases react during cooling to form a new solid phase, called a *peritectoid.*

permanent emulsion
An *emulsion* in which the droplets of the dispersed phase retain their individual identities for a long time, owing to the admixture of an *emulsifying agent.*

permanent gas
A gas the temperature of which is higher than its *critical temperature,* and which consequently cannot be liquefied by pressure alone.

permanent hardness
With reference to water, the hardness which remains in water after prolonged boiling, owing to the continued presence of calcium and magnesium chlorides or sulfates.

permanent magnet
A *magnet* which retains its magnetism after it has been removed from an external magnetic field.

permanent magnetism
Magnetism independent of the presence of any external magnetic field.

permanent set
The condition which a solid body assumes under the effect of a stress which reaches its *elastic limit.*

permeability
(1) The ability to penetrate or pass through a given substance or medium.— (2) Permitting the passage of a given substance.— (3) See *magnetic permeability.*

permeance
The reciprocal of magnetic *reluctance.*

permittivity
The ratio of the electric displacement of a medium to the electric force which produces it. (The term *relative permittivity* used as a synonym for *dielectric constant.*)

permutation
(1) Any of the different arrangements which can be formed of a number, n, of things taken r at a time. The number of such permutations of n things taken r at a time is denoted by the symbol P(n,r).— (2) The process of producing such permutations.

peroxide
(1) An oxide which yields hydrogen peroxide when treated with acid. (2) A compound of an element with oxygen in which the oxygen atoms are singly linked.

perpendicular
At right angles to a line or surface. (*Vertical* means only at right angles to the earth's surface.)

perpendicular planes
Two planes which form dihedral angles the plane angles of which are right angles.

persistent spectrum
The spectrum of a substance which results from only the most moderate excitation. The most persistent lines which remain when all others have subsided or when the quantity of the substance is diminished to a trace.

personal equation
A systematic error in observations due to the personal characteristics of the observer; this error consists of the difference between the true reading and that made by the observer.

perspective
Appearance in terms of distance. When lines joining points

on one figure to corresponding points on another are concurrent in one point, the two figures are in perspective.

perspective center

The point of origin or termination of *perspective rays.*

perspective drawing

The representation of an object on a plane surface as it appears to the eye; it affords a better pictorial effect than other types of drawing.

perspective plane

Any plane which contains the *perspective center.*

perspective ray

A line joining the *perspective center* and a point object.

perturbation

In crystallography, the influence of thermal agitation on the relations between atoms in a crystal or in a molecule.

perturbation

In spectroscopy, (1) an irregularity in the spacing of the lines of a band spectrum; (2) a deviation of the values of terms in a spectral series from the appropriate series formula.

pervaporation

The evaporation of a liquid, or of the *continuous phase* of a colloidal system, by suspending it in a bag of collodion or parchment or some other semipermeable material and heating it; any crystalloids contained in the liquid deposit then on the outside of the bag.

petrifaction

The gradual changing of organic substances into a stone-like or mineral structure.

petrography

The systematic study of rocks based on their mineralogical and textural characteristics as observed on field specimens and on microscopic sections.

petroleum

(1) An oily, inflammable, liquid mixture of gaseous, liquid and solid hydrocarbons, chiefly of the paraffin series, found in the earth.—(2) The name commonly given to the liquid oils found in the earth.

petroleum geology

That branch of geology which treats of the geologic factors involved in the origin, accumulation and exploitation of oil deposits. (Also called *oil geology*.)

petrology

The study of the origin and formation of rocks.

Ph

A chemical symbol for *phenyl.*

pH

Short for *pH value.*

pH value

The logarithm to the base 10 of the reciprocal of the concentration of the hydrogen ions in gram molecules per liter in an aqueous solution. Thus, water has a *p*H value of 7; acidic solutions are less than 7, and basic solutions are higher than 7. It affords a method of expressing small differences in the acidity or alkalinity of nearly neutral solutions.

phase

In chemistry, a solid, liquid or gaseous homogeneous substance that exists as a distinct and mechanically separate portion in a heterogeneous system; any homogeneous parts of a system that are separated from one another by definite physical boundaries.

phase

In physics, (1) the stage or point in a period of a periodic motion (vibration, rotation, oscillation), expressed usually in angular measure with reference to point or moment of starting or neutral position; (2) the succession of electrical impulses of an alternating current.

phase angle

In physics, the time difference, expressed in electrical degrees, between the occurrence of the positive peaks of two sine waves of the same frequency.

phase delay

In the transfer of a single-frequency wave from one point to another in a system, the time of delay of a part of the wave identifying its *phase;* it equals the ratio of the phase shift to the angular frequency.

phase diagram

A graphical representation of the equilibrium temperatures and composition limits of *phase fields* and phase reactions in an alloy system; it is constructed from thermal and other data showing the limits of temperature and composition within which the different constituents and phases of the alloy system in question are stable.

phase difference

The magnitude by which one wave leads or lags with respect to another.

phase field

The area in a constitutional or equilibrium diagram occupied by a single phase or a mixture of two phases. A field in which two phases are present usually separates two single-phase fields occupied by each of these separate phases, respectively.

phase inversion

A phase difference of 180° between two similar wave-shapes of the same frequency.

phase-lag angle

That fractional part of a complete oscillation by which an alternating current lags behind the alternating *electromotive force* which produces it. This lag is usually expressed in angular measure such that a complete oscillation is designated by 360 degrees.

phase of a wave motion

An angular measure of the stage which the vibration has reached in its progress through its cycle. Waves travelling in the same path reinforce each other if they are of the same phase (called *in phase*), whereas they tend to nullify or destroy each other wholly or partly if they are of opposite or different phases (*out of phase*).

phase rule

The statement of the relationship between the number of phases, components and degrees of freedom of a system, viz.: $F = C + 2 - P$, where F is the *degrees of freedom*, i.e., the independent variables (temperature, pressure and concentration) which may be altered without producing the disappearance of any phase or the formation of a new phase, C = the number of components, and P = the number of phases. (Also called *Gibbs' phase rule*.)

phase space

A multidimensional space (hyperspace) in which the co-ordinates represent the variables required to specify the state of a given system.

phase velocity

The velocity of the travel of wave crests and troughs through a medium; the velocity of the propagation of a given phase in homogeneous wave train.

phase wave

See *de Broglie wave*.

phenol

(1) A generic designation of the members of a class of organic compounds in the aromatic or benzene series.– (2) Specifically, the common name of carbolic acid, C_6H_5OH, a corrosive, poi-

sonous, water-soluble substance, used as a disinfectant and for industrial purposes.

phenyl

The monovalent radical C_6H_5—; symbol: Ph.

phon

A unit of loudness based on the auditory ability of the average human ear; it is equal numerically to the sound pressure level in *decibels* of a sound of a frequency of 1000 cycles per second which the listener judges to be equivalent in loudness.

phosphor bronze

An alloy of 70% to 97% copper, 3% to 13% tin, 0% to 16% lead and 0.1% to 1.0% phosphorus, and sometimes a little zinc.

phosphorescence

The re-emission by a substance of radiant energy absorbed by it, the wavelengths of the emission being different from the wavelengths absorbed, persisting even after the removal of the excitant incident radiation.

phosphoroscope

An instrument for measuring the duration of *phosphorescence* in different substances.

phosphorus

An element in Group V of the periodic table, discovered by Brandt in 1669. Atomic number 15, atomic weight 30.975; it has two principal allotropic forms, yellow phosphorus (ignition temperature in air 34°C., melting point 44.1°C., boiling-point 280°C., specific gravity 1.82) and red phosphorus (ignition temperature in air 34°C., melting point 44.1°C., boiling-point gravity 2.2), valence 3 or 5. Symbol: P.

phot

The unit of illumination; one phot is equal to one lumen per square centimeter.

photocell

See *photoelectric cell.*

photochemical equivalent

See *Einstein's law of the photochemical equivalent.*

photochemistry

That branch of chemistry which deals with chemical changes produced by light and radiant energy.

photoconductive effect

The increase of the magnitude of the electrical conductivity of certain substances (especially selenium) with the increase in the intensity of the incident light.

photoelastic dynamometer
An apparatus for the measurement of the magnitude of rapidly varying forces.

photoelastic stress
See *photoelasticity*.

photoelasticity
A technique for the investigation of stresses in engineering structures. The general method consists in examining the patterns produced by passing polarized light through a stressed model of the structure, made in a plastic material having suitable optical properties. It is then possible to calculate the magnitude and directions of stresses at all points in the model.

photoelectric
Pertaining to the electrical effects of light or other radiation; these effects can be the emission of electrons, the generation of a voltage, or a change in electrical resistance upon exposure to light.

photoelectric atcion
An ionizing action due to the loss of a negative charge, or the acquisition of a positive charge as a result of exposure to light.

photoelectric cell
A general term for any cell the electrical properties of which are effected by illumination. An electric cell capable of emitting electrons when actuated by electromagnetic radiation in the form of visible light; it transforms light energy into electrical energy. It is an instrument for detecting and measuring light electrically. (Also referred to as *photocell.*)

photoelectric cell pyrometer
An instrument for measuring high temperatures, utilizing for the purpose the light radiation from the hot body.

photoelectric effect
Electromagnetic radiation (light) incident on the surface of a body may cause the emission of electrons. For a body of a given material, the emission occurs only for a particular band of wavelengths of the incident radiation; and for a given wavelength, the rate of emission of electrons is proportional to the incident radiant flux.

photoelectric emission
The emission of electrons by certain substances when exposed to light or other radiation. For a body of given material the emission occurs only for a particular band of wavelengths of the

incident radiation; and for a given wavelength, the rate of emission of electrons is proportional to the incident radiant flux.

photoelectric equation

An equation, developed by Einstein, giving the kinetic energy of an *electron* ejected from a system in the *photoelectric effect,* where the electron is ejected by an incident *photon,* absorbing all the energy of the latter. This equation is $Ek = h\nu - \omega$ where *Ek* equals the kinetic energy of the ejected electron, *h* is the *Planck constant,* ν is the frequency associated with the absorbed photon and ω is the energy necessary to remove the electron from the system.

photoelectric pyrometer

A photoelectric cell used to measure high temperatures.

photoelectric tube

A vacuum tube in which electron emission is produced by light energy as it impinges on a photosensitive cathode.

photoelectron

An electron ejected from a surface by photoelectric emission.

photographic pyrometer

An apparatus for determining apparent surface temperatures by comparing them with the variable density of the photographic negatives taken of the hot surface under study.

photomacrograph

A photographic reproduction of any object which has not been magnified more than ten diameters.

photometer

An instrument to measure the relative brightness of objects. The brightness of the object is matched with that of a screen illuminated by a lamp of known power.

photometric density

In optics, the ability of a surface to reflect or transmit light, equal to the logarithm of the ratio of the incident light to the reflected or transmitted light. (Usually referred to simply as *density,* especially in photography where it is the measure of the relative opacity of the negative.)

photometry

The science and technique of measuring light intensity.

photomicrograph

A photographic reproduction of any object magnified more than ten diameters (usually 200 or more).

photon

A quantum or unit of visible light, and in general of all radiant energy, which may be regarded either as a wave or as a particle having the mass of $\frac{h\nu}{c^2}$, where h is Planck's constant, ν is the frequency of vibration, and c is the speed of light in a vacuum. According to the *quantum theory*, visible light and all other forms of electromagnetic radiation consist of photons moving with the speed of light.

photosensitive

Exhibiting a photoelectric effect when irradiated by visible light. Also called *light-sensitive*.

photosensitized reaction

A photochemical reaction in which an atom or molecule acts as intermediary (called the *sensitizer*) by absorbing the radiation and passing a part or all of the acquired excess energy on to the reacting molecule.

photosynthesis

A process found in green plants in which carbohydrates are formed under the influence of light with chlorophyl serving as a catalyst.

photovoltaic cell

A cell which maintains an electric current by means of an electromotive force produced by light.

photovoltaic effect

The generation of an electromotive force by the incidence of light upon the boundary between certain pairs of dissimilar materials.

physical change

The alteration of the physical properties of a substance without any accompanying change in its chemical composition.

physical chemistry

That branch of chemistry which deals with the study of the physical properties of matter, the structure of matter, and the laws and theories regarding chemical interaction and the interrelation of chemical and energy changes.

physical climatology

That branch of *climatology* which endeavors to study and explain the causes and effects of climate on the basis of climatic data, physical geography, the laws of physics, using observation, experimentation and mathematical analysis for its purposes.

physical constants

The values or characteristics of chemical substances such as the *atomic weight, melting point* and *coefficient of expansion.*

physical metallurgy

The science concerned with the physical and mechanical characteristics of metals and alloys.

physical meteorology

That branch of meteorology which endeavors to explain all atmospheric phenomena by the accepted principles of physics and thus deals with the mechanics and thermodynamics of the atmosphere, and also with the explanation of optical, acoustical and electrical phenomena.

physical pendulum

See *simple pendulum.*

physical properties

Those properties of a body which do not involve any change in its chemical composition, e.g. density, electrical conductivity, magnetic properties, and coefficient of thermal expansion.

physical solution

A solution in which the *solute* and the *solvent* mix, but without chemical reaction, so that the solvent can be separated from the solute by distillation.

physics

The study of the physical world, its phenomena and its laws, and in particular the properties of matter and energy.

physiography

The study of the features of the earth and their changes. Variously defined as physical science, physical geography, physical geology, etc.

physiology

The science and study of the vital phenomena shown by living organisms, and of the organs and their functions.

pi (π)

A Greek letter, pronounced *pie.* It has been adopted by mathematicians as the symbol for the ratio $\frac{\text{circumference}}{\text{diameter}}$ which is a constant and to twenty places of decimal = 3.14159265358979-323846. It is approximately equal to 22/7 or more accurately 355/113. Its reciprocal $\frac{1}{\pi}$ = .318309886183790671 53........

pico-

A prefix synonymous with *micromicro-*, i.e., meaning 10^{-12}. Symbol *p* or $\mu\mu$.

picture plane

A plane on which a system of lines or rays from an object is conceived to be projected to form an image or picture. In perspective drawing, the system of rays is understood to converge to a single point.

piercing point

In orthographic projection, the point in which a straight line intersects a plane.

piezoelectric axis

One of the directions in a crystal in which either tension or compression will cause it to develop piezoelectric charges.

piezoelectric effect

The interaction of electrical and mechanical stress-strain functions. A crystalline dielectric which is anisotropic and has no center of symmetry becomes electrically polarized when it is mechanically strained, the direction and magnitude of the polarization depending on the nature and amount of the strain, as well as on the direction of the strain with reference to the axis of the crystal.

piezoelectric sources

Crystals which transform the energy of compression into electrical energy through the *piezoelectric effect.*

piezoelectricity

The interconversion of mechanical and electrical energy, in which electric charges are produced as the result of properly directed mechanical stresses on a crystal. (See *piezoelectric effect.*) The term is used also to designate the reverse effect in which a mechanical strain is produced in a crystal by an applied electric field. (See *inverse piezoelectric effect.*)

piezometer

An instrument used for determining the compressibility of a substance.

pigment

A finely divided, insoluble coloring substance, used as coloring agent in paints, for cosmetics, for ceramic articles, etc. Pigments color the surface, but, unlike dyes, do not penetrate fibers, tissue, etc. beneath it.

pile

A fission chain reactor in which various isotopes are prepared for research and industry. (See *reactor.*)

pi-meson (π meson)
See *meson.*

pinch effect
An electrical phenomenon in which magnetic lines of force circling a gas compress it by their tension. It is used in an effort to confine *plasmas* for the production of controlled *fusion* of hydrogen.

pion
A *pi-meson.*

pitch
In acoustics, the attribute of auditory sensations whereby sounds are judged to be relatively high, middle or low. It is associated with the physical characteristic called the *frequency* of the vibration, being determined by the number of vibrations per second.

pitot tube
An instrument for measuring velocities of currents or moving streams of gases or liquids.

plan view
See *top view.*

Planck's constant (*h*)
A universal constant representing the ratio of the energy of a quantum of radiation to its frequency. It has the dimension of action (energy $\times$ time). Its numerical value is 6.624×10^{-27} erg-second. Also called *(Planck's) quantum of action.*

Planck's law
The fundamental law of the quantum theory, which expresses the principle that the energy of the quanta or radiation of every frequency is proportional to the frequency, viz.: $q = h\nu$, where q is the value of the quantum in energy units, h is Planck's constant, and ν is the frequency.

Planck's radiation formula
An equation which expresses the variation of the intensity of black-body radiation with wavelength at a given temperature.

It may be written as $$J\lambda = \frac{c_1\lambda^{-5}}{e^{c_2/\lambda T} - 1}$$

where $J\lambda$ is the energy emitted per unit area per unit time at a certain wavelength λ, T is the absolute temperature, e is the base of natural logarithms, and c_1 and c_2 are constants.

plane
A surface such that a straight line joining any two points of the surface lies entirely in the surface.

plane

In crystallography, the possible face of a crystal.

plane angle of a dihedral angle

An angle formed by two lines which do not lie in the same face of the dihedral angle and both of which are perpendicular to the edge of the dihedral angle where the vertex of the plane angle lies.

plane curve

A curve all points of which are in a plane.

plane figure

Any two-dimensional figure, i.e., any figure that lies completely in a plane.

plane geometry

That branch of geometry which treats of plane figures, their properties and relations.

plane of polarization

The plane which contains the waves of a plane-polarized beam of light.

plane of projection

See *projection plane.*

plane of symmetry

An imaginary plane that divides a solid (specifically, a crystal) into two parts which are the mirror images of each other. This is the plane in which a reflection, with or without a translation parallel to the plane, replaces every element (point, line and plane) of the crystal by an equivalent element.

plane-polarized light

Light vibrating in one plane along the path of propagation. It is usually obtained by passing ordinary light through polarizing crystals.

plane-polarized wave

A wave of radiant energy, especially light, confined to one plane, i.e., all vibrations of which are made to take place in one direction, which is perpendicular to the direction of propagation of the wave.

plane surface

A *plane.*

plane trigonometry

The science which deals with the measurement of plane triangles; that is, it determines the numerical values of three parts (sides or angles) of a triangle when the other three parts which must include a side are numerically given.

planetary electrons
The electrons which are found in the various subshells outside of the nucleus in *K, L, M shells,* etc.

plasma
A fourth state of matter described as "electrified gas." The hottest and swiftest of the magnetic fluids, plasma is a region in a gas discharge which contains nearly equal numbers of *positive ions* and *electrons,* and hence is nearly neutral. Believed to be the cause of *cosmic rays.*

plasma jet
A stream of charged hydrogen atoms injected into an evacuated glass tube and accelerated to produce *fusion.*

plastic *(noun)*
A generic term for a class of synthetic or natural materials which can be shaped when soft and hardened afterwards. Synthetic plastics consist of a resin or some other organic binder.

plastic body
An elastic body.

plastic deformation
Permanent deformation of a solid.

plastic flow
The deformation of a metal by the mechanism of *slip* along definite geometric planes within the crystals.

plastic strain
The permanent deformation after removal of load; synonymous with *permanent set.*

plasticarve
A low-melting compound used in modelling or duplicating patterns.

plasticity
That property of a body by virtue of which it tends to retain its deformation without rupture after removal of the applied deforming stress. The term defines the ease of deformation.

plasticizer
A substance which when added to a mixture keeps the latter viscous.

plastometer
A machine for the determination of the stress/strain properties of metals at high temperatures and at various rates of strain.

platinum
A metallic element in Group VIII of the periodic table.

Atomic number 78, atomic weight 195.23, melting-point 1773° C., specific gravity 21.4, valence 2 or 4. Symbol: Pl.

platinum family

The elements osmium, iridium, platinum, gold and mercury (atomic numbers 76-80). Also called *platinum group*.

platinum metals

The six rarer elements in Group VIII of the periodic system which normally occur together in nature and resemble each other in properties. They comprise *ruthenium, rhodium, palladium, osmium, iridium* and *platinum*.

Plexiglas

Trade name for an acrylic plastic which is colorless, perfectly transparent and thermoplastic.

plus

The word denoting the positive sign (+). It means that the quantity following this sign must be added to other quantities in any expression.

plutonium

An artificially produced radioactive element, second member of the group of transuranic elements, not occurring in nature except for minute traces (about 1 part in 10^{14}) in pitchblende and carnotite, where it forms apparently due to the accidental absorption of neutrons by U^{238} nuclei contained in these ores. First produced artificially by E. Sergé, G. T. Seaborg, J. W. Kennedy, A. C. Wahl and McMillan, in Berkeley, Calif., in 1940. Its atomic number is 94, the mass number of its most stable isotope is 242; it has the valences 3, 4 or 6. Symbol: Pu.

plutonium reactor, plutonium pile

A nuclear reactor utilizing the fission chain reaction of plutonium as source of energy.

Pm

The symbol for the element *promethium*.

pneumatics

That branch of physics which deals with the dynamic and mechanical properties of gases.

Po

The symbol for the element *polonium*.

pOH

A term, analogous to pH, sometimes used to express directly the degree of alkalinity of a solution.

point

A geometrical element which is zero-dimensional, i.e., has

position which can be defined by coordinates, but has no extension or dimensions (no length, breadth or thickness) or size.

point of concurrence
Any point at which three or more lines meet.

point of contact
The point at which two curves, or a straight line and curve which have no other common point, meet.

point of intersection
The common point of two or more lines or curves, i.e., the point in which they meet or intercept each other.

point of osculation
See *tacnode.*

point source of radiation
Any source of radiation conceived of as viewed from a distance sufficiently great in comparison to the linear extension of the source to make it negligible and to permit treating it as a dimensionless point.

poise
The unit of viscosity. It is the force (one dyne) which, when exerted on a unit area between two parallel planes one square cm. in area and one cm. apart, produces a difference in streaming between the two planes of one cm. velocity per second.

Poiseuille flow
The steady *laminar flow* of a viscous liquid through a pipe of circular cross-section.

poison
(1) In general, any substance that can cause a disease or death in a living organism. (2) In chemistry, any foreign substance which impairs or nullifies the action of a catalyst. (3) In metallurgy, any substance the admixture of which to an alloy impairs its quality.

polar
A line drawn through a circle is the *polar* of the point of intersection of the tangent to the circle, drawn from the points of intersection of the line and the circumference of the circle. (The point is called the *pole.*)

polar chart
A *great circle chart* the point of tangency of which is one of the poles of the earth.

polar coordinates
A system of coordinates for describing the position of a point

in space with respect to an arbitrarily selected point of origin by means of two directions and a distance, that is, the *vectorial angle* and the *radius vector.* From a fixed point 0 called the *origin* or *pole* on a straight line OX called the *initial line* a straight line OP called the radius vector is drawn to any fixed point P in a plane. The angle XOP enclosed by the initial line and the radius vector is called the *vectorial angle.* The radius vector and vectorial angle together are called the polar coordinates of the point P.

polar distance of a circle of a sphere

The *spherical distance* from any point on the circle to the nearer *pole.*

polar form of a complex number

The form of a complex number as expressed in *polar coordinates,* viz.: $r\ (\cos\theta + i \sin\theta)$, where r is the radius vector and θ the vectorial angle of the point which represents the complex number in question.

polar triangle

A spherical triangle the *poles* of the sides of which are the *vertices* of a given spherical triangle. (Cf. *self-polar triangle.*)

polarimeter

An instrument used to measure the number of degrees an optically active substance rotates the plane of polarized light. It consists of two prisms so arranged that polarized light from the first prism can pass through the second when the instrument is set at zero. The rotation of a substance is determined by placing it between the two prisms and rotating the second to compensate for the change that the substance has caused in the plane of the light emerging from the first.

polariscope

An instrument for studying polarized light and determining its properties. It consists of a combination of two Nicol prisms (the *polarizer* and the *analyzer*) or some other polarizing substance, such as Polaroid, between which a crystal or crystal sections may be examined in polarized light. If the two Nicol prisms or the two objects replacing them have their vibration planes at right angles, they are said to be *crossed,* and in that case *extinction* of the light occurs, since the light polarized by the polarizer is extinguished by the analyzer, so that darkness results.

polarity

(1) A term applied to electrical machinery or apparatus when

it is desired to indicate which terminal is positive and which is negative. (2) The quality of an electrical charge, i.e. whether it is positive or negative.

polarization of an electric battery

As applied to an electric battery polarization occurs if ions produced at the cathode do not migrate rapidly enough through the electrolyte. By accumulating at the cathode they set up a potential opposing that of the battery, thereby reducing or stopping the output of current.

polarization of radiant energy

The process of confining the vibrations of the wave to definite pattern or direction, as in one plane, in an ellipse, or some other definite curve. Also the state of the radiant energy when its vibrations have been so affected.

polarized light

Light in which the transverse vibrations occur in a specific plane circularly or elliptically.

polarizer

(1) Special material such as a *Nicol prism* and Polaroid sheet which transmits light vibrations predominantly in one direction. (2) In a *polariscope* or *polarimeter,* the first Nicol prism or Polaroid, which polarizes the light arriving from the outside system.

polarizer

In electricity, a substance which when added to an electrolyte increases the polarization.

polarizing angle

That angle of incidence of light upon a given substance at which the light undergoes maximum polarization.

polarograph

An instrument used in chemical analysis. It measures the current which flows when a predetermined potential is applied to two electrodes immersed in the solution being analyzed.

Polaroid

The trade name for a type of glass consisting of two ordinary sheets of glass with a layer of a special cellulose acetate fiber between them; this fiber contains crystals of alkaloidal iodide. This product has the property of polarizing light.

pole

Either of the two points on the surface of a sphere cut by the diameter of the sphere which is perpendicular to a great circle or small circle of that sphere. On the Earth the north and south poles are those points on the surface which are pierced by a diagonal at right angles to the Equator. The point of origin of *polar coordinates* is also called the pole.

pole of a circle of a sphere

Either of the two extremities of the axis of the circle.

pole of a magnet

See *magnetic pole.*

pole of the celestial sphere

Either of the two points in which the earth's axis, extended, cuts the *celestial sphere.*

polonium

A radioactive element in Group VI of the periodic table, discovered and first isolated by Pierre and Marie Curie in 1898. Atomic number 84, mass number of the longest-lived isotope 210. Symbol: Po.

poly-

A prefix meaning *many.*

polybasic acid

Any acid each molecule of which yields more than one hydrogen ion in an aqueous solution.

polycrystalline

Composed of several or many crystals, in contradistinction to a substance that is a single crystal.

polygon

A totally enclosed *geometrical* plane figure bounded by straight lines having as many angles as it has sides. The terms sides, angles, vertices, diagonal are employed as usual.

polygon method of adding vectors

A method of adding vectors graphically: the vectors are placed "head to tail," and the resultant is drawn from the origin to the head of the last vector, thus closing the polygon.

polygon of forces

The polygon obtained when applying the *polygon method of adding vectors* to three or more vectors acting on the same point and which are in equilibrium. Forces may be represented as vectors. (See also *parallelogram* of forces.)

polygonal number

Designation of a type of numbers occurring in the summation of certain series. They begin with unity and obtain the continued summation of the successive terms of an arithmetic progression whose common difference is a whole number. Thus the arithmetic progression 1, 2, 3, 4, 5, 6,........ (common difference one) gives by summation of successive terms (1, 1 + 2, 1 + 2 + 3, etc.) the series of the triangular numbers 1, 3, 6, 10, 15........ The polygonal numbers constitute the first order of the *figurate numbers.*

polyhedral angle

An angle formed by three or more planes (the *faces* of the polyhedral angle) which meet at a point (the *vertex*). Each two successive faces form a *dihedral angle* of the polyhedral angle, and the *edges* of these dihedral angles are the edges of the polyhedral angle; the angles formed by any two consecutive edges are referred to as the *face angles* of the polyhedral angle.

polyhedron

A solid which is bounded by planes. (See *convex polyhedron, regular polyhedron, similar polyhedrons.*)

polymer

Any of two or more *polymeric* compounds; specifically, one of higher molecular weight, in particular one produced from another by *polymerization.*

polymeric

Consisting of the same chemical elements in the same proportion by weight, but differing in molecular weight.

polymerism

A change in the properties of a compound whereby the elements remain in the same percentage but assume different molecular weights.

polymerization

A chemical reaction in which two or more molecules of the same substance combine to form a compound from which the original substance may or may not be regenerated, the new molecular weight being a multiple of that of the original compound.

polymerize

(1) To undergo polymerization. (2) To produce polymerization.

polymorphic

Possessing the property of being able to crystallize in two (*dimorphic*), three (*trimorphic*), four (*tetramorphic*) or more forms.

polymorphism

The condition of being able to have more than one form of

crystallization. (If the substance is able to crystallize in either of two forms, the condition is called *dimorphism;* if there are three possible forms, it is a case of *trimorphism;* the existence of four forms is *tetramorphism.*)

polymorphous

See *polymorphic.*

polynomial

An algebraic expression of the form $a_0x^n + a_1x^{n-1} + a_2x^{n-2} + \dots\dots + a_{n-2}x^2 + a_{n-1}x + a_n$, where the *a's* are constants, with $a_0 \neq 0$, and n is a positive integer (called the *degree* of the polynomial). If this expression (which is also called a *rational integral function* of degree n) is equated to zero, the resulting equation is referred to as a *polynomial equation* of degree n.

polynomial equation

A *polynomial* (q.v.) equated to zero.

polynomial expression

See *polynomial.*

polynomial in one variable x

A function made up of a sum of terms each of the form Ax^N where A is a constant and N is a positive integer or zero.

polynomial in two variables x and y

A function F where $F(x,y) = a_0(y)x^n + a_1(y)x^{n-1} + \dots\dots a_{n-1}(y)x + a_ny$ where the coefficients are themselves polynomials in y.

polysaccharides

The collective designation of a large class of organic compounds; every polysaccharide is a natural carbohydrate and on hydrolysis yields more than three monosaccharides. (The starches and cellulose are important representatives of this class.)

porcelain

A hard, white material, impermeable to liquids and unaffected by acids, made of kaolin (a decomposed granite rock material) with admixtures of materials containing silica.

porous

Containing minute cavities or open spaces.

positive

Plus, to be added. Any number not preceded by a minus sign is regarded as a positive quantity.

positive angle

In trigonometry, an angle formed by a counterclockwise revolution of the *terminal line* about the vertex. (Cf. *standard position of an angle.*)

positive charge

The type of electric charge which is present in a substance the atoms of which have less than their normal number of electrons.

positive electricity

The kind of electricity which predominates on a body composed of glass after it has undergone electrification by rubbing with silk. It is due to a deficiency of electrons.

positive eyepiece

An *eyepiece* consisting of two lenses of equal size, flat on one side and convex on the other, with their curved sides facing each other.

positive ion

An atom which has less than its normal number of electrons.

positive lens

See *convergent lens.*

positive quantity

Any quantity preceded by a plus sign.

positive rays

Streams of positively charged ions, created from the rarefied gases of the discharge space of a discharge tube, or in other vessels containing a rarefied gas, by the loss of electrons by the atoms of the gas. These beams of ions are deflected by electric and magnetic fields, and are attracted by negatively charged bodies. The positive ray particles are several thousand times heavier than an electron, and their velocity is much less than the velocity of an electron at the same voltage. Also called *canal rays.*

positive trigonometrical angle

Cf. *negative trigonometrical angle.*

positive valence

The valence of an atom which yields an electron or electrons. Its numerical value equals the number of electrons which the atom is able to give up.

positron ($_1e^0$, e, $\beta+$)

A sub-atomic particle, discovered by C. D. Anderson in 1932. The positron is an elementary particle, identified as a positively charged electron. It is exactly identical with an electron in atomic weight (5.4862×10^{-4} a.m.u.), theoretical rest mass

(9.1066×10^{-28} gram) and energy (0.515 Mev, equal to 8.18×10^{-17} erg), and its positive charge is equal to the numerical value of the negative charge of the electron (4.8025×10^{-10} e.s.u., or 1.6023×10^{-20} e.m.u.).

positron emitters (β+-emitters)

Certain unstable radioisotopes which emit positrons, thus reducing their atomic number by 1.

postulate

A position assumed without proof and taken for granted to be correct. A self-evident problem. A statement, in geometry, concerned with points, lines, angles and other geometrical forms which, in general application is correct. E.g., two straight lines can intersect in one point only.

potassium

A monovalent metallic element, one of the alkali metals, in Group I of the periodic table, discovered by Davy in 1807. Atomic number 19, atomic weight 39.100, melting point 62.3° C., boiling-point 760-762°C., specific gravity 0.865. One of the very few elements below atomic number 82 which have a naturally occurring radioactive isotope. Symbol: K.

potential

In *dynamics,* the sum of the products of all the pairs of particles of two systems (one from each system), each product divided by the distance between the pair–in the limit, a double integral; the work that would be done by a system of attracting and repelling masses in moving from positions infinitely remote from one another to their actual positions.

potential

In *electrostatics,* at any point near or within an electrified body, the quantity of work necessary to bring a unit of positive electricity from infinite distance to that point, the given distribution of electricity remaining unaltered. It has a meaning very similar to that of *voltage,* and is measured in volts.

potential

In *magnetostatics,* at any point in a magnetic field the quantity of work expended in bringing a positive unit magnetic pole from a fixed given distance to that point.

potential difference

A difference in the electrical states existing at two points which causes a current to tend to flow between them. The electric potential difference between two points is equal to the work per unit charge, i.e., to the work associated with the transfer of a unit quantity of positive electricity from one point to the other. If outside energy is required to transfer a unit quantity of positive electricity from point A to point B, the point A is at a higher potential than the point B.

potential energy

Energy due to position. The potential energy of a body or of a system of bodies in a given configuration is the work required to bring the system from an arbitrarily chosen reference configuration to the given configuration without change in the kinetic energy of the system.

potential series

A synonym for *electrochemical series.*

potential temperature

In meteorology, the temperature that a specimen of air or other gas would assume if brought adiabatically to a standard pressure, usually selected as 1,000 millibars.

potentiometer

An instrument for the measurement of electric quantities which depends for its operation upon the balancing of an unknown *potential difference* against a known potential difference obtained by the passage of a fixed current through an adjustable resistor or of an adjustable current through a fixed resistor.

pound

The mass of the *imperial standard pound,* a platinum cylinder kept at the Board of Trade Standards Office. 1 lb. = 453.592 gm.

pound calorie

A synonym for *centigrade heat unit.*

pound degree

The heat required to raise one pound of water through one degree of temperature. A pound degree centigrade, also known as the *centigrade heat unit* or the *pound calorie,* is the heat required to raise 1 lb. of water through 1°C., while a pound degree Fahrenheit is the heat required to raise 1 lb. of water through 1°F.

poundal

The absolute unit of force in the foot-pound-second system; 1 poundal is defined as the force which produces an acceleration of 1 foot per second per second in a mass of 1 lb.

powder

Solid matter in the form of an aggregation of loose particles of very small size, usually between 1 and 1000 microns.

powder metallurgy

A metallurgical technique for the production of metallic articles of various shapes without resorting to operations such as melting, casting, etc. It consists essentially of the sintering of pulverized metals at high temperatures in molds of specified shape. (It is somewhat similar to the technique employed in making *plastics*.)

powder method

A method of X-ray or crystal analysis, using a loose aggregate of small crystals with chaotic orientation, applied to methods using solid metals composed of crystalline aggregates. (See *Debye-Scherrer method,* and *Hull method.*)

power

In physics, the work accomplished in a given time. It is force times the distance through which the force acts divided by the time during which the force is acting, or to be transferred to another source of power supply.

power

In mathematics, the result of multiplying a number by itself a given number of times; the number of the times it is so multiplied, called the *exponent,* is occasionally also referred to as the "power" of the number. Thus $b \times b \times b \times b$ is called b to the "fourth power" or "the power of four," being b multiplied by itself four times. $b \times b$ is b to the "second power" or "power of two" but it is usually called "b squared" and $b \times b \times b$ is b to the "third power" or "power of three," usually called "b cubed."

power factor

In electricity, the ratio of active power to apparent power; it is equal to the cosine of the angle by which the current is out of phase with the voltage.

power gas

See *producer gas.*

power of a point with reference to a circle

$PA \times PB$ when A and B are points at which any line through P meets the circle. $PA \times PB$ is considered positive or negative according as P is outside or inside the circle.

power of a point with reference to a sphere

The power of the point with reference to any of the circles

that are formed by a plane passing through the point under consideration and the center of the sphere.

power series

A series in which the terms contain positive integral powers of a variable, arranged in ascending order, as well as constant coefficients; e.g., $a_0 + a_1x + a_2x^2 + \dots\dots + a_nx^n + \dots\dots$, called a power series in *x*, or $a_0 + a_1\ (x-2) + a_2\ (x-a_2)^2 + \dots\dots + a_n(x-a)^n \dots\dots$ which is called a power series in $x-a$.

Pr

(1) The symbol for the element *praseodymium.* (2) The chemical symbol for the radical *propyl.*

praseodymium

A metallic element, one of the rare-earth group of the periodic table, discovered by Welsbach in 1885. Atomic number 59, atomic weight 140.92, melting-point 940°C., specific gravity 6.475, valence 3 or 4. Symbol: Pr.

precession

A motion of the axis of a symmetrical rotating body, which slowly describes a cone with the *centroid* as vertex, owing to the action of a small *torque* tending to change the direction of the axis, as in a spinning top or a gyroscope.

precession axis

See *axis of precession.*

precipitate

A solid or liquid made insoluble by a chemical agent or by a physical process, and which separates from a solution and settles at the bottom of the vessel containing it.

precipitate

(1) To become insoluble and to separate from a solution. (2) To cause to become insoluble and to separate and settle from a solution.

precipitation

The formation of an insoluble compound called the *precipitate,* by the addition of a suitable precipitant to a substance in solution, followed by the deposition of such precipitate.

precision

Degree of refinement in the performance of an operation or in the statement of a result, as exemplified by the number of decimal places to which a computation is carried out. The term relates to the quality of execution, and is distinct from *accuracy* which relates to the quality of the result. The use of the term *precision* has been extended by custom to methods and instruments employed in obtaining results of a high order of accuracy.

pressure

A force, per unit area, exerted over the surface of a body. Expressed as dynes per square centimeter.

pressure filtration

Filtration by applying pressure to the liquid to force it through the filter.

primary carbon atom

In organic chemistry, the designation applied to a carbon atom which has only one of its *valence bonds* satisfied by another carbon atom.

primary colors

The so-called primary colors are red, green, and blue. It was demonstrated almost a hundred years ago that all colors and shades of color can be duplicated by the admixture in various proportions of the light of these three colors. The principle is now applied in color photography.

primary cosmic rays

See *cosmic rays.*

primary dark space

See *Aston dark space.*

primary electron

An electron originating from a *primary emission.*

primary emission

The emission of subatomic particles or electromagnetic radiation, capable of exciting the substance receiving it to a *secondary emission* of particles or radiation.

primary radiation

(1) See *primary emission.* (2) With reference to x-rays, the rays which are emitted directly from the anticathode in the x-ray tube.

prime factors

The *prime numbers* of which a given number is the product. Thus 2, 3, and 7 are the prime factors of 42 and 2, 2, 3 and 5 are the prime factors of 60.

prime meridian

In geography, the meridian used as an origin for the measurement of longitude. The meridian of Greenwich, England, is generally accepted as the prime meridian.

prime number

A whole number greater than 1 which is divisible only by itself and by 1.

principal axis of a lens
The straight line which joins the centers of curvature of the spherical surfaces of the lens; if one of the surfaces is a plane, the principal axis passes through the center of curvature of the spherical surface and is perpendicular to the plane surface.

principal focus of a lens
In optics, the point through which the rays of light parallel to the principal axis pass.

principal parameter of a conic
The *latus rectum,* i.e., a line drawn through the focus point at right angles to the principal axis; it is equal in length to the distance from the focus point to the directrix.

principal value of an inverse trigonometric function
For the arc-sine, arc- secant, act-tangent and arc-cotangent, the principal value is the value which is the *smallest numerically.* For the arc-cosine and arc-cosecant, it is the *smallest positive value* of the angle.

principle
In chemistry, a substance which endows a compound of vegetable origin with its characteristic property or effect.

principle of least time
See *Fermat's principle.*

principle of proportional parts
If a number is increased by a very small part of itself, the increase in the logarithm of that number is approximately proportional.

principle of stationary time
See *Fermat's principle.*

prism
A solid the bases or ends of which are congruent and parallel polygons and the sides of which are parallelograms, each of which has one side in common with a boundary line (edge) of the base. It is formed by two parallel lines which intersect a closed *prismatic surface* and cut all the *elements.* The two congruent parallel faces of the prism are called its bases, and the other faces are referred to as *lateral faces.* A prism that has a base perpendicular to an edge is called *right* prism, all other prisms are *oblique prisms.* A right prism having a regular polygon for a base is a *regular prism.* Prisms are frequently classified according to the shape of the base (e.g., a prism having a triangle for a base is called a *triangular prism,* whereas a *quadrangular prism* has a quadrilateral base, etc.).

prismatic

Pertaining to, resembling or having the properties of a prism; pertaining to the effects produced by prisms.

prismatic surface

A surface generated by the movement of a straight line (the *generatrix*) which always intersects a broken line (the *directrix*) in a plane and is always parallel to a fixed line situated outside the plane. (Any position assumed by the generatrix is called an *element* of the surface.

prismatoid

A polyhedron all the vertices of which are situated in two parallel planes.

probability

The number of ways in which an event can occur divided by the total number of ways it can occur and fail to occur. If an event can occur in *a* ways and the chance of it not occurring is in *b* ways the probability of its occurring is $\frac{a}{a + b}$ and of its failing is $\frac{b}{a + b}$. Thus, if in a class of 29 boys 5 can get tickets for a football match and 24 cannot, then the probability of any one particular boy getting a ticket is 5/29 and of his not getting a ticket is 24/29.

probable error

An allowance based on average errors made in a large number of analogous experiments or calculations.

problem

A question proposed for solution.

producer

An apparatus for manufacturing gas for fuel and for lighting, called *producer gas*.

producer gas

A combustible gas obtained from solid fuel in a *gas producer*. It is usually made by passing air mixed with steam over red-hot carbon, e.g. coke, anthracite, or bituminous coal. The gas so obtained contains about 30% carbon monoxide, 13% hydrogen, 2% carbon dioxide, the remainder being principally nitrogen. Water gas and semi-water gas are modifications of producer gas. (Also referred to as *generator gas, power gas,* and occasionally also as *Siemens gas.*)

product

In mathematics, the result of multiplying two or more expressions, terms, numbers or quantities together. Thus $4 \times 5 = 20$, where 20 is the product of 4×5; or $ab \times bc = ab^2c$, where ab^2c is the product of $ab \times bc$.

pro-eutectoid

The constituent which separates out of a solid solution before the formation of the *eutectoid.*

profile projection plane

In orthographic projection, a projection plane that is perpendicular to the vertical as well as to the horizontal projection plane.

profilometer

An instrument for measuring the roughness of surfaces, including internal and external surfaces, straight, tapered and circular.

profiloscope

A simplified form of the *profilometer* developed for use in the workshop.

progression

A series of numbers or quantities or magnitudes arranged according to some definite, constant rule or principle. (Cf. *arithmetic progression, geometric progression* and *harmonic progression.*)

projectile

Any object set into motion with an initial velocity and then permitted to move under the action of no other force or influence than that of gravity.

projection

The method or technique of representing an object, an area, or a portion of space on a two-dimensional surface or in a three-dimensional scale model; also the resulting representation. In geometry, the extension of lines or planes to intersect a given surface; the foot of the perpendicular from a point to a line or plane. The transfer of a point from one surface to a corresponding position on another surface by graphical or analytical methods.

projection plane

A plane surface onto which a point or object in space is projected.

projectors

In descriptive geometry, lines drawn from a point or object in space to the projection plane; in *orthographic projection,* these lines are perpendicular to the projection plane.

prolate cycloid
A *trochoid* with loops.

promethium
A metallic element, one of the rare-earth group of the periodic table, not occurring in nature, first announced by Harris, Yntema and Hopkins in 1926. The first positive chemical identification of radioisotopes of this element was made by Marinsky, Glendenin and Coryell, who showed in 1945 that the radioisotopes of mass numbers 147 and 149 are by-products of the neutron-induced nuclear fission of uranium. The atomic number of this element is 61, the mass number of its longest-lived isotope is 145. Symbol: Pm.

promoter
In chemistry, a substance which is added to a *catalyst* to accelerate the action of the latter.

proof resilience
The greatest amount of energy that can be stored up by a body strained only elastically. (See *resilience*.)

propagation
(1) The travel or spreading of an effect or motion, especially of a vibration or wave motion, through a medium. (2) The transmission of an effect or motion, especially of a vibration or wave motion, by a medium.

propagation factor
See *propagation ratio.*

propagation ratio
In a wave, from one to another, the ratio of the displacement at the second point to that at the first point; it is expressed as a vector involving both amplitude and phase angle. (Also called *propagation factor, transfer ratio* and *transfer factor*.)

proper fraction
A fraction the numerator of which is smaller than its denominator; e.g., $3/4$, 117/120, etc. The value of a proper fraction is greater than 0 and less than 1.

proportion
An equation each side of which consists of a ratio of two numbers. Four numbers are said to be in proportion when the ratio of the first to the second is equal to the ratio of the third to the fourth. Thus 3 is to 4 as 15 is to 20; or 3 : 4 :: 15 : 20, or $3/4 = 15/20$.

proportional
Unless otherwise specified, a term synonymous with *directly proportional.*

proportional equation
Two equal ratios joined by an equality sign. (E.g., $1/3 = 3/9$.)

proportional parts
See *principle of proportional parts.*

proportionality
See *limit of proportionality, range of proportionality.*

proposition
A theorem or a problem. A formal statement of a truth to be demonstrated or of an operation to be performed. (A proposition that is to be taken for granted is called an *axiom;* one that is to be proven is referred to as a *theorem.*)

propyl
The monovalent radical C_3H_7—; symbol: Pr.

protactinium
An extremely rare metallic radioactive element in Group V of the periodic table, discovered by O. Hahn and L. Meitner in 1917, formerly called also *protoactinium.* Atomic weight 231, valence 5. Symbol: Pa.

protein
A complex nitrogenous organic compound occurring in organic matter (vegetable as well as animal), which is a necessary food for human beings and animals.

protium
The lightest isotope of hydrogen ($_1H^1$). It constitutes 99.98% of ordinary hydrogen.

proton
A sub-atomic particle, recognized as one of the fundamental nuclear particles, discovered by Rutherford in 1920, and identified as the nucleus of the common hydrogen isotope $_1H^1$. The mass of a proton is about 1847 times the mass of an electron, slightly less than the mass of a neutron, viz. 1.00758 a.m.u., or 1.67248×10^{-24} gram, corresponding to an energy of 938.2 Mev. Protons are positively charged; the numerical value of the positive charge of each proton is equal to the electronic charge, i.e. 4.8025×10^{-10} e.s.u., or 1.60203×10^{-20} e.m.u. The nuclei of atoms are considered to consist of protons and neutrons (collectively called *nucleons*) ; the number of the protons in the nucleus always equals the atomic number of the isotope, while the combined total of all the nucleons is the mass number.

proton acceptor
In chemistry, a substance which gains a hydrogen ion (H^+).

proton donor

In chemistry, a substance which yields a hydrogen ion (H^+).

proton microscope

An optical apparatus built at the Collège de France in Paris, France, capable of magnifications of 600,000 or more, and thus six times as powerful as the strongest *electron microscope.*

proton synchrotron

A device which accelerates protons to high velocities and high energy levels.

protoxide

An oxide which contains the least number of oxygen atoms in a series of oxides.

protractor

An instrument or scale for measuring angles.

proximate analysis

Cf. *ultimate analysis.*

proximity effect

The phenomenon of nonuniform current distribution over the cross-section of a conductor caused by the variation of the current in a neighboring conductor.

pseudocrystalline

Not crystalline, i.e., having a structure in which no true crystals are present, and yet possessing regularities which produce a diffraction pattern different from those of liquids or amorphous solids. (E.g., stretched rubber.)

pseudomorph

A crystal which has assumed a geometric form other than its own, the original mineral having been subjected to molecular displacement by another substance.

Pt

The symbol for the element *platinum.*

Pu

The symbol for the element *plutonium.*

pulsatance

In a vibrating or oscillating quantity, 2π times the frequency.

pulsating current

In electricity, a periodic current which is the sum of a direct current and an alternating current; it changes in value, but not in direction.

pulse

A transient disturbance. A wave in which the displacement at each point of the medium is an aperiodic function of the time. A pulse produces, at any point, a single displacement, subsequent to which the medium returns to a state of passive equilibrium.

pulse polarizer

An apparatus for studying electrochemical corrosion phenomena in metals, making use of the principles and phenomena of *polarization.*

pure mathematics

The study and development of the principles of mathematics as such (for their own sake and possible future usefulness) rather than for their immediate practical applications. (Also called *abstract mathematics.*)

putrefaction

Progressive decomposition of organic matter, including dead animal tissue, owing to the action of bacteria.

pyknometer

A device for measuring the density of a substance, usually a liquid, by determining the weight of a known volume.

pyramid

A solid the base surface of which is a rectilinear polygon and the side surfaces of which are triangles, each triangle having a side line of the base for its base line, and the vertices of the triangles meeting in one point which is called the *apex* or *summit* of the pyramid. It is formed by one *nappe* of a closed *pyramidal surface* cut off by a plane that cuts all the *elements.* If the *base* is a regular polygon and the perpendicular line from the *vertex* to the base (called the *altitude*) meets the base at the center point, the solid is called a *regular pyramid.*

pyramidal surface

The surface generated by a straight line (the *generatrix*) as it moves so that it always intersects a fixed broken line (the *directrix*) in a plane and passes always through a fixed point (the *vertex*) which does not lie in the plane of the fixed broken line. (Each position assumed by the generatrix in this movement is called an *element* of the surface.)

pyranometer

An instrument for measuring the transparency of the sky and so determining the amount of solar radiation.

pyrheliometer

An instrument for measuring the intensity of the sun's radiation.

pyrite
A generic designation of a group of sulfide minerals, e.g., iron pyrite (FeS_2), copper pyrite ($CuFeS_2$).

pyro-
A prefix indicating a connection with fire or heat.

pyroconductivity
Electrical conductivity which develops with rising temperature, and notably upon fusion, in solids which are practically non-conductive at atmospheric temperatures.

pyroelectric
Possessing or exhibiting the property of becoming electrically polarized and developing electric charges when exposed to changes in temperature. This property is present in certain asymmetric crystals.

pyroelectricity
Electricity developed in *pyroelectric* substances (certain asymmetric crystals) when exposed to changes in temperature.

pyrolysis
The decomposition of organic substances by heat.

pyrometer
An instrument for measuring temperatures. The main types are: (1) *Resistance pyrometers,* making use of the property of change of resistance of a substance with temperature. (2) *Thermoelectric pyrometers,* using the principle of the *thermocouple.* (3) *Optical pyrometers,* in which the temperature is obtained by comparison of the color of brilliance of the hot body with a known value. (4) *Radiation pyrometers,* based on the measurement of heat radiation. (5) *Photoelectric cell pyrometers,* using the light radiation from a hot body.

pyrometric cone
A small, pyramid-shaped piece of clay or ceramic materials, the compositions of which are so adjusted as to melt at a definite temperature under standard conditions; used to indicate furnace temperatures.

pyrometry
The science and technique of temperature measurement.

pyrophoric alloys
Alloys, usually of iron and cerium, which emit sparks when submitted to friction. Such alloys are used for cigarette lighters.

pyrostat
An automatic device for regulating high temperatures.

pyro-total radiation pyrometer

An instrument for measuring high temperatures; it consists, essentially, of a *thermocouple* enclosed in an evacuated bulb, and a quartz lens which focuses the radiant heat emitted by the hot body onto the thermocouple. (See *total radiation pyrometer.*)

Pythagorean theorem

The square on the hypotenuse of a right triangle is equal to the sum of the squares on the other two sides.

Q

Q

The symbol for *Q-factor*.

Q-factor

A measure of the efficiency of coils and capacitors and resonant circuits; it expresses the ratio between the total energy in the circuit and the energy which is dissipated. (Symbol: Q.)

quadrant

One-fourth of the area of a circle, bounded by two radii at right angles to one another, one usually being horizontal and the other vertical, and the arc being the quarter of the circumference of the circle which lies between the two radii. The arc subtends a right angle at the center of the circle.

quadrant rule

In any spherical triangle, one half of the sum of two angles is in the same quadrant as one half of the sum of the opposite sides.

quadrantal

(1) Containing the fourth part of a measure. (2) Relating to a *quadrant.*

quadrantal triangle

In spherical trigonometry, a spherical triangle one side of which equals $90°$.

quadratic

Of the second degree.

quadratic equation

An equation having two roots and only one unknown which appears with the exponent 2. Example: $x^2 - x = 42$. Also known as an *equation of the second degree.*

quadratic surd

The indicated square root of a number which is not a perfect square.

quadratic trinomial

An expression of three terms which contains the second power of an unknown quantity as the highest power of that quantity. Thus $4x^2 - 5x + 6$.

quadric

In mathematics, of the second degree or order.

quadric surface

Any surface that can be expressed by an algebraic Cartesian equation of the second degree. Quadric surfaces can be classified in two general groups: (1) *non-degenerate quadric surfaces* (divided into *central quadrics,* i.e., the *ellipsoid* and the *hyperboloids of one sheet* and *of two sheets,* and the *non-central quadrics,* i.e., the *elliptic paraboloid* and the *hyperbolic paraboloid*), and (2) the *degenerate quadric surfaces* (q.v.).

quadrilateral *(noun)*

A four-sided polygon, i.e., any geometric plane figure bounded by four straight lines and having four interior angles.

quadrilateral *(adj.)*

Four-sided, bounded by four lines.

quadrimolecular

See *tetramolecular.*

quadrinomial

A *polynomial* in four terms. E.g., $(a + x + b + y)$.

qualitative analysis

The determination of the constituents of a substance regardless of the respective percentages of their presence.

quantic

A homogeneous algebraic function of two or more variables in general, containing only positive integral powers of the variables, and called *quadric, cubic, quartic, quintic,* etc., according as it is of second, third, fourth, fifth, etc., degree.

quantitative analysis

The determination of the relative proportions of the constituents of a substance (by weight or by volume).

quantity

In mathematics, any arithmetic, algebraic or analytic expression of value or number.

quantized

(1) Composed of or relating to quanta; (2) expressed in terms of the quantum theory or quantum mechanics.

quantometer

An electronically operated form of direct reading spectrograph.

quantum

A discrete "packet" or unit of energy, angular momentum, or of other physical quantities, representing a minimum, indivisible quantity thereof. The value of a quantum of radiant energy is $h\nu$ where h is *Planck's constant* and ν is the *frequency* of the radiation. (The plural of *quantum* is *quanta.*)

quantum jump

A sudden change in the energy level of an atom or molecule, accompanied by the emission or absorption of a quantum or quanta of radiant energy.

quantum mechanics

A physical theory dealing with atomic structure and phenomena subject to quantum conditions. It is based largely on Heisenberg's uncertainty principle, embraces also his matrix mechanics, Schrödinger's wave mechanics and also the transformation theory of Dirac and Jordan.

quantum number

A number (an integer) characteristic for a quantum condition and associated with each given quantum state, or energy level, of the atom. Quantum numbers appear in the formulas of series terms of atomic spectra. A change of a given quantum number for another generally corresponds to a quantum of radiant energy or light photon of a certain frequency.

quantum of action

A synonym for *Planck's constant.*

quantum theory

The theory, formulated by Max Planck in 1900, stating that all forms of radiant energy are emitted in discrete, indivisible "packets" of energy, called *quanta* (the plural of the noun *quantum*), and that the size of each such unit or quantum of energy is proportional to the frequency of the radiation ($q = h\nu$ where h is *Planck's constant* and ν is the frequency)—in other words, that energy is atomistic in nature, variable only in multiples of this elementary quantum.

quarter-wave plate

A plate of mica or some other birefringent crystal material of such thickness as to introduce a phase difference of one-quarter cycle between the *ordinary* and *extraordinary* rays resulting from the *double refraction* of light.

quartic

In mathematics, of the fourth degree or order.

quartz

A natural mineral consisting of crystallized silica (silicon dioxide, SiO_2); in its pure native form it occurs in colorless hexagonal crystals, having a hardness of 7 on the *Mohs scale,* a density of 2.66, and a mean *index of refraction* of 1.55.

quaternary
In general, composed of four elements or components.

quaternary alloy
An alloy containing four principal elements.

quaternary compound
In general, a combination of four different elements, i.e., a compound each molecule of which consists of four different atoms.

quinary
In general, composed of five elements or components.

quinary alloy
An alloy containing five principal elements.

quinquevalent
Having a *valence* of five.

quintic
In mathematics, of the fifth degree or order.

quotient
The number of times a divisor will go into a dividend. It is the result of a process of division.

R

R

A chemical symbol indicating any *radical* in general.

r

See *roentgen.*

r unit

See *roentgen.*

Ra.

The symbol for the element *radium.*

RaA, RaB, RaC, RaC′ (RaC_1), RaC″ (RaC_2), RaD, RaE, RaF, RaG

The symbols for *radium A, radium B, radium C,* etc.

rabatment

A plane C is said to be rabatted on to another plane B when C is turned about its line of intersection with B until it coincides with B.

racemate

A salt of a *racemic* acid.

racemation

See *racemization.*

raceme

A compound which is the combination of dextrorotatory and levorotatory forms, and is therefore optically inactive.

racemism

The combination of dextrorotatory and levorotatory substances or molecules, resulting in an optically inactive compound or substance.

racemization

The process of changing an optically active substance to an optically inactive one. (Also referred to as *racemation.*)

radar

A method of electronic radio detection and ranging, employing microwaves and ultrahigh frequencies. It is based on the principle that ultrahigh frequency radio waves radiating through space are reflected from the surface of objects within the field of radiation. The direction of such an object in space can be determined by a directional antenna array, and its distance measured by the time required for the reflected waves (the *echo*) to travel back to their point of origin. (The word *radar* is a combination of the initial letters of the words *ra*dio *d*etection *a*nd *r*anging.)

radial velocity

The velocity of approach or recession between two bodies, as between an observer and a source of radiation, in a line connecting the two.

radiamatic pyrometer

An instrument for measuring temperatures in the range between 30° and 300°C.

radian

The angle (also called *unit angle*) subtended at the center of a circle, which angle is opposite to an arc of the circle equal in length to the radius of the circle. It is 57° 17′ 44″.8062740964 for every circle. This is the unit of circular measurement, and π times this angle equals two right angles; i.e., $3.14159 \times 57°\ 17'\ 44''.806 = 180°$ or half the circumference of a circle. $\pi \times$ radian = 2 right angles.

radiancy

The radiant flux density at an element of the surface of a source of radiant energy.

radianlength

In a sinusoidal wave, the distance between phases which differ by an angle of one radian; it equals the wavelength divided by 2π.

radiant energy

Energy emitted and propagated in the form of electromagnetic waves.

radiant flux

The time rate of flow of radiant energy. It is measured and expressed preferably in *watts,* or in *ergs* per second.

radiant heat

(1) Heat in transit in the form of electromagnetic radiation. (2) Heat resulting from the transformation of radiant energy into heat energy (molecular vibration).

radiant intensity

The *radiant flux* emitted per unit solid angle *(steradian)* in a specified direction.

radiant power

The *radiant flux,* i.e., the amount of radiant energy emitted, per second.

radiate

(1) To emit energy in the form of waves. (2) To emit beams of particles (e.g., alpha radiation).

radiation

(1) Radiant energy. (2) The transmission of heat or light by waves unaffected by the medium through which they pass. The term radiation includes energy in the form of the gamma rays of radium, x-rays, ultraviolet radiation, visible light, infrared radiation or heat, and electric or radio waves. (3) The emission and propagation of radiant energy in the form of *quanta*. The shorter the wavelength of the radiation, the more energy is contained in each quantum of radiation. The term "radiation" is generally extended to include not only the various forms of *electromagnetic radiation* but also streams of sub-atomic particles ("alpha radiation," "beta radiation") and cosmic rays ("cosmic radiation"), as well as the quanta of energy and the sub-atomic particles themselves.

radiation counter

A device that responds to the passage of nuclear radiation, by electronic amplification producing an audible click or a count registered through some mechanical recorder.

radiation laws

The physical laws expressing certain general basic principles obeyed by phenomena or radiation. (For the most important of these laws see *Kirchhoff's law, Stefan-Boltzmann law, Planck's radiation formula,* and *Wien's law*.)

radiation pressure

The pressure acting on a surface exposed to radiation. Its magnitude is proportional to the density of the radiant energy in the space in which that surface is exposed to the radiation. (This pressure is exerted, to varying degrees, by all forms of electromagnetic radiation, including light.)

radiation pyrometer

An instrument for determining the temperature of a distant source of heat. The heat radiated from a hot body is focused either by lens or mirror onto a *thermocouple* or a *thermopile,* thus giving rise to an *e.m.f.* proportional to the temperature of the hot body. Optical filters may be employed to render the pyrometer responsive to selected wavelengths, but in all cases other than *black body* conditions appropriate *emissivity* corrections are required for accurate measurement. (See also *total radiation pyrometer*.)

radiation survey meters

Instruments for the detection and measurement of radiation, operating principally on the ionization-chamber or electroscope principle. These instruments are not only valuable tools of theoretic scientific research, but also essential aids for the establishment and maintenance of *radiologic safety* for the protection of the personnel working in the vicinity of sources of radiation.

radical

In mathematics, the indicated *root* of a quantity. Also, a synonym for *radical sign.*

radical

In chemistry, a group of atoms, having an unsatisfied *valence,* acting as a single element in a chemical reaction, but normally incapable of separate existence.

radical axis

In mathematics, a fixed line from any point on which the tangents drawn to two fixed circles are equal. It is perpendicular to the line joining the centers of the two circles. It is the locus of points whose powers with respect to the two circles are equal. If the circles cut one another, the radical axis passes through their points of intersection.

radical center

The point at which the *radical axes* of three circles, taken in pairs, meet.

radical sign

The sign $\surd$ or $\sqrt{\ }$ which when prefixed to or placed above a quantity, called the *radicand,* indicates that the root (of the order indicated by a small figure or letter, the *index,* placed above the sign) is to be extracted. Thus the notation $\sqrt[n]{a}$ indicates that the *n*th root of *a* is to be determined. (When no index appears, the radical sign refers to the square root of the radicand.) Also called simply *radical,* and, occasionally, *root sign.*

radicand

The numeral or quantity appearing under a *radical sign.*

radio

The science, technique and process of communication by means of radio waves.

radio-

A prefix customarily used to replace the adjective "radioactive" in terms such as "radioelement," "radioisotope," etc., and in particular in the designation of artificially produced radioisotopes, e.g., radio-iodine, radiophosphorus, etc.

radio astronomy

That branch of astronomy which is concerned with the study of the electromagnetic impulses of radio energy emitted by celestial bodies and regions and the study of the universe by using radio waves instead of visible light frequencies.

radio frequency

A frequency at which electromagnetic radiation of energy can be used for purposes of communication. The range of these frequencies, referred to as the *radio spectrum,* lies generally above the frequencies of normal audible sound and below the frequencies of heat and light waves. Radio frequencies have been classified by the Federal Communications Commission into seven bands, which are as follows:

Designation	*Abbreviation*	*Frequency (kc = kilocycles mc = megacycles)*
Very low frequency	vlf	10–30 kc
Low frequency	l-f	30–300 kc
Medium frequency	m-f	300–3,000 kc
High frequency	h-f	3–30 mc
Very high frequency	vhf	30–300 mc
Ultrahigh frequency	uhf	300–3,000 mc
Superhigh frequency	shf	3,000–30,000 mc

radio noise

In radio astronomy, this term is used to denote the emission of electrical energy with a spectrum that is essentially a continuous one, in contradistinction to the discrete frequencies of radio signals.

radio source

A source of interstellar or intergalactic emission in the radio wavelengths. Today it is considered certain that at least the majority of the strong radio sources are not stars; they correspond to numerous classes of galactic and extragalactic objects. At least four broad types: (1) Remnants of supernovae, (2) galactic nebulosities of a new type, (3) normal external galaxies, and (4) peculiar external galaxies.

radio spectrometer

An instrument for the measuring of radio wavelengths. It is

used in radio astronomy to investigate a specific atomic spectral line as well as for the survey of the continuous radiation over a wide frequency range.

radio spectrum
The entire range of frequencies in which useful radio waves can be produced. (See *radio frequency*.)

radio storm
In radio astronomy, a prolonged period of disturbed emission, i.e., one that lasts for hours, at times even for days.

radio telescope
An apparatus designed to intercept radio waves emitted from that region of the celestial sphere toward which it is aimed, and to measure the radiation received, i.e., to act as a *radiometer.*

radio wave
A wave of radiant energy the frequency of which is a *radio frequency.*

radio window
The "transparent" section of the atmosphere through which radio telescopes can scan the universe.

radioactinium (RdAc)
A radioactive thorium isotope ($_{90}Th^{227}$).

radioactive
Capable of emitting spontaneously rays consisting (at least in part) of material particles traveling at high velocities.

radioactive constant
See *decay constant.*

radioactive decay
The gradual diminishing of the quantity of a radioactive substance due to the spontaneous disintegration of nuclei by the emission of α or β particles, γ-rays or positrons.

radioactive deposit
A radioactive film forming on an object or surface exposed to a *radioactive emanation.*

radioactive emanation
A gaseous product of the spontaneous disintegration of a radioactive substance. The most important radioactive emanations are *actinon, radon* and *thoron.*

radioactive equilibrium
This term is applied to the state which prevails when the rate of the formation of the atoms of a radioactive product equals the rate of the radioactive decay of the atoms of the parent

isotope, so that the amount of the radioisotope remains constant. In the case of a parent isotope with a long half-life, the radioactive equilibrium is said to be *secular,* in the case of a rapidly decaying substance it is referred to as *transient.*

radioactive family

See *radioactive series.*

radioactive product

The daughter isotope produced by the radioactive decay of another isotope (called the *parent*).

radioactive series

A radioactive series, also called radioactive family, is a group of radioactive isotopes, each of which is the product of the radioactive decay of the preceding one; the last member (called *end product*) of a radioactive series is a stable isotope without any trace of radiation. At the present time, four radioactive series are known, three of them natural and one artificial, viz.: the *uranium-actinium series* (also called *actinium series*), the *uranium-radium series* (also called *radium series*), the *thorium series,* and the artificially produced *neptunium series.*

radioactive tracer technique

See *tracer.*

radioactivity

The continuous emission of electromagnetic or corpuscular radiation by the nuclei of the atoms of certain isotopes, resulting in the transformation of the isotope into another isotope or another element. Radioactivity may be natural or artificially induced.

radioautograph

The photographic record, consisting of a blackened area, produced by a radioactive material by mere contact for a certain period of time with a photographic plate or film. Radioautographs are used principally to detect the location or progress of radioactive *tracers* in the human body.

radiochemistry

That branch of chemistry which deals with radioactive elements and radioisotopes.

radioelement

A radioactive element, with atomic number above 82, which has no stable isotope. Nuclear physicists usually speak of the various isotopes of these elements (e.g. ionium, mesothorium, etc.) as different elements.

radiogenic
A product of radioactivity or radioactive decay.

radiogram
See *radiograph.*

radiograph
A shadow picture produced by passing x-rays or gamma-rays through an object and recording the variations in the intensity of the emergent rays on a suitable photographic or sensitized film. (Also referred to as *radiogram.*)

radiographic contrast
The difference in blackening on a radiograph resulting from variations in the thickness or density of the material under examination.

radiographic density
The degree of blackening of a *radiograph.*

radiography
The science and technique of making photographic records of x-rays or gamma radiation of radioactive substances, and of the studying and interpretation of such records.

radioisotope
An artificially produced radioactive isotope of an element. Induced radioactivity was discovered in 1934 by F. Joliot and I. Curie. Radioisotopes can be and are used for literally hundreds of practical purposes, and the field of their employment is expanding continually. They are used not only in nuclear physics and chemistry as sources of ionizing radiation and to study nuclear characteristics and properties, such as decay schemes, the nature, spectra and absorption coefficients of radiation, nuclear spin, etc., but also as *tracers* in biology and medicine, both for diagnosis and therapy, animal and plant physiology and agriculture, industry, metallurgy, etc.

radiologic safety
Protection of workers in radiological establishments—determined by radiation survey meters or photographic film.

radiology
The study of radioactivity, x-rays and cosmic rays. In medicine, the diagnosis and treatment of injuries and diseases caused by radioactivity. (The term, in both senses, includes the making, study and interpretation of *radiographs.*)

radioluminescence
The luminescence evoked by radioactive emissions or x-rays. If the luminescence is caused by exposure to β or γ radiation

and subsequently to light, it is called *radio-photoluminescence;* if after exposure to β or γ radiation the material is exposed to heat, the luminescence is called *radio-thermoluminescence.*

radiometer

An instrument for measuring radiant energy.

radiomicrography

See *microradiography.*

radiomicrometer

An extremely sensitive radiation detector.

radiopaque

Opaque to x-rays; blocking the passage of radiant energy.

radiophotoluminescence

Luminescence evoked by radioactive emission on x-rays.

radiosensitivity

The relative susceptibility of substances, especially of organic tissue and organisms, to the damaging effect of radiation.

radiosonde

An instrument equipped with elements for determining the pressure, temperature, relative humidity and other conditions prevailing in the upper strata of the atmosphere, as well as with radio units for automatically transmitting the measurements to ground stations. Radiosondes are carried aloft by balloons to elevations of 10-12 miles, and thus are effective tools in high-altitude meteorological and other physical research.

radiothermoluminescence

After exposure to beta or gamma radiation, the material is exposed to heat. The resulting luminescence is called radiothermoluminescence.

radiothorium (RdTh)

A radioactive thorium isotope ($_{90}Th^{228}$).

radium

A very rare, divalent, metallic, radioactive element in Group II of the periodic table, one of the alkaline earth metals, discovered and first isolated by Pierre and Marie Curie in 1898. Atomic number 88, atomic weight 226.05, melting-point about 700°C., resembles barium in chemical properties; its chief sources are pitchblende and carnotite, and it occurs also in bröggerite, cleveite and sea water. Symbol: Ra.

radium A (RaA)

A radioactive isotope of polonium ($_{84}Po^{218}$).

radium B (RaB)

A radioactive isotope of lead ($_{82}Pb^{214}$).

radium C (RaC)
A radioactive isotope of bismuth ($_{83}Bi^{214}$).
radium C′ (RaC′)
A radioactive isotope of polonium ($_{84}Po^{214}$).
radium C″ (RaC″)
A radioactive isotope of thallium ($_{81}Tl^{210}$).
radium C_1 (RaC_1)
See *radium C′*.
radium C_2 (RaC_2)
See *radium C″*.
radium D (RaD)
A radioactive isotope of lead ($_{82}Pb^{210}$).
radium E (RaE)
A radioactive isotope of bismuth ($_{83}Bi^{210}$).
radium E″ (RaE″)
A designation occasionally used for the radioactive thallium isotope $_{81}Tl^{206}$.
radium F (RaF)
A radioactive isotope of polonium ($_{84}Po^{210}$).
radium G (RaG)
The stable lead isotope ($_{82}Pb^{206}$) which is the end product of the *uranium-radium series*.
radium lead
See *radium G*.
radium series
See *uranium-radium series*.
radius of a circle
Any straight line drawn from the center to the circumference.
radius of a regular polygon
A straight line that joins the center of the polygon and any of its vertices.
radius of a sphere
A line that joins the center of the sphere with any point of its surface.
radius of curvature
The radius of the *circle of curvature*.
radius of curvature of a circle
The absolute value of the reciprocal of the curvature.
radius of gyration
The perpendicular distance between the axis of rotation and the point in which the mass of a rotating body may be assumed to be concentrated without any change in its *moment of inertia*.

radius vector

The straight line drawn from the intersection point (zero point) of two coordinates to a point lying in their plane.

radix of a system of logarithms

See *base of a logarithm.*

radix of a system of numeration

See *base of a number system.*

radon

A zero-valent, gaseous, radioactive element in Group 0 of the periodic table, discovered by Dorn in 1900. Atomic number 86, atomic weight 222, melting-point —110°C., boiling-point 65°C., density 9.725 grams per liter. Symbol: Rn. This element was formerly called also *niton* (*Nt*) and *emanation* or *emanium* (*Em*); the latter name is still used, especially in German literature.

rainbow

A circular arc of concentric spectral-colored bands, seen on a sheet of water drops, as rain, fog, spray, etc., owing to the reflection and refraction of light on the droplets.

Raman effect

A phenomenon observed when light is scattered by molecules of transparent substances; namely, the appearance of additional spectral lines close to each prominent line in the spectrum of the incident light, owing to a difference between the wavelength of the scattered light and that of the incident light. These extra spectral lines are referred to as *anti-Stokes lines.*

random orientation

The condition existing when the arrangement of the crystal grains is irregular, i.e. when the principle axes of the grains do not conform or tend to conform to any particular direction.

range of a relation or function

The set of all second elements of the set ordered pairs forming the relation or function. (Cf. *domain of a relation or function.*)

range of proportionality

In a tensile test, that range of loading in which the strain produced is directly proportional to the applied stress. The *law of proportionality* between the stress and strain in this range is known as *Hooke's law.* (Also called *elastic range.*)

range of stress

The complete range of stress—the mean of all these stresses could be called the mean stress.

Rankine scale

A temperature scale in which the zero is *absolute zero* and the size of the degrees is that of the Fahrenheit scale, i.e. the temperatures in degrees Rankine are equivalent to degrees Fahrenheit plus 459.6. Thus the freezing point of water is 491.69°R., and its boiling point is 671.69°R.

Raoult's law

The fractional decrease of vapor pressure of a solution, as compared with vapor pressure of pure solvent, is equal to the mol fraction of solute in the solution.

rare earth elements

See *rare earth metals.*

rare earth metals

The 15 metallic elements of atomic numbers 57 to 71 (lanthanum, cerium, praseodymium, neodymium, promethium, samarium, europium, gadolinium, terbium, dysprosium, holmium, erbium, thulium, ytterbium and lutetium). They are so similar in their chemical properties that they are considered to occupy the position of a single element in Group III of the periodic table. Chemically, scandium (atomic number 21) and yttrium (atomic number 39) while not members of this group, are generally included because of their chemical similarity to these 15 elements. The rare earth elements are also called *lanthanides* or *lanthanide group,* after the first member of the group.

rare earths

The oxides of the *rare earth metals.*

rare gases

The six gaseous elements, helium, neon, argon, krypton, xenon and radon, which make up Group 0 of the periodic table. Each of these six elements is a colorless, odorless gas. They are chemically inactive, zero-valent, practically without any tendency to combine with each other or any other element. They possess the highest excitation and ionization energies. Except for radon, every rare gas can be obtained from the atmosphere. Also called *inert gases* or *noble gases.*

rare metals

The metallic elements rubidium, rhodium, palladium, osmium, iridium and platinum, in Group VIII of the periodic table, so called because they occur in small quantities only. All these elements are *noble metals.*

rarefaction
The process or act of making or becoming less dense; usually said with reference to gases.

rating
A rating of an electrical machine, apparatus or device is a designated limit of operating characteristics based on definite conditions. Such operating characteristics as load, voltage, frequency etc., may be given in the rating.

rating curve
A curve which is the graphical expression of the relation between mutually dependent quantities.

ratio
The statement of the quotient of two quantities. The ratio of two numbers or quantities is the result of dividing one by the other. Thus the ratio of b to c is written: $\frac{b}{c}$; b:c; b ÷ c or *as b is to c*. It is the relation which one quantity bears to another of the same kind. Every ratio is an abstract quantity and merely expresses the number of times one quantity contains another.

ratio of equality
A ratio in which the *antecedent* (or *numerator*) equals the *consequent* or (*denominator*). Thus 2 : 2 or 9/9.

ratio of greater inequality
A ratio in which the antecedent (or numerator) is greater than the consequent (or denominator). Thus 5 : 3 or 5/3.

ratio of lesser inequality
A ratio in which the antecedent (or numerator) is less than the consequent (or denominator). Thus 4 : 7, or 5 : 9; or 4/7, or 5/9.

rational function
The quotient of two polynomials.

rational integral function of degree *n*
See *polynomial.*

rational number
An integer or a fraction in which both the numerator and the denominator are integers, and which can be expressed without the use of the *radical sign*. E.g., 3, 12, ¾, etc.

raw ore
Ore as mined.

ray of light
The geometrical conception of a single element of light propa-

gated in a straight line and of infinitesimal cross-section, used in tracing analytically the path of light through an optical system.

Rb

The symbol for the element *rubidium.*

Re

The symbol for the element *rhenium.*

reactance

Impedance offered to the flow of alternating current by the inductance or capacitance of a circuit or a part thereof. It is the imaginary component of impedance. The reactance of a portion of a circuit for a sinusoidal current and potential difference of the same frequency is the product of the sine of the angular phase difference between the current and potential difference times the ratio of the effective potential difference to the effective current, there being no source of power in the portion of the circuit under consideration. The reactance of a circuit is different for each component of an alternating current.

reactant

Any of the substances or molecules which enter into a chemical reaction with each other and thus produce the *resultant* of the reaction.

reacting force

According to *Newton's third law of motion,* when a body exerts a force on a second body, the second body exerts an equal and opposite force on the first. The force exerted by the first body on the second is the *acting force,* and that exerted by the second one on the first is the *reacting force.*

reaction

In physics, the force which acts in opposition to a given force (cf. *Newton's law of motion—3*).

reaction

In chemistry, any chemical change or any transformation of molecules of one kind into molecules of another kind.

reactions of the first, second, third and fourth order

(1) A classification of chemical reactions according as they involve and produce a change in one, two, three or four molecules. Also called *monomolecular* or *unimolecular, dimolecular* or *bimolecular, trimolecular* or *termolecular,* and *tetramolecular* or *quadrimolecular reactions,* respectively. (2) The terms are used also to express the number of the different

substances (molecular or atomic species) the concentration of which determines the rate of the reaction. In this sense there is also a reaction of zero order (see *zero-order reaction.*)

reactive

Readily susceptible to chemical or nuclear change or transformation.

reactor

The term "reactor" has been adopted by the Atomic Energy Commission to replace the formerly used designation *"pile"* or *"atomic pile."* A reactor (frequently referred to also as a *nuclear reactor,* or *fission chain reactor*) is an apparatus designed and operated for the purpose of initiating and maintaining a nuclear fission chain reaction in a fissile material, either for the generation of heat for power purposes or primarily for the production of plutonium from uranium, in which case the heat generated is considered merely as a by-product of the operation of the reactor. According as whether uranium or plutonium is used as reactor "fuel," i.e. the fissile material in which the fission chain reaction is started and maintained, we speak of *uranium reactors* and *plutonium reactors.*

reagent

Any substance which, when added to another, causes a chemical reaction.

real axis

In the *Argand diagram,* measured along the *x-axis.*

real gas

A term often used to denote any actually existing gas, in contradistinction to the *ideal gas.*

real image

An optical counterpart of an object, produced on the surface at which light rays converge after passing through a lens. The real image of an object-point is the point at which light originating in the object-point is finally converged after traversing the optical system.

real number

Any number, rational or irrational, other than an imaginary or complex number.

Réaumur scale of temperature

A temperature scale in which the freezing-point of water is 0° and its boiling-point 80°. The abbreviation of "Réaumur" is R.

reciprocal ***(adj.)***
Mutual; affecting both elements alike.

reciprocal ***(noun)***
The reciprocal of any number, *a*, is unity divided by the number, *a*. Thus, the reciprocal of 5 is 1/5, the reciprocal of $\frac{2}{3}$ is 3/2. The formation of the reciprocal of a fraction merely inverts the fraction.

reciprocal proportions
See *law of constant proportions, law of equivalent proportions, law of multiple proportions.*

reciprocity
See *duality*.

recoil electron
The recoil of an electron when struck by gamma rays. Also known as *Compton electron.*

recorder
In *telemetry*, that part of the telemetering system which receives the impulses from the *measurand transmitter* through the *telemetering link* and sorts them and displays them visually (hence also called a *display*), records them electronically, or reproduces them electronically.

recrystallization
In chemistry, the process of dissolving crystals and concentrating the solution, thus allowing the crystals to re-form, usually for the purpose of purification of the substance.

rectangle
A four-sided plane figure every angle of which is a right angle.

rectangular
At right angles; resembling a rectangle.

rectangular components of a vector
The projections of the vector on a set of rectangular axes.

rectangular coordinate system
The system of coordinates (also called *Cartesian coordinate system*) consisting of two perpendicular lines, each of which bears a number scale, the zero point being the point of intersection of the two lines (the *axes* of the system).

rectangular hyperbola
A hyperbola the transverse and conjugate axes of which are of equal length, and the asymptotes of which are at right angles to each other. Also called *equilateral hyperbola.*

rectangular solid
A parallalepiped, i.e., a prism bounded by six parallelograms,

the opposite sides being equal and parallel, and the lateral edges perpendicular to the base.

rectification

In chemistry, the repeated distillation of a liquid for the purpose of purification.

rectification

In electricity, the conversion of an alternating current into a direct current.

rectifier

A device which converts alternating current into direct current by virtue of a characteristic permitting appreciable flow of current in only one direction.

rectifier anode

The electrode of the *rectifier* from which the current flows into the arc.

rectifier cathode

The electrode of the rectifier into which the current flows from the arc.

rectilinear

Straight-lined; having the characteristics of a straight line.

rectilinear figure

Any closed plane figure that is bounded by straight lines.

rectilinear motion

Motion in a straight line.

recuperative furnace

A *reverberatory furnace* in which the air for combustion (under a slight pressure) is preheated by entering through pipes in the chimney flue, around which the products of combustion pass, the gases always passing through in the same direction without reversal.

recuperator

An apparatus for recovering the heat of hot spent gases and using it for the preheating of the incoming fuel or air, in a continuous operation.

recurring decimals

Decimals in which the numbers after (i.e., to the right of) the decimal point repeat themselves in fixed periods of figures, usually indefinitely. Thus $5/11 = 0.454545$........, usually written 0.45.

recurring series

Any series in which, from and after a certain term, each term equals the sum of a fixed number of the preceding terms multiplied respectively by certain constants. Thus in the series $1 + 2x + 3x^2 + 4x^3 + 5x^4 + \ldots\ldots$ a term can be obtained by multiplying its two preceding terms by $(-x^2)$ and $2x$ respectively and adding the two products together.

red hardness

The hardness of a metal, commonly high-speed tool steel, which is retained even at red heat.

red heat

The temperature at which a metal appears red hot when viewed in a dark room, i.e. about 550°C.

red shift

(1) In general, a shift or displacement of the lines of a spectrum toward the red end, i.e., toward the greater wavelengths, owing to the *Doppler effect* or to gravitational attraction (see *Einstein shift*). (2) Specifically, the apparent increase of wavelengths observed in the spectra of distant nebulae and galaxies, manifesting in a displacement of the spectral lines toward the red end of the spectrum, the extent of the displacement increasing progressively with the increase in distance. (See *expanding universe*.)

red shortness

Brittleness or low shear resistance in steel at red heat, due usually to a high content of sulfur and a low manganese content.

redox

The phenomenon of mutual reduction and oxidation; oxidation-reduction.

reduce

In arithmetic, to change from one denominator to another. Thus 2 tons = 40 cwt. is tons reduced to cwt.

reduce

In chemistry the act of adding one or more electrons to an atom (hence also referred to by the synonym, *to electronate*). The term thus covers three acts or processes: (1) the increase of the valence number of an atom by a corresponding increase in the valence number of another atom (which undergoes *oxidation*); (2) the deprivation of oxygen from a substance or the production of the metallic state; (3) the addition of hydrogen to an element or compound.

reducing agent

A substance capable of supplying electrons to another substance and thus bringing about *reduction*.

reducing atmosphere

An atmosphere containing elements or compounds which react with or neutralize the effect of oxygen. Such atmospheres prevent oxidation or bring about chemical change involving a decrease in the state of oxidation.

reducing flame

A flame containing a carbonaceous gas in excess of that required to produce a neutral flame and thus effecting *reduction.* (Also referred to as *carburizing flame.*)

reduction

The gain, by the substance being reduced, of one or more electrons. Invariably accompanied by oxidation of some other member of the system under consideration. The term covers also the addition of hydrogen atoms and the loss of oxygen atoms. It is the reverse of *oxidation.*

reduction to lowest terms

With reference to fractions or expressions involving fractions, the elimination of all common factors of the numerator and denominator by division.

redundant equation

A derived equation that has more solutions than the given equation from which it is obtained.

re-entrant angle

An angle that is greater than 180° and is an interior angle of a plane polygon.

re-entrant polygon

A polygon one or more of the interior angles of which is a *re-entrant angle* (greater than 180°, but less than 360°). Also called *reflex polygon.*

reference line

See *folding line.*

reflectance

The measure of the ability of a body to reflect a given form of electromagnetic radiation, specifically light; it is the ratio of the intensity of the reflected radiant flux to that of the incident radiant flux.

reflected rays

The incident rays after they are reflected from a reflecting surface.

reflecting microscope

A microscope in which a pair of concave and convex mirrors replaces the objective lens system.

reflection

The process by which a part of a radiation impinging on a surface which separates two media of different densities is turned back into the medium from which it impinges on that boundary surface. The rays impinging on the surface are called *incident rays,* those rebounding from the surface are referred to as *reflected* rays. The angles formed by these rays with the normal at the point of incidence (*angle of incidence,* and *angle of reflection*) are generally equal and lie in the same plane.

reflectivity

The ratio of the reflected portion of the incident radiant energy to the total amount of that energy that incides on the reflecting surface.

reflector

(1) A device to redirect radiation in a desired direction or directions. (2) A reflecting telescope.

reflex angle

An angle greater than 180° but less than 360°.

reflex polygon

See *re-entrant polygon.*

reflux condenser

A form of chemical apparatus, consisting of a cooling coil mounted on a vessel containing a boiling solution, which continuously condenses the escaping vapor and returns it to the vessel.

refluxed

A term applied in testing corrosion resistance to boiling solutions in which the escaping vapor is condensed and returned to the solution containing the metal under test.

refracted ray

A ray which has been bent as the source of light passed into a medium of different optical density.

refraction

A change of direction of rays (as of light) which are obliquely incident upon and pass through a surface bounding two media in which the ray has different velocities (as light going from air to water). It is found that for isotropic media: (1) When passing into a denser medium, the ray is refracted toward a perpendicular to the surface, and when passing into a less dense medium, it is bent away from the perpendicular; (2) the incident and refracted rays are in the same plane;

(3) the sines of the angle of incidence and the angle of refraction bear a constant ratio to each other for any two given media (the *angles of incidence* and *refraction* are the angles which the incident and refracted ray make, respectively, with the perpendicular to the surface). If the first medium is air, this ratio is called the *index of refraction* or the *refractive index* of the second medium.

refractive index
See *index of refraction.*

refractivity
The property of refraction, equal to $(n - 1)$ where n is the *index of refraction.*

refractometer
An instrument for direct measurement of *indices of refraction* to the fourth or fifth decimal place.

refractor
A device, usually of prismatic glass, which redirects the light of a lamp in desired directions, principally by refraction.

refractoriness
The index to the heat-resisting properties of a *refractory.*

refractory *(adj.)*
Resistant to heat.

refractory *(noun)*
A material, usually ceramic, which resists the action of heat and will not fuse at high temperatures (e.g., fireclay, silica or dolomite). Used for furnace linings and similar purposes.

refrangible
Capable of being refracted.

refrigerant
A substance, such as a liquid of low boiling temperature, suitable for use to produce low temperatures.

regelation
The melting of ice under pressure acting on it, followed by refreezing after the pressure is removed.

regenerative chambers
See *regenerators.*

regenerative furnace
A type of *reverberatory furnace* in which the air and gaseous fuel are preheated by passing through *regenerators.*

regenerators
Two pairs of chambers (*regenerative chambers*), one for gas and one for air in each pair, each filled with a checkerwork of

refractory bricks. The object of the checkerwork is that the bricks may absorb the heat from the hot waste gases as they pass through on their way from the furnace to the chimney. When the checkerwork of one pair of chambers is heated, the flow of the heated gases is reversed, to pass through the other pair of cool chambers, while the hot checkerwork gives up its heat to the incoming air and *producer gas* on their way to the furnace.

regular crystal
See *cubic crystal, cubic system.*

regular decagon
A plane figure having ten equal sides and ten equal angles.

regular dodecahedron
A polyhedron with twelve faces, all of which are regular pentagons, and with congruent polyhedral angles.

regular hexahedron
The *cube.*

regular icosahedron
A polyhedron bounded by twenty congruent equilateral triangles and having congruent polyhedral angles.

regular octahedron
A polyhedron bounded by eight congruent equilateral triangles and having congruent polyhedral angles.

regular polygon
A polygon in which all sides are of equal length and all angles are equal.

regular polyhedron
A *polyhedron,* all faces of which are congruent *regular polygons,* so arranged that all dihedral angles are equal, and the same number of faces meet at each vertex. The five regular polyhedrons known are: the *tetrahedron,* the *cube,* the *octahedron,* the *dodecahedron,* and the *icosahedron.*

regular prism
A *prism* the *base* of which is a regular polygon and is perpendicular to an edge.

regular pyramid
A pyramid the base of which is a regular polygon and the *altitude* (the line extending from the center of the base to the apex) is perpendicular to the base.

regular reflection
See *specular reflection.*

regular solid

A *regular polyhedron* (q.v.).

regular system

The cubic system of crystal with all right angles.

related numbers

Any numbers whose relation to each other is expressed by a formula.

relation

A set of ordered pairs.

relative density

(1) The ratio of the density of a gas to that of hydrogen under similar conditions of pressure and temperature. Also called *density ratio.*— (2) *Specific gravity.*

relative error

The ratio of the absolute error ($\pm m$) to the true value (x) or $\frac{\pm m}{x}$. The absolute error of plus or minus one inch in measuring a board 38 inches long gives a relative value of $\frac{\pm 1}{38}$.

relative frequency

In frequency distributions, the frequency of a specific class mark divided by the total number of figures in the data.

relative humidity

In meteorology, (1) the ratio of the amount of moisture in a given volume of space to the amount of moisture which could be contained in that volume in the state of *saturation;* (2) the ratio of the actual vapor pressure to the saturation vapor pressure.

relative permittivity

A synonym for *dielectric constant.*

relative valency effect

The tendency of a metal of higher *valency* to dissolve one of low valency rather than vice versa.

relative viscosity

The viscosity of a substance as compared with that of water at a temperature of 20°C, i.e., 1.005 centipoises.

relatively prime

Said of two or more expressions which have no factors in common. (The factor unity is disregarded since unity is a factor of every number.)

relativity

The theory of relativity is a modern system of natural philosophy, developed by Dr. Albert Einstein on the basis of a mathematical theory of the *space-time continuum,* and comprising his Special Theory of Relativity (1905) and General Theory of Relativity (1915). Certain concepts and magnitudes which the classical theory considered to be absolute (e.g. the simultaneity of spatially distant events not coincident in space-time, the length of an extended solid body, etc.) are considered by the theory of relativity to be relative to the choice of co-ordinate system, or frame of reference, and thus relative to the observer. On the other hand, certain concepts and magnitudes which the classical theory considered to be relative (e.g. the velocity of light in a vacuum) are regarded as absolute by the theory of relativity. The theory of relativity rejects the separation of the four-dimensional space-time into a three-dimensional space and a one-dimensional time. It considers mass and energy to be interdependent and interchangeable, merely two different aspects of the same fundamental concept, as if mass were merely congealed energy, and energy attenuated mass. This principle of the equivalence of mass and energy and its equation, $E = mc^2$ was the chief foundation of the researches which eventually resulted in the actual release of nuclear energy. Two basic postulates of the theory of relativity are: (1) the laws of mechanics are not affected by a uniform rectilinear motion of the system of co-ordinates which serve as their frame of reference, and if two systems of co-ordinates are moving with uniform but different linear velocities, observers in either system cannot possibly learn more about the motion than that a relative motion does exist; (2) the velocity of light is constant and independent of the motion of the observer, so that all measurements of the velocity of light will give the same numerical value in either of the above cited two moving systems.

reluctance

The opposition of a magnetic path to the establishment of *magnetic flux* in the material. Thus it is the ratio of the *magnetomotive force* to the magnetic flux, i.e., the reciprocal of the *permeance,* and is measured in gilberts.

reluctivity

The reciprocal of the *magnetic permeability.* Also called *specific reluctance.*

r.e.m., rem, Rem

See *roentgen equivalent, man.*

remainder
The number or quantity left over after one number or quantity has been divided by, or separated into equal parts by another. Thus 22 divided by 5 gives 4 equal parts and a remainder of 2.

remainder theorem
If a polynomial, $f(x)$, is divided by $(x - r)$, the remainder is $f(r)$.

remanence
The *magnetic induction* which remains in a magnetic circuit after the removal of an applied *electromagnetomotive force.* (Also called *remanent induction.*)

remanent induction
See *remanence.*

remote velocity
The velocity of an object taken as a whole relative to the surrounding fluid, as distinguished from the local velocity of any of its parts.

r.e.p., rep, REP
See *roentgen equivalent, physical.*

replaceable hydrogen
The hydrogen *atoms* in a molecule of an acid which are replaceable by metal atoms when the acid is neutralized with a base or replaced by an active metal.

replacement
In chemistry, a reaction in which an element, molecule or radical in a compound sets the replaced element, molecule or radical free. (See also *double replacement.*)

replacement series
A list of the metals arranged in the order of the magnitudes of their oxidation potentials.

reproduction factor
See *multiplication factor.*

residual electricity
The quantity of electricity which remains in a condenser after discharge.

residual ionization
The ionization of air or another gaseous medium in a vessel without any demonstrable influence of any ionizing agency, such as a radioactive substance or a source of x-rays in the vicinity, etc.; residual ionization is at the present time generally attributed to the effects of cosmic rays.

residual magnetism

Remanence.

residual stress

The stress which exists in an elastic solid body in the absence of, or in addition to, the stresses caused by an external load. (Also called *locked-up stress.*)

resilience

A general term for the capacity of an elastically strained body to spring back on removal of the load. This property is conferred by the potential energy stored up in an elastically strained body and given out when the load is removed. It is usually expressed in meter-kilograms per square centimeter or foot-pounds per cubic inch.

resin

A solid or semi-solid amorphous organic compound or mixture of such compounds, with no definite melting point and no tendency to crystallize. Resins may be of vegetable or animal origin (e.g. *shellac*), or may be synthetic. Natural resins are distinguished from gums in that the resins are insoluble in water, but certain synthetic water-soluble materials are referred to as resins.

resistance

In electricity, the ratio (expressed in ohms) of the voltage (expressed in volts) to the current (expressed in amperes). It is the scalar property of an electric circuit, or any body that may be used as part of an electric circuit, which determines for a given current the rate at which electric energy is converted into heat or radiant energy and which has a value such that the product of the resistance and the square of the current gives the rate of conversion of energy.

resistance pyrometer

An instrument for measuring high temperatures which functions on the principle that the electrical resistance of a metal changes with temperature; the change in resistance of a coil of fine wire, usually platinum, is calibrated with temperature.

resistivity

See *specific resistance.*

resistor

A device, the primary purpose of which is to introduce *resistance* into an electric circuit.

resolution

The separation of spectral lines, vectors, etc. into their components (Cf. *resolving power.*)

resolving power

In optics, in general, the smallest angular distance at which a lens permits the images of two point sources of light to be perceived as separate points. In reference to microscopes, the number of lines per millimeter which can be clearly separated by its lenses.

resonance

A quality describing an atom which emits rays of the same wavelength as it has previously absorbed. In physics, the condition existing between one coordinate of a system which is executing oscillations or vibrations and a periodic agency which maintains the oscillations or vibrations when a small amplitude of the periodic agency produces in the system a relatively large amplitude of oscillations or vibrations.

resonance in an electric circuit

Resonance may exist in an electric circuit possessing inductance, resistance and capacitance between the quantity of electricity which oscillates and a periodically applied electromotive force which sustains the oscillations. Any one of the three kinds of resonance, viz., *amplitude resonance, period resonance, or phase resonance,* may exist, depending on the constants of the circuit.

resonator

An apparatus or system in which some physical quantity is capable of being put into a state of oscillation by oscillations of another system.

rest mass

According to modern, relativistic physics, the mass which a body has when at absolute rest, and which increases when the body is in motion, according to the following formula:

$$m_1 = \frac{m_0}{\sqrt{1 - \frac{v^2}{c^2}}}$$

where m_1 is the mass in motion, m_0 is the rest mass, v the velocity of the motion, and c is the speed of light. The classical, Newtonian physics makes no distinction between rest mass and mass in general.

restitution

In chemistry, the opposite of *substitution,* i.e., the restoration of the original compounds.

resultant

In mathematics and physics, the sum of two *vectors*. Defined as a force which is a combination of two or more concurrent forces. This force is the equivalent of the concurrent forces and can replace them. It has magnitude, sense and direction based on the original concurrent forces. The resultant is equal in value to the *equilibrant,* but its direction is opposite to that of the latter. See also *eliminant.*

resultant

In chemistry, the designation of the molecules produced by a chemical reaction between the molecules called the *reactants.*

retardation

Deceleration.

retort

(1) A spherical glass vessel with an extended side tube, used in chemical laboratories for the distillation of liquids.– (2) A metallic or refractory-lined vessel for distilling metals, alloys and ores.

retro-

A prefix meaning *backward.*

reverberation

The persistent continuation of sound within an enclosed space in which the sound waves keep being reflected again and again.

reverberatory furnace

A furnace in which the ore or metal is exposed to the action of flame, but is not in contact with the fuel. The principle of the furnace is the deflection or reverberation of the heated gases from the sloping roof on to the metal.

reversible process

In physics, a process in which the opposing force is infinitesimally smaller than the driving force. Such a process can be reversed by an infinitesimal increase in the opposing force.

reversible reaction

A chemical change in which the products of reaction may be made to return to the original substances by slightly changing the conditions of, for example, temperature, pressure and concentration; an equilibrium condition in which the extent of reaction in either direction is determined by pressure and temperature.

rhenium

A metallic element in Group VII of the periodic table, discovered by Noddack and Tacke in 1925. Atomic number 75, atomic weight 186.31, melting-point above 3000°C., specific gravity 10.5, valence 2, 3, 4, 6 and 7. Its naturally occurring isotopes are: Re^{183} (stable, natural abundance 38.2%) and Re^{187} (radioactive, half-life 4×10^{12} years, natural abundance 61.8%). (Cf. *electron structure of atoms, isotope, periodic table.*) Symbol: Re.

rheology

The science of the deformation and flow of matter.

rheostat

An adjustable *resistor* so constructed that its resistance may be changed without opening the circuit in which it may be connected, thus providing the means to vary readily the amount of resistance present in the circuit.

rhodium

A metallic element in Group VIII of the periodic table, one of the rare metals, discovered by Wollaston in 1804. Atomic number 45, atomic weight 102.91, melting-point 1966°C., specific gravity 12.5. Symbol: Rh.

rhombic system

See *orthorhombic system.*

rhombohedral system

The crystallographic system comprising crystals which have parallel 6-fold rotary-reflection axes, or a single set of parallel 3-fold rotation axes, in either instance without planes of symmetry perpendicular thereto, and with or without other elements of symmetry. The typical rhombohedral crystal has three equal axes mutually inclined at an angle not usually equal to 90 degrees. (Also called *trigonal system.*)

rhombohedron

A prism bounded by six parallelograms.

rhomboid

A parallelogram all the angles of which are oblique, and in which the adjacent sides are not equal (the opposite sides, of course, are equal).

rhombus

A four-sided plane figure bounded by straight lines all of equal length. Opposite angles are equal but none are right angles. The diagonals are unequal, but bisect one another and are at right angles to one another.

rhumb line

A line on the surface of the earth, or a sphere in general, which intersects all meridians at the same angle.

rider

In mathematics, a subsidiary problem. In chemistry, a device for providing small weight units on a balance.

right angle

The angle formed by two straight lines perpendicular to one another. It is the angle at the centre of a circle subtended by a quarter of the circumference of the circle. In the *sexagesimal system* which is the universally used system, a right angle is $90°$ (ninety degrees). In the *centesimal system* a right angle is 100^g (one hundred grades).

right-angle triangle

A plane triangle in which one angle is a right angle.

right ascension

The arc of the celestial equator, or the angle at the celestial pole, measured eastward from the hour circle of the vernal equinox to the hour circle of a given celestial body, either through 24 hours or 360 degrees.

right circular cone

A *cone of revolution,* i.e., a cone the *base* of which is a circle and the *axis* of which is perpendicular to that base.

right circular cylinder

A cylinder the *bases* of which are perpendicular to an *element,* and the right sections of which are circles. Also called *cylinder of revolution.*

right-hand rule

See *left-hand rule.*

right-handed elliptically polarized wave

An elliptically polarized wave in which the rotation of the direction of displacement is clockwise for an observer looking in the direction in which the wave is traveling.

right-handed quartz

Quartz which rotates the plane of polarization to the right, i.e., clockwise, as seen when looking back toward the source of light.

right line

A straight line.

right parallelepiped

A *rectangular solid.*

right prism

A *prism* the *base* of which is perpendicular to an edge. If the base is a regular polygon, the right prism is called a *regular prism.*

right section

With reference to a cylinder or a prism, the *section* produced by a plane that cuts all the *elements* or lateral edges and is perpendicular to one of them.

ring

In chemistry, (1) a closed chain of atoms; (2) a circular structure. (See *ring structure, ring system.*)

ring structure

The arrangement of atoms typical of aromatic compounds, the molecules of which contain one or more rings (closed chains of atoms) which are connected by a single atom or by a simple or double *valency bond.*

ring system

The graphic representation of the *ring structure* of a compound.

R. M. S.

The abbreviation for *root-mean-square value.*

Rn

The symbol for the element *radon* (q.v.).

roasting

Heating an ore in contact with air.

roentgen (r)

The quantity of x-rays or gamma radiation such that the associated corpuscular emission per 0.001293 gram of air produces, in air, ions carrying 1 electrostatic unit of electricity of either sign.

roentgen equivalent, man

A radiological unit, also referred to as *r.e.m. unit* (abbreviated *rem* or *REM*), equal to that quantity of radiation which when absorbed by a human being produces the same effect as the absorption of 1 r (one roentgen) of high-voltage x-rays.

roentgen equivalent, physical

A radiological unit, also referred to as *r.e.p. unit* (abbreviated *rep* or *REP*), defined as that dose of ionizing radiation other than x-rays or gamma-rays which produces energy absorption of 93 ergs per gram of tissue (defined also as such radiation producing energy absorption of 83 ergs per gram of tissue).

Roentgen rays

Properly spelled "Röntgen rays," and commonly known as x-rays; penetrating radiations similar to light, having wave lengths of the order of a thousandth to a millionth of those of light in the visible spectrum (10^{-7} to 10^{-10} cm). In practice they are usually generated by allowing a stream of high-speed electrons to impinge on a metal target. The term "Roentgen rays" is preferred by medical authorities, but "x-rays" is in more general use by physicists.

roentgenometer

A device for measuring the intensity of ionizing radiations.

rolling

The process used in metallurgy which consists of rolling slabs of metal into sheets or bars.

rolling friction

The friction set up when a body rolls on a surface, as opposed to sliding over it.

roman numerals

A system of writing integers, used by the Romans, in which I denotes 1; V, 5; X, 10; L, 50; C, 100; D, 500; M, 1000. All integers are then written using the following rules: (1) when a letter is repeated or immediately followed by a letter of lesser value, the values are added; (2) when a letter is immediately followed by a letter of greater value, the smaller is subtracted from the larger. The integers from 1 to 10 are written: I, II, III, IV, V, VI, VII, VIII, IX, X. The tens are written: X, XX, XXX, XL, L, LX, LXX, LXXX, XC, C. Hundreds are written: C, CC, CCC, CD, D, DC, DCC, DCCC, CM, M.

root-mean-square value (R.M.S.)

The square root of the mean value of the squares of the instantaneous values of a periodic quantity during one complete cycle.

root-mean-square velocity (R.M.S. velocity)

The square root of the *mean square velocity.*

root of a number or quantity

The number or quantity which raised to the indicated power, i.e., multiplied by itself the number of times indicated by the index of the radical, equals the given quantity. Thus, the nth root of a (written $\sqrt[n]{a}$) is the number the nth power of which equals a.

root of an equation

A number which satisfies the equation, by providing its solution, in that when it is substituted for the unknown in the equation, it makes the latter an *identity.*

root sign

See *radical sign.*

rosin

A vegetable *resin;* the molten resinous residue remaining in the still after the distillation of spirits of turpentine from crude turpentine obtained from living pine trees. The molten resin is strained, and solidifies to form a solid rosin.

rotation

Motion in which the system moves about a fixed, stationary point constituting a *center* or *focus,* or about a fixed, stationary line forming an *axis* of the rotation, all particles or points of the system moving in circles about the center or focus with the same angular velocity.

rotation axis

(1) See *axis of rotation.*— (2) The *axis of symmetry of a crystal.*

rotation diagram

A special type of *Laue pattern,* viz., the photographic record of diffracted beams produced when a slender beam of x-rays is directed on a rotating single crystal.

rotation spectrum

An x-ray spectrum or diffraction pattern, produced by a crystal which is rotated while it is traversed by an x-ray beam.

rotator

An optical apparatus, as a quartz plate cut perpendicular to the *optic axis,* which rotates the plane of polarization of light traversing it.

rotatory

(1) In general, pertaining to, possessing or producing rotation.— (2) Optically active, i.e., able to rotate the plane of polarization of light. (See *dextrorotatory* and *levorotatory.*)

rotatory motion

See *rotation.*

rotor

The rotating member of a machine.

round off a number to n places

To write the approximate value of the number to n significant digits.

r. p. m.

The abbreviation for *revolutions per minute.*

Ru

The symbol for the element *ruthenium* (q.v.).

rubber

An elastic solid substance obtained by the coagulation of the milky juice (*latex*) of certain tropical plants, notably the tree *Hevea brasiliensis.* Natural rubber in the raw state is weak, plastic, inelastic, and has the general empirical formula C_5H_8; vulcanization, a process which adds sulfur to the raw rubber, turns it into a strong, nonplastic, elastic, insoluble material.

rubidium

A monovalent metallic element in Group I of the periodic table, one of the alkali metals, discovered by Bunsen in 1860. Atomic number 37, atomic weight 85.48, melting-point 38.5°C., boiling-point 690-700°C., specific gravity 1.53. Symbol: Rb.

ruled surface

A surface through any point of which there passes a line that lies entirely on that surface.

ruthenium

A metallic element in Group VIII of the periodic table, discovered by Claus in 1845. Atomic number 44, atomic weight 101.1, melting-point about 2450°C., specific gravity 12.06, valence 3, 4 or 8. Symbol: Ru.

Rydberg constant

A physical constant, relating to the wave number of spectral lines. Its smallest numerical value is: for *hydrogen* (R_H), 109,-677.58 ($\pm$0.01) cm^{-1}; for *infinite mass* (R_∞), 109,737.30 cm^{-1}.

S

S

The symbol for the element *sulphur.*

S pole

In a magnet, the *south-seeking pole,* i.e., the pole which always points toward the magnetic south.

Sa

An alternative symbol for the element *samarium.*

saccharimeter

A *polariscope* used for determining the quality of sugar, making use of the property of sugar of rotating the plane of planc-polarized light.

saccharometer

A fermentation tube used for the determination of the concentration of sugar solutions by ascertaining their density.

sag

See *sagitta.*

sagitta

The maximum elevation of a convex surface or the maximum depression of a concave surface. (Also called *sag.*)

salt

(1) In chemistry, any substance other than water produced by a reaction between an acid and a base; a compound of a metal or a positive radical and a nonmetal or a negative radical; the product of the displacement of the hydrogen of an acid by a metal or an electropositive radical, or of the neutralization of an acid with an alkali, or of a double replacement. (2) Common or table salt, i.e., sodium chloride (Na^+Cl^-).

salting out

The addition of a solid salt, or a concentrated solution of a salt to a solution or a colloid, causing the solute (liquid or solid) or the disperse phase to separate out.

samarium

A trivalent metallic element, one of the rare earth group of the periodic system, discovered by Boisbaudran in 1879. Atomic number 62, atomic weight 150.43, melting-point above 1300°C., specific gravity 7.7. Symbol: Sm or Sa.

sample
That portion of a substance or material selected for testing.

saponification
(1) The hydrolysis of fats by alkalies, forming glycerol and the alkali salts of the fatty acids. (2) Sometimes the definition is extended to include any ester hydrolysis, whether by alkali, acid, steam, or enzyme.

saponification number
The number of milligrams of KOH needed to saponify one gram of a fat or oil.

saturated compound
A compound in which all the valencies of the elements are satisfied. Most commonly used to indicate a single bond between adjacent carbon atoms.

saturated magnetization
That condition of a body when an increase in the magnetizing force produces no change in the intensity of magnetization.

saturated solution
A solution containing the maximum amount of a particular substance in solution that it can dissolve at that temperature.

saturated vapor
A vapor which is in equilibrium with its liquid at a given temperature, so that it cannot hold any more substance in the gaseous phase unless the temperature is increased.

saturated vapor pressure
The pressure exerted by a *saturated vapor* at the given temperature.

saturation
In chemistry, (1) maximum or total absorption of a substance by a solvent; (2) complete neutralization of an acid or of a base; (3) total and full satisfaction of all the valency bonds within a molecule.

saturation
In meteorology, that state of the atmosphere in which it can contain no more moisture.

saturation of magnetization
See *magnetic saturation.*

saturation of a color
See *colorimetric purity.*

Sb
The symbol for the element *antimony.*

Sc
The symbol for the element *scandium.*

scalar *(noun)*
Short for *scalar quantity.*

scalar *(adj.)*
Following the laws of arithmetical algebra.

scalar function
A scalar function of a variable scalar quantity is a scalar quantity which has one or more definite values for every value, within assigned limits, of the independent variable.

scalar product
The scalar product of two vectors is the scalar quantity of the magnitudes of the two vectors by the cosine of the angle between them. The scalar product of two vectors A and B may be indicated by means of a dot, A•B. Another common method is to enclose A and B in parentheses, (AB). If the values of the two vectors are given in terms of their regular components, then $A \cdot B = (AB) = A_xB_x + A_yB_y + A_zB_z$. Example: Work is the scalar product of force times displacement. Also called *dot product.*

scalar quantity
Any quantity which has magnitude only. The relation between any physical quantity and the unit used to measure it is completely described by a real number. Examples of scalar quantities for which the relationship to the unit can be represented by a positive number: mass, energy, electric resistance. Examples of scalar quantities for which the relationship can be represented by either a positive or a negative number: time, temperature, quantity of electricity.

scale error
An error in the reading of an instrument, owing to an imperfection or imperfections of the graduations of its scale.

scale hardness
See *Mohs' scale.*

scale of a drawing or map
The ratio of the distance between any two points in the drawing or map to the length between the two points which they represent.

scale of notation
Any recognized system of *numeration,* as the *denary* or *decimal* system which is now in general use in the western world.

scale of relation

The coefficients with which the preceding terms of a *recurring series* are multiplied in order to obtain the following term.

scalene triangle

A triangle in which no two sides and no two angles are equal.

scandium

A trivalent, metallic element in Group III of the periodic table, discovered by Nilson in 1879. Atomic number 21, atomic weight 44.96. Although not a member of the group of rare-earth elements, it is usually classed with them chemically because of its similar chemical properties. Symbol: Sc.

scattering

Dispersion or diffusion in various directions.

scattering media and surfaces

Media and surfaces which redirect the radiation impingent on them into a multiplicity of separate rays or pencils by reflection or transmission, as, e.g., rippled glass redirects light into separate pencils of light rays.

Schrödinger equation

A wave equation formulated by Schrödinger to represent the de Broglie wave, giving the relation between the wave function, the particle mass, the total energy, the potential energy, and Planck's constant. It is the basic equation of *wave mechanics.*

science

The ordered arrangement of ascertained knowledge, including methods by which such knowledge is extended and the criteria by which its truth is tested. The older term, natural philosophy, implied the contemplation of natural processes per se, but modern science includes such study and control of nature as is, or might be, useful to mankind.

scientific notation

The method of expressing very large or very small numbers as a number between 1 and 10 multiplied by 10 to a whole-number power. The exponent is the number of the zeros omitted after the number between 1 and 10; a negative exponent indicates the number of zeros omitted before the number which is regarded as following a decimal point. (E.g., 6,570,000,000,000 is written as 6.57×10^{12} and .00000000000657 as 6.57×10^{-12}.)

scintillation

(1) The production of a brilliant flash or spark. (2) The

minute flash of bright light made by an alpha particle or proton as it strikes a fluorescent screen in the dark at high speed.

scintillation counter

Radiation on zinc sulfide causes luminescence which provides individual flashes. This light, which is produced in a single scintillation, falls on the cathode of a photomultiplier tube and amplification results. This is now a very sensitive device.

scintilloscope

A radiation survey instrument, built on the principle of visual counting of the scintillations produced by a bombardment by alpha particles.

sclerometer

An instrument for measuring the hardness of substances according to the pressure that has to be applied to a diamond-pointed tool to produce a scratch.

scleroscope

An instrument for measuring the hardness of metals, according to the extent of the rebound of a steel ball dropped from a certain height.

scratch hardness

See *Mohs scale.*

Se

The symbol for the element *selenium.*

-scope

A suffix which identifies an instrument designed for observing or watching, usually as distinct from measuring.

scoria

See *slag.*

sec

Short for *secant.*

secant

One of the trigonometrical functions or ratios. In the right-angle triangle it is the hypotenuse divided by the adjacent side to the chosen angle. It is the reciprocal of the cosine.

secant curve

A graphic representation of the secant function or ratio; i.e., the graph of $y = \sec x$; it is one of the *trigonometric curves.*

secant of a circle

A straight line which passes through the circle in two points.

second

Unit of time. Universally used in all systems. Equal to 1/86,400 of the mean solar day.

second mean value theorem
See *extended law of the mean.*

second of arc
A measure of an angle, 1/60 of a minute.

secondary axis of a lens
The straight line which passes through the optical center of the lens and forms an angle with the *principal axis.*

secondary battery
See *storage battery.*

secondary cosmic rays
See *cosmic rays.*

secondary electron
An electron originating from a *secondary emission.*

secondary emission
The emission of beams of particles or electromagnetic radiation by a substance excited to such activity by the incidence of such beams or radiations emitted by another source (the *primary radiation*).

secondary ionization
Ionization produced by electrons dislodged by γ radiation.

secondary unit
A multiple or submultiple of a *fundamental unit.*

secondary wave
(1) See *secondary emission.* (2) A wavelet emitted by a point or an advancing wave front according to *Huyghens' principle.*

section
The plane figure produced by the intersection of a plane and a solid.

sector of a circle
That area of a circle bounded by an arc of the circumference of the circle and the radii of the circle drawn to the end points of the arc. The portion on either side of the radii is a sector.

sedative
A chemical compound which counteracts stimulation and irritation or nervous tension and produces a calming, soothing effect.

sediment
A deposit of insoluble material settled or precipitated from a liquid where it was held in suspension.

sedimentary rock
Rock formed by the accumulation of sediment which may include the remnants of plants, animals, products of climatic

actions or chemical reactions, and also fragments and remains of previously existing rocks.

sedimentation

The precipitation and settling of insoluble materials from a suspension, either spontaneously by free fall, or induced artificially by means of a *centrifuge.*

Seebeck effect

See *thermoelectric effect.*

segment

A portion of a line or figure cut off by a line or plane. (See *line segment.*)

segment of a circle

That part of the area of a circle bounded by any straight line (chord) drawn from one point on the circumference to another point on the circumference, and that part of the circumference between the two points. The portion on either side of the straight line (chord) is a segment. (See *angle in a segment of a circle.*)

seismograph

An instrument for registering and recording the time, direction and intensity of earth tremors.

seismology

The study of earthquakes and of the elastic properties of the earth.

selective absorption

The process in which a substance absorbs, either completely or partly, only certain wavelengths of an incident radiation, while it reflects or transmits the other regions of the radiation spectrum. For example, a substance that is seen as green, has absorbed all the colors of the spectrum except the green which it reflects.

selective reflection

The property of a substance owing to which it reflects certain colors of the spectrum while absorbing others. (Cf. *selective absorption.*) (2) The process or the phenomenon produced by this property.

selective transmission

(1) The property of a substance owing to which it transmits freely certain colors of the spectrum while it absorbs all the others. (Cf. *selective absorption.*) (2) The process or the phenomenon produced by this property.

selenium

A non-metallic element in Group VI of the periodic table,

discovered by Berzelius in 1817. Atomic number 34, atomic weight 78.96. Its principal allotropic forms are: red selenium (melting-point 180°C., specific gravity 4.45) and gray selenium (melting-point 220°C., boiling-point 690°C., specific gravity 4.8). Valence: 2, 4 or 6. Symbol: Se.

self-absorption

The absorption of radiation by the emitting substance itself.

self-conjugate triangle

The *harmonic triangle* when a conic passes through the four apexes of a quadrilateral. (Cf. *self-polar triangle.*)

self-diffusion

The movement of atoms through a *lattice* in which all atoms are identical, migration occurring simply by exchange of position.

self-inductance

The property of an electric circuit which causes it to oppose any change of current in it, and which determines for a given rate of change of current in it the electromotive force induced in the same circuit.

self-induction

The production of a counter electromotive force in a conductor when its own magnetic field collapses or expands with a change in current in the conductor.

self-polar triangle

If a conic passes through the four apexes of a quadrilateral, the *harmonic triangle* is said to be self-polar with respect to the conic. (Cf. *self-conjugate triangle.*)

self-propagating reaction

A chemical reaction which, once it has been started, is able to proceed without any further application of extraneous energy. (See *chain reaction.*)

self-stifling reaction

A chemical or nuclear reaction the effects or products of which are such as to prevent it from becoming *self-propagating.*

semi-

A prefix, used in Latin compounds, meaning *half.* (Analogous with the prefix *hemi-* which is used in compounds of Greek derivation.

semicircle

One-half of a circle; a segment of a circle bounded by the diameter and half the circumference. (See *angle in a semicircle.*)

semiconductor
A material intermediate between metals and non-metals, and capable of passing an electric current under certain conditions.

semiconjugate axis of a hyperbola
One-half of the conjugate axis, measured from the point where the latter intersects the transverse axis.

semicubical parabola
The curve representing the locus of the equation $y^2 = ax^3$.

"semi-hot laboratory"
Premises where the β and γ radiation is more than 10 millicuries but less than 1 curie.

semimetal
A term, now obsolete, for nonmalleable metals as for example, *bismuth* and *arsenic.*

semipermeable
Permeable to certain kinds of molecules, but stopping the passage of others.

semisteel
The name sometimes given to the metal resulting from the addition of mild steel to the charge of pig iron in the cupola. The term is a misnomer, for the product is still cast iron although the total carbon content is lower than usual and the graphite is in a finer state of division.

semitransverse axis of a hyperbola
One-half of the transverse axis, measured from either vertex.

semiwater gas
A mixture of carbon monoxide, carbon dioxide, hydrogen and nitrogen obtained by passing a mixture of air and steam continuously through incandescent coke. Its *calorific value* is low, about 125 B.T.U. per cu. ft.

sensation unit
The former name of the *decibel.*

sense
In mathematics, a synonym for *direction.*

sense of inequalities
If the signs of inequality in two or more inequalities point in the same direction (e.g., $a > b$ and $x > y$), the inequalities are said to have the *same sense;* if the signs point in opposite directions (e.g., $a > b$ and $x < y$), the inequalities are said to have *opposite sense.*

sensed line
See *directed line.*

sensible atmosphere
That part of the atmosphere that may be felt, i.e., that offers resistance. (Cf. *effective atmosphere.*)

sensible heat
The heat absorbed or emitted by a body or gas on heating or cooling through a certain range of temperature, no change of state or of allotropic form being involved.

sensitivity
(1) The speed of the response of an instrument or machine to unit change in the input. (2) The change in reading per unit of the quantity measured.

sensitizer
The intermediary upon which an atom or molecule acts in a *photochemical reaction.*

septagon
A polygon bounded by seven straight lines.

sequence
An ordered set of elements, called *terms.*

series
A succession of terms or quantities, which, after the first term or quantity, are formed according to some common law. The various quantities are called the *terms* of the series. Thus in the series 2, 5, 8, 11........each term is formed by adding 3 to the preceding term; and in the series 3, 12, 48, 192........each term is formed by multiplying the preceding term by 4.

series connection
See *in series.*

servo mechanism
Apparatus for exercising accurate and more or less remote control.

sesqui-
A prefix meaning *one and one-half,* used in chemistry to indicate a proportion of two to three, as, e.g., in manganese sesquioxide (Mn_2O_3).

set
A collection of objects.

settling
Separation of a denser solid from a liquid by gravity.

sexagesimal system of angular measure
The system in which the right angle is divided into ninety equal parts, each called a *degree* (symbol = °); the degree is divided into sixty equal parts each called a *minute* (symbol

= ′) and the minute is divided into sixty equal parts each called a *second* (symbol = ″) ; angles smaller than seconds are usually measured as decimals of a second. Example: 105 degrees, 23 minutes, 37.216 seconds is written 105° 23′ 37″ .216.

sextic

In mathematics, of the sixth degree or order.

shade

The part of the surface of an object from which light is blocked by the object itself.

shear

A deformation of a solid produced by the shearing stress.

shear modulus

The ratio of shearing stress to shearing strain. (Also called *coefficient of rigidity* or *modulus of rigidity*.)

shearing strength

The resistance offered by a material to a force that tends to cause one layer of the material to slide over an adjacent layer.

shearing stress

A stress that tends to slide one part of a body upon another.

sheath

This term is used to designate an incomplete electron shell, i.e. a *shell* (q.v.) which does not contain the full number of electrons which it would be capable of containing. According to I. Langmuir (*Phenomena, Atoms and Molecules,* Philosophical Library, New York, 1950), "The sheath of an atom consists of all the electrons in the outside shell provided that this layer is incomplete when the atom is neutral. The rare gases are the only elements whose neutral atoms have no sheaths, i.e. have all their electrons arranged in complete shells." (Cf. *kernel.*)

shell

According to *Pauli's exclusion principle,* the extranuclear electrons do not circle around the nucleus all in orbits of the same radius, but are arranged in orbits at various distances from the nucleus. The extranuclear orbital electrons are thus assumed to be arranged in a series of concentric spheres, called *shells,* which are designated, in the order of increasing distance from the nucleus, as *K, L, M, N, O, P* and *Q* shells. The number of the electrons which each of these shells can contain is limited.

shellac

An animal *resin,* produced by insects parasitic on certain trees in India and Southern Asia. It is used in lacquers and varnishes and as a modifier for synthetic resins.

shielding

In nuclear physics and engineering, the material interposed between a source of radiation and an irradiated object for protection against the dangers of the radiation; also, the process of so reducing the radiation hazard.

shock wave

A wave involving a sudden change in pressure and particle velocity in the propagating medium. Specifically, a sound wave set up by an object moving at supersonic speeds, causing increased energy losses.

short-circuit

An abnormal connection of relatively low resistance, whether made accidentally or intentionally, between two points of different potential in a circuit.

short wave

In radio terminology, waves of shorter wavelengths and higher frequencies than those currently used for commercial broadcasting, i.e., the range of frequencies between and 1500 and 30,000 kilocycles.

shrinkage

Loss of volume or bulk.

shunt

A resistor of low value which provides an electrical bypath; used for the measurement of electric current by means of a potentiometer, or of an ammeter, through which only a fraction of the total current passes.

Si

The symbol for the element *silicon.*

side of a polyhedron

Any of the lines which bound the polygon.

side of a polyhedron

Any of the planes or surfaces which bound the polyhedron.

side of an angle

Either of the two lines which form the angle.

side reaction

A chemical reaction which occurs simultaneously with the principal reaction, involving the same reactants, but resulting in different reaction products. Side reactions are very common in organic chemistry in particular.

sidereal year

The period of time during which the sun performs a complete revolution with reference to the fixed stars, i.e., 365.256 mean solar days.

sides of an equation

The term or group of terms appearing on either side of the sign of equality (=).

Siemens gas

See *producer gas.*

signed number

See *directed number.*

significant digits

In *rounding off* a number, the digits beginning with the first one, reading from left to right, which is not zero and ending with the last definitely specified one.

silex

Ferro-silicon mixed and bonded with chemical agents to furnish extra heat by means of exothermic reaction when the material is added to the bath.

silica (SiO_2)

Silicon dioxide. Silica forms 60% of the earth's crust and is the chief constituent of sand and silicate clays. It is extremely *refractory,* melting point 1750°C.

silica gel

A colloidal form of silica prepared by treating sodium silicate with acetic acid or hydrochloric acid, washing and drying. It is highly absorbent and its various applications include that of a drying agent both in the laboratory and industry.

silicate

A salt formed from silica or silicic acids.

siliceous

Containing *silica* or *silicates.*

silicon

A non-metallic element in Group IV of the periodic table, discovered by Berzelius in 1823. Atomic number 14, atomic weight 28.09, melting-point 1420°C., boiling-point 2600°C., specific gravity 2.4, valence 4; the second most abundant element, constitutes 27.75% of the earth's crust. Symbol: Si.

silicones

Heat-stable compounds of silicon in which the silicon atoms are linked up by oxygen atoms, the remaining valencies of the silicon atoms being saturated with hydrogen or organic radicals. Silicones with widely varying properties can be produced. They are chemically inert.

silver

A monovalent metallic element in Group I of the periodic table. Atomic number 47, atomic weight 107.880, melting-point 960.5°C., boiling-point 1955°C., specific gravity 10.5. Symbol: Ag.

silver brazing alloys

These are usually ternary alloys of silver, copper and zinc, e.g. a soft silver solder contains 67% silver, 20% copper and 13% zinc. A typical hard silver solder consists of 80% silver and 20% copper, the silver being added to lower the melting point of the solder and increase its fluidity, while still retaining high strength. Certain silver brazing alloys may contain other metals such as cadmium or tin.

silver family

The elements ruthenium, rhodium, palladium, silver and cadmium (atomic numbers 44-48). Also called *silver period.*

silver solder

See *silver brazing alloys.*

similar

Said of triangles or polygons all corresponding angles of which are equal, and all corresponding sides of which are proportional.

similar cones of revolution

Right circular cones generated by causing right triangles to revolve about their corresponding legs as axes of the revolution.

similar cylinders of revolution

Right circular cylinders generated by causing similar rectangles to revolve about corresponding sides.

similar figures

Figures the corresponding angles of which are equal and the ratio or proportion of whose corresponding sides is equal.

similar polyhedrons

Two or more *polyhedrons* the corresponding *faces* of which are equal in number and are similar each to each, as well as placed similarly, and the corresponding *polyhedral angles* of which are equal.

similitude

The state of being similar. (See *center of similitude.*)

simple closed curve
See *closed circuit.*

simple equation
An equation having only one unknown quantity which does not appear as a denominator of any fraction nor with any exponent higher than 1. Also known as an *equation of the first degree.*

simple expression
In mathematics, an expression that consists of one term only; e.g., 4*a* or merely *4* or *a*.

simple fraction
A single fraction that has both its terms whole numbers. Thus ⅔ or 8/5.

simple harmonic motion
A vibratory to-and-fro linear motion about an equilibrium position such that the acceleration is always proportional to the displacement from the equilibrium position and directed towards it. It is often referred to as the projection of uniform curvilinear motion upon a diameter of the circle of reference.

simple microscope
The ordinary microscope with one lens.

simple pendulum
A particle suspended by a rod or cord, the weight of which is negligible, the body being treated as if it were concentrated at its center of gravity. Also called *physical pendulum.*

simple quantity
In mathematics, a quantity that consists of only one quantity or term; e.g., 2*a,* or merely *2* or *a.*

simple sugar
See *monosaccharides.*

simplex
More or less simple, plain, elementary, unadorned, not complicated.

simultaneous equations
Two or more equations satisfied by the same set of values of the unknown quantities. Thus, in the equations $3x + 7y = 27$ and $3x + 2y = 12$, the values of $x = 2$ and $y = 3$ satisfy both equations.

sin
Short for *sine.*

sine
One of the trigonometrical functions or ratios. In the right-

angle triangle it equals the perpendicular (or opposite) side divided by the hypotenuse or $\sin A = \frac{a}{c}$. It is the reciprocal of the secant or $\sec A = \frac{1}{\sin A} = \frac{c}{a}$; $\frac{\sin A}{\cos A} = \tan A$. Taking the hypotenuse as $= 1$, versed sine of $A = 1 - \cos A = \text{vers } A$; $\sin^2 A + \cos^2 A = 1$.

sine curve

A graphic representation of the sine ratio, i.e., the graph of $y = \sin x$; it is one of the *trigonometric curves.*

sine law

See *law of sines.*

sintering

(1) The technique of heating a mass of crystalline or amorphous particles for a predetermined time and then cooling it in order to produce allotropic crystals and to cause them to coalesce into a solid mass. (2) The process of such coalescence of particles.

sinusoidal

Varying in proportion to the sine of an angle or time function.

sinusoidal wave

A wave the displacement of which varies as the sine (or cosine) of an angle that is proportional to time or distance, or both.

siphon

A U-shaped tube, the two legs of which are unequal in length; used to transfer liquid from one level to a lower one. The short leg of the tube is immersed into the vessel in which the liquid level is higher, and atmospheric pressure causes the liquid to flow through the tube and through the longer leg into the vessel in which the level of liquid is lower.

skew lines

Lines that do not lie in the same plane, are not parallel and do not meet or intersect.

skiagraph, skiagram

Synonymous terms for *radiograph.*

skin effect

The phenomenon of nonuniform current distribution over the cross-section of a conductor, caused by the variation of the current in the conductor itself, resulting in a tendency of alternating currents to flow near the surface of the conductor.

This effect increases the resistance and becomes more marked as the frequency rises.

slag

A molten or solidified substance consisting of oxides in combination with the fluxing material used in smelting operations. It is the nonmetallic fused by-product of a furnace operation. Also called *cinder* or *scoria.*

slant height of a cone of revolution (or right circular cone)

Any of the *elements* of the cone.

slant height of a frustum of a cone of revolution (or right circular cone)

That part of an *element* which lies between the two *bases.*

slant height of a frustum of a regular pyramid

The common *altitude* of its lateral faces.

slant height of a regular pyramid

The common *altitude* of its lateral faces.

slide rule

A mechanical device to aid in calculating by the use of logarithms. It consists essentially of two rules, one sliding in a groove of the other, containing logarithmic scales by means of which products and quotients are calculated by adding and subtracting logarithms.

sliding friction

The resistance offered to the motion of a solid surface which just slides over another.

slime

(1) In general, a fine powder in suspension. (2) A finely divided insoluble metal or compound forming on the surface of an anode or in the solution during electrolysis. (3) In hydrometallurgy and wet concentration, the finely divided portion of ore which settles very slowly from aqueous suspension.

slip

The mechanism of cold deformation in a crystal, in that one part of the crystal glides over another along certain atomic planes, known as *slip planes.*

slip plane

An atomic plane in a crystal along which the deformation known as *slip* occurs.

slit

In spectroscopic and other optical apparatus, a narrow slit with parallel sides, used to permit radiation to pass through or to produce diffraction, and occasionally also as a line source of radiation.

slitless-spectrograph

Beams of rays are limited by a wedge brought into contact with the face of a crystal.

slope

In analytic geometry, the tangent of the angle referred to as the *inclination* (q.v.) of a line. It is usually designated by the letter *m*.

slow neutrons

See *thermoneutrons*.

slug

The unit of mass in the FPS system of units; 1 slug is defined as the mass to which a force of 1 pound will impart an acceleration of 1 foot per second per second.

Sm

The symbol for the element *samarium*.

small calorie

See *calorie*.

small circle of a sphere

Any section produced by a plane that passes through a sphere, without passing through its center.

smelting

A metallurgical process, or series of processes, whereby a metal or compound is separated in a state of fusion from its ore or other material with which it is chemically combined or physically mixed. Care should be taken to distinguish between the terms smelt and *melt*.

smoke

In general, finely divided solid particles suspended in a gas; it usually issues forth from a burning substance and consists of solid particles of carbon dispersed in air.

smoothing

The process of eliminating, graphically or by arithmetical formulas, from tabular data, certain irregularities which are of no importance or significance for the immediate object.

Sn

The symbol for the element *tin*.

Snell's law

The experimental law of the refraction of light, stating that the ratio of the sine of the angle of incidence to the sine of the angle of refraction is a constant which is independent of the magnitude of the angle of incidence. This constant ratio is called the *index of refraction*. The law states, furthermore, that the incident ray, the refracted ray and the perpendicular to the point of incidence lie in one and the same plane. (Many authorities refer to this statement as a distinct law of Snell.)

soda

The common designation of sodium carbonate, one of the most important chemicals for industrial purposes, manufactured from common salt; when causticized with quicklime, it is converted into caustic soda.

soda ash

A commercial sodium carbonate, 99% pure Na_2CO_3.

Soddy Fajans displacement law

The emission of an alpha particle by a substance produces a displacement to the left by two places in the periodic table, i.e., a reduction of the atomic number of the substance by 2, whereas the emission of a beta particle produces a displacement to the right by one place in the periodic table, i.e., an increase of the atomic number by one.

sodium

A monovalent metallic element in Group I of the periodic table, one of the alkali metals, discovered by Davy in 1807. Atomic number 11, atomic weight 22.997, melting-point 97.7° C., boiling-point 880°C., specific gravity 0.97. Symbol: Na.

soft coal

See *bituminous coal.*

soft radiation

A term applied to radiation of limited penetrating power.

soft soldering

Soldering with a low temperature jointing alloy. The most widely used alloy contains about 60% lead and 40% tin, and has a melting point of 180°C. Tin is added to lower the melting point as required.

soft water

Water relatively free of salts of calcium and magnesium.

sol

The liquid phase of a colloidal solution.

solar constant

The rate at which solar radiant energy impinges on a surface perpendicular to the radiation, outside the atmosphere of the earth, at the mean distance of the earth from the sun; the mean value of this constant is 1.94 gram calories per minute per square centimeter.

solar energy

The energy radiated from the sun.

solar engine

An engine designed to convert the radiant energy of sunlight to work.

solar furnace

An apparatus for producing intense heat by concentrating the rays of the sun in a vacuum chamber in order to vaporize metals for spectrographic study.

solar noise

In radio astronomy, radio waves emanating from the sun. The term, "noise," is used here to denote electrical energy with a spectrum that is essentially a continuous one, in contradistinction to the discrete frequencies of radio signals.

solar phoenix

A picturesque reference to the *carbon cycle* which supplies the vast amounts of energy radiated by the sun; the origin of this expression is the restoration of the carbon isotope ${}_6C^{12}$ in the series of thermonuclear reactions reminiscent of the rise of legendary phoenix from its own ashes.

solar radiation

Radiation received directly from the sun.

solar radio astronomy

That subdivision of radio astronomy which studies the sun and its emission of electromagnetic energy manifesting as solar radio waves and solar noise.

solar spectrum

The band of colors produced by passing sunlight through a prism. The colors of the visible spectrum are violet, blue, green, yellow, orange and red.

solder

A general term for alloys used for joining metals by the process called *soldering.*

soldering

Uniting two pieces of metal by means of a third of lower melting point. For satisfactory union, the surfaces to be joined must be metallic, i.e. free from oxide films, and this is assured by using a *flux* which melts at a lower temperature than that of the *solder.* The molten solder and the flux must wet the two surfaces, i.e. make metallic contact with them. For use with soft solder, zinc chloride is a suitable flux. A higher temperature or *hard solder,* which is much stronger, is known as *silver solder,* for which borax is a suitable flux. Whatever the process used, the solder metal is non-ferrous and has a melting point well below that of the parent metal.

soldering acid
A workshop term for *hydrochloric acid* (HCl).

solenoid
An electromagnet having an energizing coil approximately cylindrical in form, wound as a helix with a small pitch or as two or more coaxial helices, and operating on a movable core or plunger.

solid
(1) Physically, a mass or body that exists in all three dimensions, i.e., has length, breadth and thickness. (2) Chemically, a substance in a state of aggregation in which it has a definite shape and volume and a higher degree of cohesion among its molecules than in any other state of aggregation.

solid angle
The ratio of the area of the surface of the portion of a sphere enclosed by the conical surface forming the angle, to the square of the radius of the sphere.

solid geometry
That branch of geometry which treats of three-dimensional objects, their properties and relations.

solid of revolution
A solid resulting from the revolution of a plane surface about a fixed line (called the *axis of revolution*).

solid shrinkage
Contraction in volume of a mass of molten metal or alloy on cooling from the freezing point to normal temperature.

solid solubility
The extent to which one metal is capable of forming solid solutions with another. This capacity varies considerably; some metals are almost insoluble in each other, while others are mutually soluble in all proportions.

solid solution
A homogeneous solidified mixture of two or more substances.

solid state
See *solid.*

solidification shrinkage
Contraction in volume of a cooling mass of molten metal or

alloy as it passes from the liquid to the solid state at the freezing point.

solidus

A line in a *phase diagram* indicating the temperature at which solidification is completed, or melting begins. (Cf. *liquidus.*)

solstices

Either of the two times during the year when the sun is at its greatest distance from the celestial equator.

solubility

The extent to which a given substance (the *solute*) will mix with a liquid (the *solvent*) to form a homogeneous system (the *solution*); measured and expressed in terms of the concentration of a saturated solution at a given temperature.

solubility curve

A line in a constitutional diagram showing the mutual solubility of the two constituents in relation to the temperature. The temperature is plotted vertically while the percentage composition is plotted horizontally.

solubility product

An equilibrium constant, which is the product of the concentration of ions of a substance in a saturated solution.

soluble

Capable of dissolving in or mixing with a liquid (the *solvent*) to form a homogeneous system (a *solution*).

soluble anode

An anode which goes into solution during an electrolytic process.

solute

A substance that dissolves in or mixes with a *solvent* to form a homogeneous system called a *solution.*

solution

(1) In chemistry, the homogeneous system produced by mixing a solid, liquid or gaseous substance (the *solute*) with a liquid (the solvent), not necessarily involving a chemical reaction. When a chemical reaction takes place, the resulting substance is called a *chemical solution;* otherwise, it is designated a *physical solution.* (2) The process of producing such homogeneous system by mixing a solute and a solvent.

solution

In mathematics, the required answer to a problem; also, the process of obtaining that answer.

solution of a triangle
The method or procedure of finding the unknown parts (sides or angles) of a triangle.

solution potential
The difference in potential of a material and of a reference electrode in a solution.

solvate
(1) A complex consisting of some particles of solvent associated with solute. (2) To form such complex particles.

solvent
The liquid in which a substance (the *solute*) is dissolved to form a *solution.*

solvus line
A solid solubility curve in a *phase diagram,* which indicates the limit of solubility of one phase in another at any temperature.

sonar
A method of measuring the thickness of deep-ocean sediments by exploding depth-charges thousands of feet below the sea surface and recording the echoes as reflected from boundary layers deep in the ocean floor.

sonic
Pertaining to or utilizing sound waves.

sonic frequencies
Sound frequencies, i.e., frequencies between 20 and 20,000 cycles.

sonic speed
The speed of an object moving at Mach 1 relative to the surrounding fluid.

soporific
(1) Possessing the property of inducing sleep. (2) As a *noun,* a generic term for all compounds or drugs used for inducing sleep.

sorption
A term which includes both (molecular) *absorption* and *adsorption.*

sound
(1) A wave phenomenon propagated through an elastic medium and capable of exciting the auditory nerve. (2) The auditory sensation produced by such wave phenomenon.

sound energy flux density
Sound energy per unit area measured as watts per sq. centimeter.

sound intensity

The sound energy transmitted per unit time in the specified direction through a unit area at right angles to that direction at the point for which the sound intensity is to be determined; measured and expressed in ergs per second per square centimeter, occasionally also in watts per square centimeter. (Also called *sound energy flux density*.)

sound wave

A traveling wave produced by vibrations in an elastic medium at a rate that can be heard.

south-seeking pole

That pole of a magnet which always points toward the magnetic south. Referred to as *S pole.*

space

The entire universe beyond the atmospheric envelope of the earth; the near-vacuum in which the solar system, stars, nebulae, galaxies exist.

space charge

The cloud of electrons existing in the space between the cathode and plate in a vacuum tube, formed by the electrons emitted from the cathode in excess of those immediately attracted to the plate.

space corner

A term occasionally used for a trihedral angle.

space group

A group of points infinitely extended and possessing the same elements of symmetry and the same translations as one of the 230 different arrangements theoretically possible for atoms in crystals.

space lattice

The dimensional geometric pattern in which the atoms of a metal arrange themselves, and upon which a crystal is built. There are several known lattice configurations but most metals crystallize in one of three types: (1) *face-centered cubic;* (2) *body-centered cubic;* and (3) *close packed hexagonal.*

space-time

A concept originated by Minkowski and developed by Einstein as a fundamental principle of the *theory of relativity.* The space-time continuum is a four-dimensional continuum, with four co-ordinates, viz. the three ordinary spatial coordinates plus the time co-ordinate.

spallation
A nuclear disintegration which causes a large electric pulse visible in an ionization chamber.

spark gap
An arrangement of two electrodes between which a disruptive discharge of electricity may take place, and such that the insulation is self-restoring after the passage of the discharge.

specific area
The total surface area of the particles in one gram of a specific material. Also called *specific surface.*

specific conductance
Conductance through a cubic centimeter of a given substance, expressed in mhos. Also called *conductivity.*

specific electronic charge
The ratio of the charge of the electron to its mass, i.e., 1.7592 $\times 10^7$ e.m.u. per gram. Symbol: e/m.

specific energy
Internal energy per unit mass.

specific gravity
The weight of a given volume of substance compared to that of an equal volume of chemically pure water at 4°C. (39.2° F). Example: pure commercial sulphuric acid has a specific gravity of 1.8 as compared to distilled water which has a specific gravity of 1.00. (Also referred to as *relative density.*)

specific heat
The specific heat of any given substance is the ratio of its *thermal capacity* and the thermal capacity of water; its numerical value equals the thermal capacity of the substance under consideration. (See also *negative specific heat.*)

specific humidity
In meteorology, the mass of water vapor in a unit mass of moist air, usually expressed in grams per gram or kilogram of moist air. (Cf. *mixing ratio.*)

specific ionization
A synonym for *ionization coefficient.*

specific magnetic susceptibility
The *magnetic susceptibility* of a substance per unit mass, i.e., the ratio of the magnetic susceptibility of the substance to its density.

specific reflectivity
Molecular reflectivity divided by the molecular weight.

specific refraction

At any given wavelength, the specific refraction of a substance equals $\frac{n^2 - 1}{n^2 - 2} \times \frac{M}{d}$, where n is its *index of refraction* at the given wavelength, M is its molecular weight, and d is its density.

specific refractivity

The *refractivity* of the substance divided by the density.

specific reluctance

The reciprocal of *magnetic permeability*. Also called *reluctivity*.

specific resistance

The electrical resistance between opposite faces of a unit cube of a given material at a given temperature. (Also called *electrical resistivity* and *volume resistivity*.)

specific surface

See *specific area*.

specific tenacity

A value, suggested by Rosenhain, representing the ratio of tensile strength to density, and enabling one to obtain a fair comparison of the strength of materials differing widely in density. The specific tenacities of widely different materials are only really comparable, however, if they have similar ductility.

specific thermal resistance

A synonym for *thermal resistivity*.

specific volume

The volume per unit mass of a substance; the reciprocal of the density.

specific weight

Weight per unit volume. See also *specific gravity*.

spectral line

One of the lines in a *line spectrum*.

spectral series

A series of frequencies in a radiation or absorption spectrum, following certain definite numerical laws.

spectrograph

The photograph of a spectrum, made with a *spectrograph*.

spectrograph

(1) A recording form of *spectroscope* for producing photographic or electronically controlled records of spectra. (2) In x-ray work, an instrument using an extended surface, i.e.

a photographic plate or film, or a fluorescent screen, for receiving the x-ray diffraction pattern.

spectrographic analysis
The qualitative and quantitative analysis of a substance, its composition and properties on the basis of its study with the spectrograph.

spectrometer
An instrument for obtaining and studying spectra and for measuring the wavelengths of spectral lines.

spectrophotometer
An apparatus for studying and measuring the distribution and relative intensities of the components of different frequencies in the spectrum of a light source.

spectroscope
An apparatus for the dispersion of radiation (spreading individual wavelengths) and analysis of the spectrum obtained.

spectrum
An array of frequencies or wavelengths resulting from the dispersion of an electromagnetic radiation. (See *solar spectrum.*)

spectrum analysis
The study and interpretation of spectra and *spectral lines,* and the investigation and qualitative analysis of substances and bodies through their spectra and spectral lines.

spectrum colors
The colors appearing in the continuous spectrum of white light, viz.: red, orange, yellow, green, blue, indigo, violet.

specular reflection
Reflection of light or other forms of radiant energy from a surface so smooth that its inequalities are small in comparison with the wavelength of the incident rays, so that each incident ray gives rise to a reflected ray in the same plane. (Also called *regular reflection.*)

speculum metal
An alloy consisting of two parts of copper to one part of tin; it is used for mirrors and reflectors.

speed
The rate of motion of a body (cf. *velocity*). [G: *Geschwindigkeit*]

speed of sound
The speed at which sound waves travel through a medium.

(See *mach.*) Sound travels at different speeds through different mediums, and at different speeds under different conditions of temperature, pressure, etc. In air under standard sea-level conditions, sound travels at about 1,100 ft/sec or 750 mph.

spelter
The commercial name of the bath of molten zinc into which metal objects are dipped in order to be galvanized.

sphere
A solid, the surface of which is the locus in space of all the points which are the same equal distance (the *radius*) from a fixed point (the *center*).

spherical
Pertaining to a sphere or spheres; shaped like a sphere.

spherical aberration
In optics, an imperfection or defect of the lens, viz., the inability of the lens to bring the marginal and central rays of light to one focus, bringing the former to a focus nearer to the lens itself than the latter, resulting in a loss of sharpness of the image.

spherical angle
An angle on a sphere formed by the arcs of two *great circles.*

spherical cone
A *zone* (a spherical segment) which has only one *base.*

spherical degree
A *spherical triangle,* two sides of which are *quadrants,* and the third one an arc of exactly 1°; it is a unit of measuring spherical surfaces.

spherical distance
The distance between two points on a sphere, measured along the *minor arc* of the *great circle* which passes through both of them.

spherical excess of a spherical polygon
The difference, measured in *spherical degrees,* between the sum of the angles of the spherical polygon and the sum of the angles of a plane polygon having the same number of sides, i.e., the number of spherical degrees by which the sum of the angles of the spherical polygon exceeds $(n - 2)\ 180°$, where n is the number of the sides of the polygon. Thus, e.g., the spherical excess of a spherical quadrilateral is the sum of its angles minus 360°.

spherical excess of a spherical triangle
The sum of the degrees of the angles of the spherical triangle less 180°.

spherical geometry
That branch of geometry which treats of spherical magnitudes.

spherical lens
A lens the surfaces of which are segments of spheres.

spherical polygon
A closed figure on a sphere, each of the boundaries of which is an arc of a *great circle.*

spherical sector
A solid generated by causing a *sector* of a semicircle to rotate about its diameter as axis.

spherical segment
The solid formed between two parallel planes which pass through a sphere. (The solid formed when one of the planes passes through the sphere and the other one is tangent to it is likewise called a *spherical segment.*)

spherical surface
A curved surface every point of which is at the same identical distance from a single point inside that surface; that point inside the curved surface is called its *center.*

spherical triangle
A closed figure on a sphere bounded by three lines, each of which is an arc of a *great circle.*

spherical trigonometry
The study of spherical triangles—finding unknown sides, angles, and areas by the use of trigonometric functions of the plane angles which measure angles and sides of the triangles.

spherical wave
A wave the wave front of which is a spherical surface.

spheroid
An *ellipsoid of revolution* which resembles a sphere.

spherometer
A generic term for instruments for measuring the curvature of spherical surfaces.

spin
The rotation of an atomic nucleus or subatomic particle about its own axis, and also the angular momentum produced by this rotation. The spin of a nucleus or particle is considered to be responsible for its magnetic moment.

spinel

The name applied to a group of minerals of general formula XO, Y_2O_3 or XY_2O_4 where X may be a divalent metal such as magnesium, nickel, iron, cadmium, copper, manganese or zinc and Y a trivalent metal such as aluminum, iron, chromium, gallium, indium or vanadium. Some other minerals, particularly those of general formula X_2YO_4 may have the same structure as spinels. Spinels belong to the isometric system, have imperfect cleavage, and often exhibit simple or complex twinning on octahedral faces.

spinthariscope

A simple apparatus, introduced by Crookes, for visually observing the scintillations produced by the impact of alpha particles on a zinc sulfide screen. The instrument, equipped with a magnifying eyepiece, can be used to measure the range of the alpha particles emitted by the various radioactive substances.

spiral

The plane locus of a point which moves about a fixed center while continually increasing its distance from it.

spiral of Archimedes

A curve which represents the locus of a point moving at a uniform speed along the radius vector as the latter moves at a uniform angular speed.

spongy

Porous.

spontaneous

Occurring owing to inherent properties, without external inducement or activation.

spontaneous decay (disintegration, transmutation) of elements

The change, by emission of alpha or beta particles, of radioactive elements into other radioactive elements, which form a *radioactive series* or *family,* the end product of which is a stable isotope (an isotope of lead in the three natural radioactive series, an isotope of bismuth in the artificial neptunium series).

square

In algebra and arithmetic, the second power of a quantity, i.e., the product obtained by multiplying that quantity by itself.

square

In geometry, a plane figure bounded by four straight lines having all its four interior angles right angles and all its sides equal lengths. Its diagonals bisect one another at right angles.

square degree

A unit for the measurement of *solid angles;* defined as the solid angle subtended by a square each side of which subtends an angle of degree. One square degree equals $\left(\frac{\pi}{180}\right)^2$.

square root

The expression, term or quantity the second power of which (i.e., which when multiplied by itself) equals the given expression, term or quantity. Thus the square root of 64 is 8, because $8 \times 8 = 64$ and is written $\sqrt{64} = 8$; or $\sqrt{x^2 - 2xy + y^2} = x - y$ because $(x - y)\ (x - y) = x^2 - 2xy + y^2$.

square wave

A wave which has equal positive and negative alterations, with respective amplitudes about equal to the length of alteration.

squaring the circle

The classical insoluble problem of mathematics which calls for the construction of a square exactly equal in area to a given circle.

Sr

The symbol for the element *strontium.*

stabilization

In chemistry, the addition of a substance (called *stabilizer*) in order (1) to increase the stability of another substance or of a compound, or (2) to counteract the effect of an *accelerator.*

stabilizer

In chemistry, a substance which increases the stability of another substance or compound, or which counteracts the effect of an *accelerator* or otherwise slows down or retards a chemical reaction.

stable substance

(1) In chemical terminology, a substance able to resist successfully the attempts to decompose it. (2) In nuclear physics, a nonradioactive substance.

standard atmosphere

The atmosphere used as an international standard presumes for mean sea level and a temperature of 15°C., a pressure of

1,013.2 millibars, lapse rate of 6.5°C. per kilometer from sea level to 11 kilometers.

standard calorie

The amount of heat required for raising the temperature of one gram of water from 3.5°C. to 4.5°C. Also called *4° calorie.*

standard cell

A voltaic cell made to conform with certain specifications in order to be used as a standard of potential difference.

standard conditions

When used in connection with experiments or measurements, especially those concerning or involving gases, this term means the temperature of OC. and the pressure of 760 millimeters of mercury. (Also referred to as *normal temperature and pressure.*)

standard cubic foot of gas

That quantity which would occupy a volume of one cubic foot at a pressure of 14.65 pounds absolute and a temperature of 60°F.

standard deviation

In the study of the variability or dispersion of a set of observations, the measure of the average of the amounts that the individual numbers deviate from the mean of all the observations.

standard electrode potential

See *standard potential.*

standard gravity

A conventional value of the acceleration of gravity, adopted by the International Bureau of Weights and Measures, viz.: 980.665 centimeters per second per second.

standard position of an angle

The position which an angle is said to occupy with respect to a *rectangular coordinate system* when its vertex is coincident with the zero point of the system and one of its two sides (called the *initial side*) with the *x axis* of the system. (The other side, the revolution of which about the vertex is regarded to have formed the angle, is called the *terminal side.*)

standard potential

Standard potential of an electrolytic reaction is the theoretical reversible potential for a system in which the products and reactants are at unit activity. Conventionally the standard potential of an $H \rightleftharpoons H^+$ electron reaction is taken as zero. (Also called *standard electrode potential.*)

standard solution

In chemistry, any solution the composition of which is known.

standard volume

The volume occupied by one *mole* of gas at the temperature of 0°C. and under the pressure of 760 millimeters of mercury. It equals 22.414 liters. (Also called *gram-molecular volume.*)

standing wave

(1) A periodic wave which at all points in space has simultaneous proportionate variations with time. (2) The resultant of the encounter of two waves identical in wavelength and amplitude, and traveling through a medium at identical velocity, but in directly opposite directions. (Also called *stationary wave.*)

standing wave

In electricity, the resulting wave when the electric variation in a circuit takes the form of periodic exchange of energy between current and potential forms without translation of energy.

starch

The generic designation of a group of *carbohydrates* found in all plants (except in fungi), the general formula of which is $(C_6H_{10}O_5)_n$.

stark effect

The effect of a strong electric field on spectral lines, usually manifesting in a resolution of lines.

stat-

A prefix to names of electrical units, indicating that they are stated in the practical system of units.

statcoulomb

The amount of electric charge which when placed in vacuum one centimeter from another charge of equal magnitude, but of opposite sign, will repel it with a force of one dyne. (Also called *electrostatic unit of charge* and abbreviated as *E.S.U.* or *e.s.u.*)

state of aggregation

The physical condition of matter, referred to as *solid, liquid* or *gaseous.*

static

Any electrical disturbance caused by atmospheric conditions, especially the natural electromagnetic disturbance of radio reception produced by them. (Also called *atmospherics,* but this synonym has fallen in disuse in the United States.)

static electricity
Non-flowing electricity, usually generated by friction.

static universe
The universe postulated by opponents of the theory of the *expanding universe:* a closed universe of a finite volume.

statics
That branch of *dynamics* which deals with bodies in equilibrium, situations in which the several forces are so related as to balance or neutralize each other so far as giving the body motion as a whole is concerned. (According to modern, relativistic physics, this is possible in relation to a certain frame of reference or system of co-ordinates only.)

stationary flow
See *steady flow.*

stationary wave
See *standing wave.*

stator
That portion of an electrical machine which contains the stationary parts of the magnetic circuit with their associated windings.

statvolt
Ergs per stat-coulomb; 1 stat-volt = 300 volts.

steady flow
In hydraulics, a motion of a liquid in which the liquid moves with a given, constant magnitude and direction at all points. (Also called *stationary flow.*)

steady state
An equilibrium between the formation and decay of a substance.

steady-state theory of the universe
The view that the major, overall features and properties of the universe remain unchanged in time. A logical conclusion of this theory is that the universe has no beginning and no end, that time is an infinite flow, and that matter is being constantly created, new galaxies and star clusters formed within an expanding universe at a rate that compensates for their mutual recession owing to the expansion.

steel
A malleable alloy of iron and carbon, the carbon content being usually less than 1.7%. It is produced in the fluid con-

dition, either by the *crucible, Bessemer, open hearth,* or *electric furnace.* Such steel is practically free from *slag. Shear steel* is an exception to this definition.

steeloscope

An instrument used for estimating the metallic elements present in a sample of steel from its arc spectrum.

Stefan-Boltzmann law

The total radiation intensity, i.e., the amount of energy radiated from a unit surface of a *black body* per unit time is proportional to the fourth power of the temperature of that black body, in that $J = \sigma T^4$, where J is the energy radiated, T is the absolute temperature of the black body, and σ is *Stefan's constant* (5.735×10^{-5} ergs per square centimeter per second per fourth power of the temperature in degrees). Also called *Stefan's law.*

Steinmetz coefficient

The term S in the expression $SB^{1.6}$ which is the formula for calculating the approximate loss of energy per unit volume per cycle due to *magnetic hysteresis,* where B is the maximum magnetic induction. This coefficient is constant for a given substance.

steradian

The unit *solid angle* subtended by surface of a sphere equivalent to the square of the radius. All of space about a point measures 4π steradians.

stereogoniometer

A device for measuring the orientation of crystals in coarse-grained polycrystalline metals.

stereogram

A model diagram giving a representation of data in a three-dimensional form.

stereographic map

A map produced by projecting a sphere on the plane of a great circle.

stereographic projection

The method of projecting a sphere on the plane of a great circle in such manner as to produce a *stereographic map.*

stereoisomer

A substance which possesses the property of *stereoisomerism.*

stereoisomerism

The property observed in certain substances which consists

in that their molecules possess an identical structure but a different spatial arrangement of their atoms.

stereomer

A synonym for *stereoisomer.*

stereoptics

The theory and practice of stereoscopy, e.g. an auto-stereoscopic photograph is one giving a three-dimensional or solid effect, without the use of a viewing device by the observer.

stereoscope

An instrument for viewing two photographs, each of which is seen simultaneously, but each with one eye only, producing an effect of perspective or three-dimensionality.

stereoscopic

Possessing the property or conveying the effect of depth, perspective or three-dimensionality.

stereoscopic fusion

The mental process by which the two perspective images produced on the retinas of the two eyes are combined in such a manner as to create the visual impression of a three-dimensional object or view.

stereoscopic image

The mental impression of a three-dimensional object which is produced by *stereoscopic fusion.*

stereoscopic vision

Vision in depth or in three dimensions owing to the spacing of the eyes which permits them to see objects from slightly different angles.

stereoscopy

The science and technique of stereoscopic effects and the methods by which they are produced.

stereovision

See *stereoscopic vision.*

stiffness

That property which enables a substance or a system to resist forces or efforts tending to produce a displacement or deviation from a course or pattern.

still

An apparatus in which a substance is changed by heat with or without chemical decomposition into a vapor, which vapor is then liquefied in a condenser and collected in another part of the apparatus.

stoichiometry

That branch of chemistry which treats of the quantitative relationships between elements or compounds, and the formulae and equations of chemical reactions.

stoke

The unit of kinematic viscosity; one stoke is the kinematic viscosity of a fluid which has a viscosity of one poise and a density of one gram per cubic centimeter.

Stokes' law

The velocity necessary to propel a spherical body of given radius at uniform speed, through a viscous medium: $V = \frac{2ga^2(d_1 - d_2)}{9\eta}$ where a is the radius of the sphere, d_1 and d_2 the densities of the sphere and the medium, respectively, and η the coefficient of viscosity. V will be in cm. per sec. if g is in cm. per sec.^{2}, a in cm., d_1 and d_2 in g per cm.^{3} and η in dyne-sec. per cm.^{2} or *poises.*

Stokes line

A spectral line which satisfies *Stokes' law.* (See also *anti-Stokes line.*)

stooping

An optical phenomenon, consisting in an apparent vertical contraction of an object owing to the fact that atmospheric refraction causes the rays from its base to curve down much more rapidly than those from the top.

storage battery

A connected group of two or more storage cells. Common usage permits this term to be applied to a single cell used independently. Also called *secondary battery* or *accumulator.*

storage cell

An electrolytic cell for the generation of electric energy in which the cell after being discharged may be restored to a charged condition by an electric current flowing in a direction opposite to the flow of current when the cell discharges.

straight angle

An angle equal to two right angles, i.e., 180°, or the angle which any segment of a straight line can be regarded as forming with either immediately adjacent, contiguous segment of the same undivided line.

strain

(1) The deformation produced in a solid body by stress.— (2)

The forces acting internally in an elastic body when it is deformed.

strain gradient

That variation in strain set up in a structure which has been stressed beyond its yield point and has thus suffered varying degrees of deformation.

stratigraphic geology

See *stratigraphy*.

stratigraphy

That branch of geology which treats of the composition, age, chronological succession and correlation of the rock strata of the earth. (Also called *stratigraphic geology*.)

stratosphere

A stratum of the atmosphere, extending upward from the upper limits of the *troposphere* to an altitude of approximately 19 miles. (Some meteorologists consider it to include the *chemosphere* and to extend upward to an altitude of approximately 50 miles.)

stray

A natural electromagnetic wave in the ether. The term is used in reference to the effect of such wave in producing erratic signals in radiotelegraphic receivers. Strays are known collectively as *static*.

streamlines

Lines which are parallel everywhere to the instantaneous direction of motion in a fluid. If *a, b* and *c* are the velocity components in cartesian coordinates, the differential equations of the streamlines are: $\frac{dx}{a}=\frac{dy}{b}=\frac{dz}{c}$.

stress

The forces exerted on, within or by a body during tension or during compression; the opposing reaction of the interior elements of a solid body against forces tending to deform them. It is measured in force per unit area. (See also *internal stress*.)

stress modulus

The stress divided by the strain of any elastic deformation.

stress range

See *range of stress*.

stroboscope

An instrument which creates the optical illusion of slowing down or stopping the motion of an object by illuminating it with flashes of intense light at regular intervals. It is used for the inspection of objects rotating at high speeds or performing periodic or aperiodic movements.

strontium

A divalent, metallic element in Group II of the periodic table, one of the alkaline earth metals, discovered by Klaproth in 1793, isolated by Davy in 1808. Atomic number 38, atomic weight 87.63, melting-point 800°C., boiling-point 1150°C., specific gravity 2.55. Symbol: Sr.

structural

Relating to the arrangement of constituents of an assembly, of atoms in a molecule, etc.

structural formula

In chemistry, a formula showing the spatial structure of the molecules of a compound, the *valence bonds* existing within it, and the atoms which they link together.

structural geology

That branch of geology which treats of the form, arrangement and internal structure of rocks. (Also called *geotectonic geology* or *tectonics*.)

structure

The sizes of the constituents of an assembly or unit, their arrangement, and the forces or linkages which hold them together. (See *atomic structure, molecular structure, nuclear structure.*)

structure-symbol

A geometrical figure symbolizing graphically the pattern in which the atoms are arranged in an organic molecule.

sub-

A prefix indicating *under* or *below.*

subatomic particles

This term is applied to all particles of less than atomic mass, i.e. the elementary particles (proton, neutron, electron, positron, neutrino, meson and γ-rays or photons), as well as the α-particles and deuterons, the charge and mass of which indicates them to be composite particles.

Elementary Particles

Generic Name	Particle	Parity	Symbol	Lifetime	Mass
Leptons	Photon	even	γ	Stable	0
	Neutrino	even	υ°	Stable	0
	Antineutrino	even	$\bar{\upsilon}^{\circ}$	Stable	0
	Electron	even	e^-	Stable	1
	Positron	even	e^+	Stable	1
L-Mesons*	Negative Muon	even	μ^-	2.2×10^{-6}	206.2
	Positive Muon	even	μ^+	2.2×10^{-6}	206.2
	Negative Pion	odd	π^-	2.56×10^{-8}	273.2
	Positive Pion	odd	π^+	2.56×10^{-8}	273.2
	Neutral Pion	odd	π°	$< 4 \times 10^{-16}$	263
K-Mesons*		odd			
	Negative K	odd	K^-	1.2×10^{-8}	967
	Positive K		K^+	1.2×10^{-8}	967
	Neutral K_1		K°_1	10^{-10}	973
	Neutral K_2		K°_2	8×10^{-8}	973
Nucleons	Proton	even	ρ^+	Stable	1836.1
	Antiproton	even	$\bar{\rho}^-$	Stable	1836.1
	Neutron	even	η°	1.04×10^3	1838.6
	Antineutron	even	$\bar{\eta}^{\circ}$	1.04×10^3	1838.6
Hyperons*	Lambda		λ°	2.77×10^{-10}	2182
	Anti Lambda		$\bar{\Lambda}^{\circ}$	2.77×10^{-10}	2182
	Negative Sigma		Σ^-	1.67×10^{-10}	2341.4
	Positive Anti Sigma		$\bar{\Sigma}^+$	1.67×10^{-10}	2341.4
	Positive Sigma		Σ^+	$.83 \times 10^{-10}$	2327.6
	Negative Anti Sigma		$\bar{\Sigma}^-$	$.83 \times 10^{-10}$	2327.6
	Neutral Sigma		Σ°	$< 10^{-11}$	2331.7
	Neutral Anti Sigma		$\bar{\Sigma}^{\circ}$	$< 10^{-11}$	2331.7
	Negative Xi		Ξ^-	$5 - 200 \times 10^{-10}$	2585
	Positive Anti Xi		$\bar{\Xi}^+$	$5 - 200 \times 10^{-10}$	2570
	Neutral Xi		Ξ°	?	?
	Neutral Anti Xi		$\bar{\Xi}^{\circ}$	?	?

* L-mesons—Light mesons: muons and pions.
K-mesons—Heavy mesons: all mesons that are heavier than pions and lighter than protons.
Hyperons—Y particles: all particles with mass intermediate between the neutron and the deuteron (nucleus of the deuterium atom).
Note: Capital Greek letters are used to designate hyperons, lower-case Greek letters are used to designate mesons.

subduplicate ratio

The ratio $a^{1/2} : b^{1/2}$ or $\sqrt{a} : \sqrt{b}$ is referred to as the subduplicate ratio of a : b.

sublimate

Solid obtained by the direct condensation of a vapor, without passing through the liquid state.

sublimation

The change of a substance from the solid state to the gaseous state without passing through the liquid state.

submicron

Any particle not visible under an ordinary microscope, but visible under the ultramicroscope, i.e., between 0.2 and 5 millimicrons in diameter. (Also called *ultramicron.*)

submultiple

A number or quantity which divides another exactly. Thus 8 is a submultiple of 72 and $a + x$ is a submultiple of $a^2 - x^2$.

submultiple angle

An angle which divides exactly in another angle. E.g., an angle of 20° is a submultiple of an angle of 60°.

subnormal

In mathematics, that distance on the axis of a conic which lies between the vertical line from a point on the curve and the intersection of the normal to the tangent at that point with the axis of the conic.

suboxide

That oxide of a given element the oxygen of which is the lowest.

subsonic speed

Any speed that is less than the velocity of sound (about 760 m.p.h. at normal sea-level conditions). In aeronautical engineering, any speed below Mach number = 0.8.

substance

A material of a definite chemical composition.

substitution

In mathematics, a method of replacing one algebraical expression for another of equal value, but different composition.

substitution

In chemistry, (1) the replacement of an atom or atom group by another; (2) a reaction in which such replacement takes place.

substitutional solid solution

A solid solution in which atoms of one element take the place of atoms of the primary, or solvent, metal.

subtangent

The distance from the intersection of the tangent at a point of a conic with the principal axis and the vertical line drawn from the point to the axis.

subtend

If P is a point outside a straight or curved line AB, then the angle APB is the angle subtended at P from A and B. An arc of a circle is said to subtend an angle ECD at the centre of a circle by drawing a radius CE and CD of the circle to each extremity E and D of the arc.

subtract

To deduct; to take away.

subtraction

The method of finding the difference between two quantities. In an expression of several terms the negative or minus terms are taken away from the positive or plus terms. Thus: + 6 pears — 2 pears + 3 pears — 4 pears = + 9 pears — 6 pears = + 3 pears. If the negative or minus terms are greater than positive terms then the answer is minus. Thus: 4 pears — 9 pears — 5 pears + 2 pears = + 6 pears — 14 pears = — 8 pears.

subtractive color

A color produced by the removal of some part of white light, as by passing it through one or more filters.

sucrose

The sugar $C_{12}H_{22}O_{11}$ (cane sugar or beet sugar), which in non-technical, colloquial terminology is usually called simply "sugar." Obtained from the sugar cane, sugar beets, and from the sap of the maple tree, it is the sugar of standard, every-day consumption.

sugar

A generic term for carbohydrates characterized by a sweet taste and having the general formula $C_nH_{2n}O_n$ or $C_nH_{2n-2}O_{n-1}$. In colloquial usage, the term denotes *sucrose*.

sulfur

An element in Group VI of the periodic table; atomic number 16, atomic weight 32.06, its two principal forms are: monoclinic (β-monoclinic) sulfur, with the melting-point 119°C., boiling-point 444.6°C., specific gravity 1.96, and rhombic (α-rhombic) sulfur, with the melting-point 112.8°C., specific gravity 2.07; valence 2, 4 or 6. Symbol: S.

summation

The act of summing or adding up two or more numbers, quantities or expressions.

summation of heat

See *Hess' law of constant heat summation.*

summation sign

The Greek letter *sigma,* written as Σ. The sum of the first to the *n*th terms of a series or set of numbers, $a_1, a_2, a_3 \ldots a_n \ldots$, is written $\sum_{i=1}^{n} a_i$ or $\sum_{i=1}^{n} a_i$, and the sum of an infinite number of terms is written $\sum_{i=1}^{n} a_i$ or $\sum_{1}^{n} a_i$ or just $\sum a_i$.

super-

A prefix meaning *above, beyond,* also *higher* or *over-*.

superconductivity

With reference to electricity, the extremely high conductivities possessed by certain substances at temperatures very close to absolute zero.

supercooling

The cooling of a liquid below its freezing point without causing solid matter to separate.

superheated steam

Steam heated above the temperature at which it can condense (100°C, i.e., 212°F.).

superheating

The process of increasing the temperature of a substance or phase above the point necessary to produce or complete a phase change. (Cf. *superheated steam.*)

supernatant liquid

The clear liquid lying above a residue that has settled to the bottom of a container.

supersaturated solution

A solution containing more solid than the quantity that can be dissolved normally at a given temperature, and which has not separated from it, although it is apt to precipitate or crystallize readily out of this unstable system. It can be produced by the rapid cooling of a saturated solution, solution conditions still obtaining.

supersaturation

(1) The state of containing more of a substance, force or prop-

erty than is required to produce the state of *saturation.*—(2) The act or process of bringing about this excess content.

supersaturation

In meteorology, the condition that exists in a given space when it contains more water vapor than is required to produce *saturation,* i.e., when its temperature is lower than that required for the phenomenon known as *condensation.*

supersonic

Of or pertaining to the speed of an object moving at a speed greater than mach 1 relative to surrounding fluid. (Technically, *supersonic* is generic and includes the concept of *hypersonic,* but some writers use it restrictively for relative speeds between mach 1 and mach 5.)

supersonic speed

Any speed exceeding the velocity of sound (about 760 m.p.h. at normal sea-level conditions). In aeronautical engineering, any speed that exceeds Mach 1.3.

supersonic testing

See *ultrasonic testing.*

supplement

The angle which has to be added to a given angle to produce a straight angle.

supplementary angles

Two angles which total 180°.

surd

The root of a number, or the sum of the roots of two or more numbers, which can merely be determined approximately. It is designated as *quadratic, cubic, quartic,* etc. surd, according as the index of the radical sign is 2, 3, 4, etc. Thus $\sqrt{2}$ is a quadratic surd, $\sqrt[3]{7}$ is a cubic surd, etc. The word is occasionally used as a synonym for *irrational number.*

surface

The two-dimensional configuration (having length and breadth, but no thickness) which bounds or encloses a solid. An area.

surface active agent

A substance added to a liquid in order to increase (occasionally to reduce) its spreading, wetting and similar properties.

surface boundaries

Straight lines or curves.

surface density of (electric) charge

The amount of charge on a unit of surface area.

surface energy
In physics, the energy per unit area that must be expended in order to increase a surface.

surface film
(1) A thin, usually monomolecular, layer of a substance on the surface of a liquid in which it is insoluble; especially a layer of oil on water.– (2) Any thin layer of a fluid on a solid produced by adsorption.

surface of revolution
The surface which is the locus of a line that revolves about a fixed axis.

surface tension
A phenomenon peculiar to the surface of liquids, in which the surface molecules appear to have a greater cohesion to one another than do the molecules in the body of the liquid, so that the surface acts like a stretched elastic film. Its magnitude depends on the nature of the liquid, its purity, its temperature, and on the kind of gas that is contiguous to it, and possibly on other factors as well. Defined as the force in dynes acting on a line one centimeter in length lying in the surface of the liquid.

surfascope
An instrument used for the optical measurement of roughness of plane and cylindrical metal surfaces.

surge
(1) A transient variation in the current and/or potential at a point in an electric circuit.– (2) In meteorology, a general change in barometric pressure apparently superposed upon cyclonic and normal diurnal changes.

surveyor's transit
An instrument for determining *altitude* and *azimuth,* consisting basically of a small telescope mounted on an axis which is in turn supported on a second axis at right angles to the first. Also called *theodolite.*

susceptibility
See *magnetic susceptibility.*

suspension
A term in chemistry, referring to the diffused, floating state of undissolved or partly dissolved particles in a liquid.

symmedians
The isogonal conjugates of the *medians* of a triangle.

symmetric figures
Two or more figures, the corresponding parts of which are respectively equal, but arranged in reverse order.

symmetrical equation
An equation is said to be symmetrical with reference to *a* and *b* if the interchanging of *a* and *b* leaves the equation unchanged.

symmetrical expression
In mathematics, an expression is said to be symmetrical in two or more variables, when two of these variables can be interchanged without altering the value of the expression. Thus $\sin(A + B)$ or $\sin(B + A)$; $\tan A + \tan B$ or $\tan B + \tan A$.

symmetry
The capability of being divided into two identical or like halves. Often qualified as symmetry with respect to a point, line or plane.

symmetry element
A general name for *axes of symmetry, planes of symmetry,* and centers of *inversion.*

sympathetic reaction
An induced reaction.

sympathetic vibration
Vibration due to resonance.

synchrocyclotron
A *cyclotron* in which the radiofrequency of the electric field is frequency modulated to permit the acceleration of particles to relativistic energies.

synchronism
The phase relationship between two or more periodic quantities of the same period when the phase difference between them is zero.

synchronize
To make two or more events, operations, reactions or processes occur at the proper time with respect to each other.

synchronous
(1) Taking place at the same time.—(2) Having the same period and phase.

synchrotron
An apparatus, originated by E. M. McMillan in 1944, to accelerate either electrons or protons. The electrons are started at a large radius, either by a preliminary acceleration in a *betatron* from which they are injected into the synchrotron,

or by direct injection from an *electron gun;* in the synchrotron they are further accelerated so as to acquire even more kinetic energy. (See also *Alternating Gradient Synchrotron.*)

syneresis

The contraction of gel accompanied by the exudation of a liquid.

synoptic

Presenting a general view. Said of a collection or set of things or parts so arranged as to exhibit the whole of the principal parts in a general view.

system

In mathematics, a set of quantities which possess some property in common, or a set of principles which embody some methodical plan, scheme or order.

system of lines

See *family of lines.*

synthesis

To produce a compound from simple elements or constituents, or from a simpler compound.

synthetic

Artificially prepared from the component elements or simpler materials; not obtained directly from natural sources.

synthetic division

A simplified method of dividing a polynomial in one variable, x, by x minus a number that may be positive or negative.

syntony

The condition in which two oscillating circuits have the same resonant frequency.

T

Ta

The symbol for the element *tantalum.*

table

A systematically compiled list of data which provides easily accessible information and a convenient source of reference.

table salt

Sodium chloride (Na^+Cl^-).

tabular differences

In *linear interpolation,* the varying differences between successsive functional values.

tabular interval

In *linear interpolation,* the constant difference between successive values of the independent variable.

tachometer

An instrument for measuring linear or angular speeds in revolutions per minute, or the rates of flow in liquids. (Also called *tachymeter.*)

tacnode

A point on a curve where the curve crosses itself to form two branches which have a common tangent and do not stop. Also called *double cusp* or *point of osculation.*

tan

Short for *tangent.*

tan-chord angle

The angle between a tangent to a circle and a chord passing through the point where the tangent touches the circle.

tangent ***(adj.)***

Touching in one point; having one point in common.

tangent ***(noun)***

One of the trigonometrical functions or ratios; it is the length of the vertical side of a right angle triangle divided by the length of the horizontal side of the right angle. If this horizontal side is made equal to the radius of a circle of radius = 1, then the trigonometrical tangent of angle A becomes

$$\frac{\text{opposite side}}{\text{adjacent side}} = \frac{\text{vertical side}}{\text{radius}} = \frac{\text{vertical side}}{1} = \text{vertical side} =$$

$$\tan A = \frac{\sin A}{\cos A.}.$$

tangent

Short for *tangent line.*

tangent curve

A graphic representation of the trigonometric tangent function or ratio, i.e., the graph of $y = \tan x$; it is one of the *trigonometric curves.*

tangent plane

A plane which contains only one *element* of the cone or cylinder, no matter how far it is extended, is said to be tangent to the cone or cylinder.

tangent spheres

Two spheres which touch a plane at the same identical point.

tangent to a circle

A straight line which has one point and one point only in common, or in contact, with the circumference of a circle. This straight line is at right angles to the radius drawn from the centre of the circle to the point of contact P on the circumference.

tangent to a sphere

A line or a plane that has only one point in common with the sphere.

tantalum

An element in Group V of the periodic table, discovered by Ekeberg in 1802. Atomic number 73, atomic weight, 180.95, melting-point above 2900°C., specific gravity 16.6, valence 5. Symbol: Ta.

target

(1) Any surface or object exposed to irradiation or to bombardment by particles.—(2) In an x-ray tube, the anode or anticathode on which the streams of electrons called *cathode rays* are focused, and which as a result of this bombardment emits the x-rays.

tautochrone

An inverted cycloid representing the force of gravity.

tautomer

A *tautomeric substance.*

tautomeric substance

A substance which has the property called *tautomerism.*

tautomerism

The property of existing in an equilibrium between two isomeric forms and being able to exhibit reactions corresponding to either form.

Tb
The symbol for the element *terbium.*

Tc
The symbol for the element *technetium.*

Te
The symbol for the element *tellurium.*

technetium
An artificially produced element, atomic number 43, mass number of the longest-lived isotope 99. This element does not occur in nature, only radioisotopes of it are known. It was first discovered by C. Perrier and E. Segré, in 1937, in their studies of molybdenum irradiated cyclotronically by 8-Mev deuterons. Later Segré and G. T. Seaborg produced Tc^{99} by deuteron and neutron bombardment of molybdenum. This isotope occurs also as a fission product of U^{235} and plutonium. Symbol: Tc.

technology
The practice and terminology of an applied science having commercial value.

tectonics
See *geotectonic geology.*

tele-
A prefix indicating *at, from* or *over a distance.*

telemeteorography
The science and technique of constructing and operating meteorographic instruments. The recording devices are at considerable distances from the recording or measuring units.

telemeter
An electronic instrument that senses and measures a quantity, as that of speed, temperature, pressure, or radiation, then transmits radio signals to a distant station, where, indicated or recorded, they are interpreted by code into the quantity provided for.

telemetering link
In telemetry, the means by which the *measurand transmitter* transmits the received electrical impulses to the *recorder;* it usually consists of a radio receiver and transmitter, but light beams or cables may be used as well.

telemetering pickup
In *telemetry,* an apparatus that converts instrument readings or mechanical actions into electrical impulses which it feeds into the *measurand transmitter.*

telemetering system

A system for the measuring, transmission and reception, recording or integration of instrument readings or data by electronic means. Such system is composed basically of *telemetering pickups,* a *measurand transmitter,* the *telemetering link,* and finally the *recorder* or *display.*

telemetry

The technique of transmitting or reading instrument recordings from a distance.

telephony

The reproduction of spoken words or other sounds at a distant point by electrical or other means, with or without connecting wires.

telephotometry

The science and technique of measuring light intensities on objects at considerable distances.

telescope

A generic term for instruments designed and employed to produce a magnified image of distant objects.

tellurium

A metallic element in Group VI of the periodic table, discovered by Reichenstein in 1782. Atomic number 52, atomic weight 127.61, melting-point 450°C., boiling-point 1390°C., specific gravity 6.25, valence 2, 4 or 6. Symbol: Te.

temperature

(1) Qualitatively, that property which determines the direction of heat flow from one body to another in contact with it. Quantitatively, the temperature of a body is defined as the average kinetic energy of random translation of its molecules.— (2) The measure of such property or thermal state on some arbitrarily chosen numerical scale.

temperature coefficient

A factor, expressed as a numerical value, which indicates the effect of changes in temperature on some given property of a substance.

temperature conversion

To convert Fahrenheit to centigrade, subtract 32, multiply by 5 and divide by 9. To convert centigrade to Fahrenheit, multiply by 9, divide by 5, and add 32.

temperature detector

Any instrument that may be used to measure the temperature of a body or of some particular part of a body, on the ground of any physical property that is dependent on temperature.

temperature gradient
(1) The difference in temperature between the two ends of a conductor of heat, per unit of length.—(2) In meteorology, the rate of change of temperature with distance in any given direction at any point.

temperature of combustion
The highest temperature reached during the combustion.

tempering
A method of removing the brittleness and internal strain resulting from hardening a metal by quenching. It is essentially a reheating process, either in air or in a liquid.

temporary emulsion
An *emulsion* which break down rapidly, owing to the high surface tension of one or both of the liquids.

temporary hardness
A form of hardness of water that can be removed by boiling. It is caused by the presence of bicarbonates of calcium and magnesium.

temporary magnet
A substance which is readily magnetized in a magnetic field, but retains very little magnetism after being removed from that field.

ten
A figure followed by one digit, denoting an integer multiple of 10.

tenacity
The maximum *tensile stress* which a material is capable of withstanding; *tensile strength.* (See also *specific tenacity.*)

tensile strain
Strain produced by tension; the ratio of the increase in the length of object acted upon to its original length.

tensile strength
The resistance of a material to a force which tends to pull it apart; it is measured in pounds per square inch and is calculated by dividing the load in pounds required to pull the material apart by its cross-sectional area expressed in square inches. (Also called *ultimate tensile stress* or *maximum stress.*)

tensile stress
Stress produced by tension; the ratio of the force to the cross-sectional area.

tensile test
In metallurgy a test in which specimens are subjected to an increasing tensile load until they fracture.

tensimeter

An instrument for measuring differences in vapor pressure.

tensiometer

An apparatus for measuring *surface tension.*

tension

Stress or strain caused by pulling.

tensor

The ratio of the lengths of two vectors.

tenth

Any fraction the denominator of which is 10. In decimal notation, the first digit following the decimal point.

ter-

A prefix meaning *three times* or *threefold.*

terbium

A trivalent metallic element, one of the rare earth group of the periodic table, discovered by Mosander in 1843. Atomic number 65, atomic weight 158.93. Symbol: Tb.

term

Each individual part of an expression or sum, which stands alone or is connected to other parts of an expression by a plus (+) or minus (−) sign. Thus in the expression $4 + ab - 3cd + \frac{a}{b}$, the terms are $+4$, $+ab$, $-3cd$, and $+\frac{a}{b}$.

terminal

A point on an electric circuit by which it can be connected to another circuit or piece of apparatus.

terminal line

In trigonometry, the line the revolution of which about a point forms an angle. (Cf. *standard position of an angle.*)

terminal side of an angle

See *standard position of an angle.*

termolecular

See *trimolecular.*

terms of a fraction

The numerator above and the denominator below the horizontal line. Thus a and b are terms of the fraction $\frac{a}{b}$.

ternary

In general, composed of three elements or components.

ternary alloys

An alloy composed of three principal elements.

ternary compound
A combination of three different elements, i.e., a compound each molecule of which consists of three different atoms.

ternary diagram
A diagram showing the constitution of a system containing three components.

terrestrial magnetism
The magnetism observable everywhere owing to the fact that the earth itself is a permanent magnet, the magnetic poles of which lie near the geographical poles.

Tesla coil
An air-core transformer having a few turns of heavy wire as primary and many turns of fine wire as secondary; it is capable of electromagnetic radiation at a very high frequency or potential.

test paper
An absorbent paper, impregnated with an indicating solution, e.g., *litmus paper,* used in chemical analysis.

test sample
A portion of a material taken for testing.

tetartohedral
Concerning crystals. Having only one-fourth of the symmetry elements possible to the crystal system in which the crystal belongs.

tetra-
A prefix meaning *four.*

tetrad
An atom or atom group possessing a valence of four.

tetragonal system
The crystallographic system in which all the forms are referred to three axes at right angles; two are equal and are taken as the horizontal axes, while the vertical axis is either longer or shorter than these.

tetrahedron
A solid bounded by four triangles. Also called *triangular pyramid* (any one of the four triangles may be regarded as the base).

tetramolecular
Pertaining to, consisting of or involving four molecules. Also called *quadrimolecular.*

tetramolecular reaction
A reaction in which four molecules participate and undergo a

change. Also called *quadrimolecular reaction* and *reaction of the fourth order.*

tetramorphic

Possessing the property of being able to crystallize in four forms or systems.

tetramorphism

The property of being able to crystallize in four systems or forms.

tetramorphous

See *tetramorphic.*

tetrode

A four-electrode vacuum tube containing an anode, a cathode, a control electrode, and one additional electrode ordinarily in the nature of a grid.

Th

The symbol for the element *thorium.*

ThA, ThB, ThC, ThC′, ThC″, ThD

The symbols for *thorium A, thorium B, thorium C,* etc.

thallium

A metallic element in Group III of the periodic table, discovered by Crookes in 1861. Atomic number 81, atomic weight 204.39, melting-point 303°C., boiling-point 1525°C., specific gravity 11.85, valence 1 or 3. Thallium is fissile, but its fission requires too high an energy input, and is not capable of sustaining a fission chain reaction; hence it cannot be considered at the present time as a practical source of nuclear energy. Symbol: Tl.

theodolite

A *surveyor's transit* (q.v.).

theorem

In mathematics a proposition to be proved; in analysis and algebra, a rule or law.

theoretical chemistry

The science of the laws of chemical changes.

theory

An explanation of certain phenomena i.e. a series of hypotheses found to be consistent with one another and with observed phenomena but not yet proved by experiment.

therm

A unit of heat equal to 100,000 B.t.u. (British thermal units).

thermal

Produced by or involving heat or temperature.

thermal analysis
A method of determining the transformation point in a substance by recording the temperature at which *endothermic* and *exothermic* effects take place on heating or cooling.

thermal capacity
The number of calories required to raise the temperature of a body 1 °C, or the quantity of heat released when the temperature of a body is lowered 1 °C, in both cases without change of state. Also called *heat capacity.*

thermal conductance
The time rate of heat flow through a unit area of a body, of given size and shape, per unit temperature difference. Also called *heat conductance.*

thermal conduction
The process of heat transfer through material medium in which heat is transmitted from particle to particle without gross displacement of the particles. Also called *heat conduction.*

thermal conductivity
The rate of transfer of heat along a body by conduction; it expresses how well a substance is able to conduct heat, and is measured in calories flowing per second across a cubic centimeter of the substance having a temperature difference of 1 °C on opposite faces (or similarly for other systems or units of measurement). Also called *heat conductivity.*

thermal conductor
A material which transmits heat readily by means of conduction. Also called *heat conductor.*

thermal contraction
The decrease in volume taking place during cooling.

thermal convection
Transmission of heat through a fluid (liquid or gas) by means of fluid currents.

thermal diffusion method
A method of the separation of various isotopes of an element, the basic principle being that when the substance is gasified, the molecules of the different isotopes, differing in mass, diffuse along different paths and accumulate at different places.

themal dissociation
A reversible chemical change induced by heat.

thermal electromotive force
The *electromotive force* generated when the junction of two dissimilar metals is heated.

thermal expansion

The increase in dimension on heating. It is usually expressed as the *coefficient of linear expansion* which is the ratio of the change in length per degree to the length at 0°C., or as the *coefficient of volume thermal expansion,* meaning the ratio of change in volume per degree to the volume at 0°C.

thermal fatigue

Failure resulting from rapid cycles of alternate heating and cooling.

thermal fission

Nuclear fission produced by *thermal neutrons.*

thermal gradient

The variation in temperature with distance.

thermal ionization

Ionization due to high temperature.

thermal neutron

A neutron slowed down by collisions with light atoms (e.g. hydrogen) to the speed of gas molecules at room temperature (about 0.025 ev at 15°C.). Also referred to as *thermal-velocity neutron.*

thermal number

In metallurgy, a value applied to measure the efficiency of metal cutting.

thermal ohm

The unit of *thermal resistance;* one thermal ohm requires a temperature difference of 1°C to drive heat at the rate of 1 *watt.*

thermal radiation

Radiation produced by the action of heat on molecules or atoms. Its frequency ranges between the extremes of infrared and ultraviolet. Commonly known as *heat.*

thermal resistance

Resistance to the flow of heat. The unit of this resistance is the *thermal ohm.*

thermal resistivity

A measure of the property of a material which resists the flow of heat within it. It is usually expressed in *thermal ohms* per cubic centimeter. (Also called *specific thermal resistance.*)

thermal shock

In metallurgy, stress developed by rapid and uneven heating of material.

thermal sintering
In metallurgy, molding by sintering, without prior *compacting.*

thermal stability
Resistance of a material to drastic changes in temperature.

thermal stresses
Stresses in metal caused by a disuniform distribution of temperature.

thermal transmission
The passage of heat by means of the combined effects of conductivity, radiation and convection.

thermal value of a chemical reaction
The amount of heat absorbed or released in an endothermic or exothermic reaction.

thermal velocity
See *thermal nutron.*

thermal-velocity neutron
See *thermal neutron.*

thermel
A thermoelectric thermometer, which may be a single *thermocouple* or a series of thermocouples (*thermopile*).

thermic
Pertaining to heat.

thermion
An ion, either positive or negative, emitted from a heated object. (Negative thermions are called *thermoelectrons.*)

thermionic current
Any current due to directed movements of *thermions,* such as the flow of emitted electrons from the cathode to the plate in a *thermionic tube.*

thermionic emission
The emission of electrons or ions due directly to the temperature of the emitter, the rate of emission increasing rapidly with the increase in temperature.

thermionic tube
A vacuum tube in which the cathode is heated in order to emit electrons (*thermoelectrons*); all or some of these electrons are collected by an anode. Additional electrodes are usually provided, to control the passage of the electrons from the cathode to the anode. Also called *thermionic vacuum tube* (in Great Britain, *thermionic valve*).

thermionic vacuum tube

See *thermionic tube.*

thermionic valve

See *thermionic tube.*

thermionics

The science dealing with the emission of electrons by hot bodies and the behavior of these electrons in a vacuum or near-vacuum.

thermistor

A temperature-sensitive resistance element of metallic oxide, whose electrical resistance decreases with increase in temperature. It has a temperature coefficient considerably greater than that of a pure metal.

thermit

A mixture of various oxides, and aluminum powder, with or without other alloying metals, all in a state of fine division used in the *aluminothermic process.*

thermit reactions

Exothermic, self-propagating processes in which finely-divided aluminum powder is used to reduce metal oxides to free metals.

thermochemical equation

An equation which sets out the changes in heat energy together with the chemical changes involved in a reaction.

thermochemistry

The science dealing with thermal changes taking place in chemical reactions.

thermocouple

A device for measuring temperature in which a pair of electric conductors are so joined as to produce a *thermoelectric effect.* When the ends of two dissimilar metals or any pair of electrical conductors, are joined, a thermoelectric effect is produced which causes an electric current to flow around the circuit formed by the two wires, the magnitude of the current being proportional to the difference in temperature between the hot and cold junctions. The points of contact are known as *thermojunctions,* the joint at the heated end as the *hot junction,* whilst the unheated joint is the *cold junction.* In general, thermocouples can be divided into two types: *precious metal,* and *base metal couples,* the former being used for high temperatures where resistance to oxidizing conditions is involved, the base metal couples being used for lower temperatures. The thermocouple transforms heat energy into electric energy.

thermodynamic potential
The measure of the energy level of a system representing the amount of work which can be obtained when the system undergoes a change.

thermodynamics
The branch of physics which deals with the quantitative and qualitative interrelations and interconvertibility of heat and other forms of energy and especially with the conversion of heat into mechanical energy.

thermoelastic
Pertaining to the thermodynamics of elastic processes.

thermoelasticity
The variation of the temperature of a material when stress is applied.

thermoelectric effect
The increase in *e.m.f.* due to a difference of temperature between two junctions of dissimilar conductors in a circuit. (Also called *Seebeck effect.*)

thermoelectric element
One of the two dissimilar conductors constituting a *thermocouple.*

thermoelectric force
The *electromotive force* due to a *thermoelectric effect.*

thermoelectric inversion
The decrease in *e.m.f.* of a *thermocouple* when the temperature increases above a certain degree.

thermoelectric method
A method of determining the *critical points* in steel by plotting the changes in electrical properties against the changes in temperature.

thermoelectric power
Quantity measured by the *electromotive force* produced by a thermocouple for unit difference of temperature between the two junctions.

thermoelectric pyrometer
An instrument consisting of a system in which there are three main components, namely: (1) the *thermocouple,* (2) the extension leads, and (3) the measuring instrument.

thermoelectric series
A series of metals arranged in the order of their thermoelectric powers.

thermoelectricity
Electricity generated by heating the junction between two dissimilar conductors. (See *thermoelectric effect.*)

thermoelectron
A negative *thermion,* i.e., a negative ion emitted from a heated object.

thermoelement
A device consisting of a thermocouple and a heating element arranged for measuring small currents.

thermogalvanic corrosion
A form of electrochemical corrosion in normally stable systems owing to temperature differences.

thermograph
A recording thermometer.

thermojunction
The point of contact of a pair of conductors forming a *thermocouple.*

thermolabile
Susceptible to destruction by heat.

thermoluminescence
Luminescence exhibited by certain substances, as fluorite, when heated, but of a wavelength distinctly shorter than that of the incandescence corresponding to the exciting high temperature.

thermomagnetic effect
The difference in magnetic properties which is caused by heat.

thermometer
An instrument for measuring temperature. Its forms and varieties in most common and general use are based on the change in volume (contraction and expansion) of mercury or some other suitable substance with the changes in temperature.

thermomilliammeter
An instrument for measuring very small electric currents.

thermoneutrons
Neutrons having kinetic energies up to about 10^2 ev.

thermonuclear reaction
A fusion of hydrogen in which the energy necessary is provided by colliding particles that have kinetic energy by virtue

of their thermal energy and agitation. This can occur only at temperatures of millions of degrees, and probably takes place in the sun.

thermopile

A number of *thermocouples* connected in series so that their electromotive forces are additive.

thermoplastic

Having the property of softening when exposed to heat.

thermoplastic solid

A substance which is adequately rigid at normal temperatures and under normal conditions of stress, but is capable of deformation under heat and pressure.

thermosetting plastic

A plastic which is made hard and inelastic by heat.

thermosphere

The *ionosphere* considered as a region of temperature variation from −28°F to several thousand degrees.

thermostat

An automatic device for maintaining an enclosure at a constant temperature. It often consists of some expansible element which automatically cuts off the heat supply (gas or electric) when the temperature exceeds the required value.

thermostress

Stress produced by differences in temperature or coefficients of thermal expansion.

thio-

A prefix indicating the presence of divalent sulfur, usually replacing oxygen.

third proportional

If three quantities a, b, c, are in continued proportion as: $\frac{a}{b} = \frac{b}{c}$ or a : b :: b : c, then the third or last quantity c is said to be a third proportional to a and b.

thixotropic fluid

A fluid which exhibits the property of *thixotropy*.

thixotropic gel

A gel which has the tendency to become liquid when it is shaken or otherwise exposed to vibrations, and to revert again to gel form after the cessation of the shaking or removal of the vibratory force.

thixotropy

(1) That property of a body by virtue of which its consistency

is reduced by previous deformation. (2) The rate of change of viscosity of a suspension as the rate of shearing strain varies. (3) See *thixotropic gel.*

Thomson coefficient

The ratio of Thomson e.m.f. in a metal to the corresponding temperature difference. (See *Thompson effect.*)

Thomson effect

The e.m.f. produced by a difference of temperature between two portions of one and the same conductor, and the liberation (or absorption) of heat taking place when a current flows from a hotter to a colder portion of the same material. (Also called *Kelvin effect.*)

thorium

A radioactive metallic element in Group IV of the periodic table, discovered by Berzelius in 1828. Atomic number 90, atomic weight 232.05, melting-point 1850°C., boiling-point about 3500°C., specific gravity 11.30-11.75, valence 4. Thorium isotopes are radioactive. Natural metallic thorium is capable of undergoing nuclear fission upon capture of a fast neutron, but yields no satisfactory chain reaction, so that thorium has not been considered as reactor "fuel" for the production of nuclear energy. Symbol: Th.

thorium A (ThA)

A radioactive isotope of polonium ($_{84}Po^{216}$).

thorium B (ThB)

A radioactive isotope of lead ($_{82}Pb^{212}$).

thorium C (ThC)

A radioactive isotope of bismuth ($_{83}Bi^{212}$).

thorium C′ (ThC′)

A radioactive isotope of polonium ($_{84}Po^{212}$).

thorium C″ (ThC″)

A radioactive isotope of thallium ($_{81}Tl^{208}$).

thorium D (ThD)

The stable lead isotope ($_{82}Pb^{208}$) which is the end product of the *thorium series.*

thorium emanation

See *thoron.*

thorium family

See *thorium series.*

thorium lead

See *thorium D.*

thorium series

A natural radioactive series, which starts with the thorium

isotope $_{90}Th^{232}$ and ends with the stable lead isotope $_{82}Pb^{208}$ (or thorium D). It is called *thorium series* after its parent which is also its longest-lived member, or $4n$ series because the mass number of each of its members is an integer multiple of 4. Also referred to as *thorium family* and as *4n series* or *family*.

thorium X (ThX)
A radioactive isotope of radium ($_{88}Ra^{224}$).

thoron (Tn)
A radioactive isotope of radon ($_{86}Rn^{220}$).

thousand
A digit followed by three digits and denoting an integer multiple of the third power of ten (i.e., 1000).

thousandth
A fraction the denominator of which is 1000; in the decimal notation, the third digit after the decimal point.

three-dimensional configuration
A solid, since it has length and width and thickness.

three-leafed rose
The graph of the equation $r = a \sin 3\theta$ or $r = a \cos 3\theta$. (Cf. *n-leafed rose.*)

three-space
A synonym for *three-dimensional.*

threshold
The point at which an effect is first produced, observable or otherwise indicated.

thrust
The force which drives a rocket missile or vehicle in a direction normally parallel to the longitudinal axis of the rocket, measured in terms of the fluid mass flow of its propellant in pounds per second multiplied by the effective exhaust velocity (relative to the airframe) in feet per second divided by the gravity of acceleration (at the equator 32.2 feet per second).

thulium
A very rare, trivalent metallic element, one of the rare earth group of the periodic system, discovered by Cleve in 1879. Atomic number 69, atomic weight 168.94. Symbol: Tm.

Ti
The symbol for the element *titanium.*

tin
A metallic element in Group IV of the periodic table. Atomic number 50, atomic weight 118.70, melting-point 231.9°C.,

boiling-point 2260°C., specific gravity 7.25 (crystalline) and 5.75 (amorphous), valence 2 or 4. Symbol: Sn.

titanium

A rare metallic element in Group IV of the periodic table, discovered by Gregor in 1790. Atomic number 22, atomic weight 47.90, melting-point 1800°C., boiling-point above 2800°C., specific gravity 4.5, valence 3 or 4. Symbol: Ti.

titer

The standard of strength of a solution, as determined by *titration.* In general the amount of one substance required to react with a given volume of another substance, as the amount of acid required to neutralize a definite amount of base. In the analysis of fats and oils, the titer refers to the temperature at which the fatty acids obtained by alkaline hydrolysis of the fat resolidify.

titration

The volumetric determination of the strength of a solution of an acid, base, oxidizing or reducing agent by delivering from a burette an appropriate reagent of known strength using a suitable indicator where necessary.

Tl

The symbol for the element *thallium.*

Tm

The symbol for the element *thulium.*

Tn

The symbol for the element *thoron.*

tolerance

The range between the upper and lower limits of a specification, e.g. as regards composition or dimensions; it is often expressed as plus or minus the specified quantity.

tomic energy

A term proposed by F. Soddy in his book, *The Story of Atomic Energy* (Nova Atlantis Publishing Co., Ltd., London, 1949), for the energy released by nuclear fission, to distinguish it explicitly from "atomic energy" as meaning the energy latent in the atom before its release. Analogously, Soddy proposes the term *"tomic bomb"* to replace the expression "atomic bomb."

tomography

A division of *radiography* dealing with the photography of a particular plane in an object while leaving out undesired detail in other planes.

tool
Any implement by means of which mechanical operations are performed.

top view
A representation of a point or object in space by projection to a horizontal projection plane, i.e., a view in which the lines of sight of the observer are vertical. Also called *plan view*.

topochemistry
The study of reactions which occur only at certain definite regions in a system.

topology
A branch of geometry which deals with topological properties, i.e., properties that are unaffected by changes in size and shape.

toric
Shaped like a *torus* or a section of one.

toroid
(1) A synonym for *torus*. (2) A synonym for *toric*.

torque
A twisting or wrenching effort. It is the product of a force and its *moment arm*, i.e., it equals the force multiplied by the center of rotation at which it is exerted. It is a vector quantity and is measured in pound-feet, centimeter-dynes, or some similar combination of units of force and of distance.

torquemeter
An instrument constructed for accurate measurement of *torque* at high speeds—up to about 17,000 r.p.m.

Torricelli tube
An early, and once universally used, name of the mercurial barometer.

torsiometer
An instrument for measuring the amount of power transmitted by a rotating shaft.

torsion
(1) Strain set up in a body by two forces acting about the same axis of rotation, in parallel planes, but in opposite directions. (2) In general, the deformation caused by twisting.

torsion test
Process of producing a strain in a material by twisting it until it "gives."

torsional balance

An instrument for measuring force fields by causing them to be opposed by a known force.

torsional moment

See *moment of torsion.*

torsional wave

A wave motion consisting of the propagation of a twist through a given medium or substance.

torsometer

An instrument for studying the elastic behavior of solids under torsion, especially by optical methods.

torus

A surface shaped like a doughnut with a hole in it, generated by the spatial rotation of a circle about an axis which is in its plane, but does not intersect it.

total base

The analytical value for the sum of the positive elements, Na, K, Ca, Mg; total mineral cationogens; sometimes called fixed base.

total internal reflection

Reflection which takes place within a substance because the *angle of incidence* of light striking the boundary surface is in excess of the *critical angle.*

total luminous flux

The total light emitted by a light source in all directions.

total radiation pyrometer

An instrument for measuring high temperatures, consisting essentially of a *thermocouple* in a cylindrical tube, the temperature of which is raised by a given temperature to a point of equilibrium, depending on the surroundings; the electromotive force of the thermocouple can be calibrated for a range of temperatures. (See also *radiation pyrometer, pyrototal radiation pyrometer.*)

total reflection

See *total internal reflection.*

trace

In chemistry, an amount of a substance too minute to be measured or present in a negligible concentration.

trace

In graphic or descriptive geometry, the point at which a straight line cuts the horizontal or vertical plane; also the line in which a plane cuts these two planes.

trace

With reference to instruments, the record made by any self-registering instrument.

trace element

In geology, geophysics, and related sciences, a general designation of the elements which are present in the earth's crust in minor amounts, and comprises all elements except for aluminum, calcium, iron, magnesium, oxygen, potassium, silicon and sodium. (Also referred to as *accessory elements* and *minor elements*.)

tracer

An atom, usually of a radioactive isotope, which can be traced and observed so as to discover the paths, sites or mechanism of a process under investigation. *Radioactive tracer technique* consists of adding to a stable element or compound a minute quantity of one of its own radioactive isotopes and subsequently introducing this *"labeled"* substance into a living organism (human, animal or plant) or into organic or inorganic compounds. The radioisotope used to "label" or "tag" the substance, being an isotope of the latter, is chemically identical with the rest and partakes in the same reactions, changes and processes as every other atom of the substance, but can be detected by its characteristic radiation, either with the aid of *radiation survey meters* or by the *radioautograph technique;* thus the behavior, progress, changes, etc., of the substance introduced can be traced, observed and studied. While in the great majority of the cases radioactive tracers are used, occasionally stable isotopes (*"non-radioactive tracers," "inactive tracers"*), too, are employed; these are not so easy or convenient to trace since they can be detected by their difference in weight (mass-spectrographically) only, but they are valuable in instances where the use of a radioactive isotope might damage the organism under study and observation. The tracer technique was first used by Hevesy in 1923 who used a natural radioactive isotope of lead for plant-physiological studies.

trans-

A prefix indicating the presence of *cis-trans isomerism* in a substance.

transcendental curve

Any plane curve except for the *algebraic curves* (e.g., the graph of a trigonometric function).

transducer

An electronic power-transforming device.

transfer factor, transfer ratio

See *propagation ratio.*

transformation

In chemistry, a change in form, structure or phase, or in the arrangement of molecules or atoms.

transformation

In mathematics, the change of an algebraic expression to one of a different form, or the replacement of a function or variable or set of variables by another.

transformation temperature

The temperature at which a chemical or physical transformation occurs, or which is produced by such transformation.

transformer

An electric device without continuously moving parts, which by electromagnetic induction transforms electric energy from one or more circuits to one or more circuits at the same frequency, usually with changed values of voltage and current.

transistor

A small, light, compact electronic device for rectification or amplification, designed to replace a vacuum tube and to perform its functions without any appreciable generation of heat and with a minimum requirement of power.

transit

An instrument used for measuring horizontal and vertical angles, for measuring distances by the stadia, and for leveling. The transit consists of a telescope attached to a vernier which may be moved around a graduated circle. There are suitable attachments for controlling the motion of the telescope and the vernier, and for enabling the graduated circle to be made horizontal.

transition

(1) The change of a substance from one phase or state of aggregation to another (e.g., melting, vaporization, etc.) or from one allotropic form to another. (2) In nuclear physics, a sudden change in the energy state of an atom or nucleus produced by the return of an electron, respectively a nucleon, from an *excited state* to a more stable state or energy level. (Cf. *quantum jump.*)

transition elements

The three triads of elements which form Group VIII of the *periodic table,* so called because each such triad is the connecting link between two periods of the system. They are: (*a*) iron, cobalt and nickel (atomic numbers 26-28), (*b*) ruthenium, rhodium and palladium (atomic numbers 44-46), (c) osmium, iridium and platinum (atomic numbers 76-78).

transition lattice

An intermediate, unstable crystallographic configuration, that forms during solid metal reactions, such as precipitation from solid solution and eutectoid decomposition.

transition metal

See *transition element.*

transition point

(1) In general, the point at which a *transition* occurs. (2) A synonym for *transition temperature.*

transition temperature

The temperature at which a substance passes from one allotropic form into another. Also called *transition point.*

translation

In mathematics, specifically in analytic geometry, the replacement of the coordinates of a point or points by coordinates in a system the axes of which are parallel with those to which the replaced coordinates were referred.

translatory motion

A motion in which all straight lines of a system retain their direction without any change.

translucent

Permitting the passage of light partially.

transmutation of elements

The change of the atoms of an element or isotope into atoms of another element or isotope. The process occurs in nature spontaneously in *radioactive* substances, and may also be induced artificially by nuclear bombardment by subatomic particles creating artificial *radioisotopes.* (The term, *transmutation of elements,* was used by the alchemists of antiquity and the Middle Ages to refer to the change of one metal into another, most notably lead into gold, and later regarded by science as an impossible quest, until the discoveries of the twentieth century proved it to be not only possible, but a very useful and practical undertaking.)

transparent

Permitting the passage of light fully and freely. Opposed to *opaque.*

transparent glass

Glass having no apparent diffusing properties. Varieties of such glass are sometimes referred to as flint, crystal, clear.

transpose

To change the sign of a term in an equation and transfer it from one side to the other.

transposition

The transferring of terms from one side of an equation to the other. The sign of the terms transferred must be changed.

transsonic speed

Any speed between *subsonic* and *supersonic* velocities; thus, in aeronautical engineering, any speed between $M = 0.8$ and $M = 1.3$ (where M is the Mach number).

transuranic elements

Those members of the *actinide group* (q.v.) of the extended periodic table which have an atomic number higher than 92. Transuranic elements do not occur in nature and have to be produced artificially from uranium, either directly or indirectly by successive steps of transmutations. Up to this date, ten transuranic elements have been produced and named, but the discovery and production of further members of the group is expected. The known transuranic elements, with their atomic numbers, are: *neptunium* (93), *plutonium* (94), *americium* (95), *curium* (96), *berkelium* (97), *californium* (98), *einsteinium* (99), *fermium* (100), *mendelevium* (101) and *nobelium* (102).

transversal

A straight line that intersects two or more straight lines.

transverse axis of a hyperbola

The line segment connecting the two vertices of the hyperbola.

transverse wave

A wave in which the direction of displacement at each point of the medium is perpendicular to the direction of propagation, i.e., the particles of the medium in which the wave propagates vibrate in paths perpendicular to the direction of the propagation of the wave.

trapezium

A quadrilateral no two of the sides of which are parallel.

trapezoid
A quadrilateral two sides of which are parallel, and the other two are not.

trapezoidal wave
A waveform of trapezoidal shape.

traveling wave
In electricity, the resulting wave when the electric variation in a circuit takes the form of translation of energy along a conductor, such energy being always equally divided between current and potential forces.

tree
In topology, a *network* which contains no *closed circuit.*

tri-
A prefix meaning *three.*

triad
(1) A group of three elements, with similar characteristics, as iron, cobalt and nickel in Group VIII of the *periodic table.* (2) A term, now obsolete, for a *trivalent* element.

triangle
A figure bounded by three lines, and having three interior angles the sum of which always equals two right angles (180°).

triangle law of vectors
See *polygon method of adding vectors.*

triangle of forces
See *polygon of forces.*

tribasic
Consisting of *tribasic molecules.*

tribasic molecule
A molecule containing three replaceable hydrogen atoms.

tribo-
A prefix indicating *friction* or *rubbing.*

tribocouple
Two chemically dissimilar metals in mutual electrical contact. The friction produced by the mechanical agitation of the two members of the couple results in the flow of an electric current. The power of a tribocouple is the magnitude of the current which it will generate under specified conditions of friction.

triboelectric
Electrifiable by friction.

triboelectric effect
The electric current produced by a *tribocouple.*
triboelectricity
Electricity induced by friction. (See *tribocouple.*)
triboluminescence
Luminescence induced by friction; a phenomenon observed especially in crystals.
triclinic system
A crystallographic system which refers to crystals with three unequal axes, which are not at right angles.
tridimensional
Having three dimensions; existing in all three dimensions.
trigger
(1) A substance which starts a chain reaction. (2) Any short pulse, positive or negative, which can be used to set into motion a chain of events.
trigonal system
See *rhombohedral system.*
trigonometric curve
The graph of a trigonometric function referred to rectangular coordinates (i.e., the *sine curve,* the *cosine curve,* the *tangent curve,* the *cotangent curve,* the *secant curve,* and the *cosecant curve*).
trigonometric equation
An equality of two expressions involving trigonometrical ratios, which is only true for certain definite values of an unknown angle.
trigonometric function
A ratio that depends on an angle, and hence is a function of that angle. The most commonly used trigonometric functions are the sine, cosine, tangent, cotangent, secant and cosecant. They are also called *circular functions,* for they arise naturally in connection with the geometry of the circle where the radius is the hypotenuse of a right angle triangle and the ratio of its sides to themselves or to the hypotenuse are the trigonometrical ratios. In any right angle triangle with the hypotenuse = 1, the sin of the chosen angle A is the length of the side opposite the angle divided by the hypotenuse to the angle. When written, it becomes $\sin A = \frac{a}{r} = \frac{a}{1} =$ the proportion of a to 1 = ratio of a to 1. Therefore the ratio of a to 1 = sin A, which is called a trigonometrical ratio. In a similar

manner the cosine, tangent, cotangent, secant, cosecant, versine, coversine, are trigonometrical ratios and in relation to the chosen angle A are:

$$\sin A = \frac{a}{r} = \frac{a}{1} = a.$$

$$\cos A = \frac{b}{r} = \frac{b}{1} = b.$$

$$\tan A = \frac{a}{b}$$

$$\cot an\ A = \frac{b}{a}$$

$$\sec A = \frac{1}{b}$$

$$\operatorname{cosec} A = \frac{1}{a}$$

vers A = the quantity by which the sine falls short of unity (one) = 1 − a.

covers A = the quantity by which the cosine falls short of unity (one) = 1 − b.

trigonometric identity

Any expression involving a trigonometric function or functions, which is equal for all values of the angles involved. Thus $\tan A = \frac{\sin A}{\cos A}$; $\sin^2 B = 1 - \cos^2 B$; $2 \sin A \cos B = \sin (A + B) + \sin (A - B)$, etc.

trigonometric negative angle

See *negative trigonometric angle.*

trigonometric ratio

See *trigonometric function.*

trigonometry

That branch of mathematics which studies the ratios between the sides of a right triangle, the relations between these ratios (the *trigonometric functions*) and their applications, especially in solving triangles in theoretical as well as practical applications. *Plane trigonometry* deals with plane triangles, *spherical trigonometry* with spherical triangles; *analytic trigonometry* investigates the trigonometric functions by algebraic operations.

trihedral angle

Corner of a polyhedron where only three plane faces meet at a point.

trillion

In the United States and France, a million million (10^{12}); in England and all over Europe, except France, a million times a million million (10^{18}).

trimolecular

Pertaining to, consisting of or involving three molecules. Also called *termolecular.*

trimolecular reaction

A reaction in which three molecules participate and undergo a change. Also called *termolecular reaction* and *reaction of the third order.*

trimorphic

Possessing the property of being able to crystallize in three systems.

trimorphism

The condition of being able to crystallize in three systems.

trimorphous

See *trimorphic.*

trinomial

An algebraic expression having three terms. For example: $x^2 - 2xy + 5y^2$.

triode

A three-electrode vacuum tube containing an anode, a cathode, and a control electrode.

triple-point

That combination of pressure and temperature conditions and magnitudes at which the solid, liquid and gaseous phases or states of a substance exist in equilibrium.

triplicate ratio

When a ratio is compounded with itself twice the result is a triplicate ratio. Thus a : b multiplied twice by a : b becomes $a^3 : b^3$.

trirectangular triangle

A *spherical triangle* containing three right angles.

trisect

To divide into three equal parts.

trisectors

Lines which divide other lines or angles into three equal parts.

tritium

The radioisotope of hydrogen with mass number 3 ($_1H^3$).

triturate

A powder produced from a solid by grinding, usually with the addition of some liquid.

triturate

To grind to a powder, usually with the addition of some liquid.

trituration

(1) The act or process of grinding a solid to a powder, usually with the addition of some liquid. (2) The powder so obtained.

trochoid

A curve which is the locus of a point on the radius of a circle that rolls along a stationary straight line; also, the locus of a point on the radius produced by such motion. A trochoid without loops is called a *curtate cycloid,* one with loops is referred to as a *prolate cycloid.* (Cf. *cycloid.*)

tropism

A directional property of a substance, especially a crystal, dependent on the vector quantities which determine the states of the constituent atoms or molecules.

tropopause

The boundary between the stratosphere and the troposphere.

troposphere

The lower layer of the earth's atmosphere, extending to about 60,000 feet at the equator and 30,000 feet at the poles. The thickness varies with the seasons and with different meteorological conditions. The temperature lapse rate in the troposphere averages 3.5°F. per 1000 feet.

true azimuth

The true azimuth of a given line is the angle measured clockwise from true north to the given line.

true solution

A term used occasionally to refer to a liquid, gaseous or solid mixture of components which are distributed uniformly throughout the entire mixture and in which the particles dissolved are molecular in size.

truncated cone, prism or pyramid

That part of a cone, prism or pyramid which lies between the base (or a plane parallel to the base) and another plane which is oblique to the base and does not intersect it within the solid itself. (If the second cutting plane is parallel to the base, the resulting solid is a *frustum of a cone, prism* or *pyramid.*)

tuberculation

A form of corrosion in which the products of corrosion appear in the form of blisters or nodules.

tungsten
A metallic element in Group VI of the periodic table, discovered by d'Elhujar in 1783. Atomic number 74, atomic weight 183.92, melting-point 3410°C., specific gravity 18.77-19.1, valence 4 or 6. Also called *wolfram* or *wolframium*. Symbol: W.

turbid
Said of a solution which has a more or less cloudy appearance owing to the presence of fine particles in a state of suspension.

turbidimeter
An instrument for the determination of matter existing as suspended particles in a liquid.

turbidimetry
A form of quantitative analysis of liquids, in which the quantity of suspended particles present in it is determined by an instrument called *turbidimeter*.

turbidity
The cloudy state of a liquid due to the presence of fine solid particles in suspension.

turbine
A machine for obtaining mechanical energy from steam or water.

turbulence
Irregular motion in a moving fluid, caused by an impediment in the stream by friction, or by vortex action.

turbulent air
Air in chaotic motion.

turbulent flow
Any part of a fluid flow in which the velocity at a given point varies more or less rapidly in magnitude and direction with time.

tweeter
A high-frequency loudspeaker used in high fidelity apparatus,

twinned crystals
Two individual crystals united symmetrically about a plane which is a possible face of the crystals of the substance or about an axis which is a possible crystal edge.

two-dimensional configuration
A plane surface, since it has length and width, but no thickness.

Tyndall beam
See *Tyndall cone*.

Tyndall cone

The band of reflected light that is visible when a beam of light is passed through a medium containing dispersed particles capable of reflecting the light. Frequently an apparently homogeneous medium is shown to have dispersed particles by the use of this test. Also called *Tyndall beam.*

Tyndall effect

The scattering of light by very small suspended particles. The smaller the particles, the greater is the polarization of the scattered light; the polarization is complete for particle diameters less than 0.1 micron.

Tyndall meter

An instrument for measuring the intensity of a *Tyndall cone.*

U

U

The symbol for the element *uranium.*

UI, UII, UX_1, UX_2, UY, UZ

The symbols for *uranium I, uranium II, uranium* X_1, etc.

ultimate analysis

Determination of the elements of which a material is composed, e.g. in coal it would show the percentages of carbon, hydrogen, oxygen, nitrogen and sulphur, whereas a *proximate analysis* of the same material would show the percentage of moisture, volatile matter, fixed carbon and ash.

ultimate tensile strength

The maximum tensile stress, computed on the original cross-section, which a solid can withstand. (See *tensile strength.*)

ultimate tensile stress

See *tensile strength.*

ultracentrifuge

A centrifuge which has very high speed and develops centrifugal forces as high as 7,000,000 times gravity.

ultrafiltration

Filtration under pressure or suction using a dialyzer or colloidal filter.

ultrafine powder

A powder composed of particles which are not greater than two microns.

ultramicron

See *submicron.*

ultramicroscope

A microscope, used largely in the study of colloids, where the source of light is a sharp beam at right angles to the line of vision. What is seen is not the particle itself, but the reflection of light from the particle.

ultrashort wave

A general term for radio waves shorter than 10 meters in wavelength (above 30 mc in frequency).

ultrasonic
Having a frequency above that of audible sound.

ultrasonic inspection
A non-destructive method of testing, based upon the fact that ultrasonic waves are reflected and refracted at the boundaries of a solid medium, from which it is possible to obtain the echoes of a wave transmitted from the surface of a test piece. (Also referred to as *supersonic testing.*)

ultrasonic waves
Sound waves of a frequency too high to be audible.

ultrasonics
The science of high frequency sound waves. The term usually refers to those frequencies that are above the normal upper limit of human audibility, about 18,000 c/s. The word is considered preferable to *supersonics* which now has a particular significance in aeronautical engineering.

ultrasonography
An application of *ultrasonics* for the detection of internal flaws in metals, using cathode-ray tube and an ultrasonic image produced by the echoes, which can be studied and interpreted like a radiograph.

ultrasonoscope
An instrument employed in *ultrasonography* for the detection of internal flaws in metals.

ultraviolet radiation
Invisible rays beyond the violet end of the solar spectrum, having a wavelength less than approximately 4000 Angstrom units.

umbra *(pl. umbrae)*
The part of the shadow of an object from which all direct light is excluded by the object itself.

unary system
A system of one component. Also called *unicomponent system*.

uncertainty principle
The theory formulated by W. Heisenberg, stating that it is impossible to determine simultaneously both the exact position and exact momentum of an electron, for if simultaneous measurements of both its position and momentum are pressed beyond a certain degree of accuracy, it becomes impossible to increase the accuracy of either measurement except at the expense of the accuracy of the other.

undamped wave
A wave which has an unchanging amplitude.
undetermined coefficient
A coefficient the value of which has not been settled, determined or found.
undecagon
A polygon bounded by eleven straight lines.
undulation theory
See *wave theory*.
undulatory
Wave-like, pertaining to or consisting of waves.
undulatory motion
See *wave motion*.
uni-
A prefix indicating *one, single*.
unicomponent system
A system of one component. Also called *unary system*.
unidirectional
Having only one direction, or taking place or exerting an effect in one direction only.
unified field theory
A theory developed by A. Einstein, treating electric, magnetic and gravitational fields and phenomena as different phases or aspects of the selfsame process.
uniform acceleration
Acceleration in which there are equal changes in the velocity in equal intervals of time. Also called *constant acceleration*.
uniform circular motion
Motion in which the speed is constant, the change in direction is uniform and continual, and the acceleration is of constant magnitude and is directed toward the center of the circle which is the path of the moving object or system.
uniformly accelerated motion
Motion in which the velocity changes at a constant rate; the direction of the motion remaining always the same, the acceleration equals the rate of the change of speed.
unimolecular
See *monomolecular*.
unit
(1) A single quantity, number or thing. (2) A standard of measurement. (3) The number *one*. (4) A number denoting

integer multiples of the number *one,* i.e., the digit immediately preceding the decimal point.

unit angle
See *radian.*

unit cell
The smallest group of *atoms, ions,* or *molecules* whose repetition at regular intervals, in three dimensions, produces the lattice of any given crystal.

unit cube
(1) A cube each edge of which measures one linear unit. (2) The *unit cell* in a cubic crystal.

unit electric charge
The *electrostatic unit* or *stat-coulomb.*

unit fraction
A simple fraction the numerator of which is 1.

unit magnetic pole
The unit of *magnetism;* defined as the magnetic pole which when separated in vacuum by one centimeter from a magnetic pole of exactly the same magnitude but opposite in sign, will repel the latter with a force of one dyne.

unit of structure
In crystallography, the group of atoms included by the *unit cell,* in particular if the arrangement of that group is such as to indicate that its members are in chemical relations to each other.

unit pole
See *unit magnetic pole.*

unit strain
In metallic objects, the change in dimensions per unit length.

unity
The number *one.*

univalent
Refers to elements or radicals which have a valence of one.

uni-univalent electrolyte
An electrolyte containing a *cation* having a valence of +1 and an *anion* having a valence of −1.

universal gas constant
The *ideal gas constant.*

universe
The entire spatial cosmos.

unknown quantity

In mathematics, the quantity or symbol the value of which is to be ascertained. Usually referred to simply as *the unknown.*

unlike terms

In mathematics, terms which differ in the power of the same literal coefficient or unknown number. Thus $4x^3$ plus $7x^2$ cannot be expressed by a single term and must be written $4x^3 + 7x^2$.

unstable

In chemistry, tending to undergo spontaneous changes or to decompose readily. (See *unstable element, unstable equilibrium.*)

unstable element

An element, the atomic nuclei of which are radioactive, emit certain radiations and change eventually into atomic nuclei of another isotope, which in turn may be also radioactive or stable.

unstable equilibrium

(1) The state of a substance which is in a process of phase change. (2) See *unstable element.*

upper air

That part of the atmosphere of the earth which consists of the *ionosphere* and the *exosphere.*

upper limit of integration

In a definite integral it represents the upper limit between which the integration is performed.

upper partial

In music, a term embracing all the tones in a compound tone (chord or discord) except the lowest *partial* or *fundamental.*

uranium

A radioactive, metallic element, one of the group of actinides in the extended periodic table, the heaviest of all the naturally occurring elements, discovered by Klaproth in 1789. Atomic number 92, atomic weight 238.07, melting-point about 1850° C., specific gravity 18.685, valence 2 or 6. The abundance of natural uranium in the earth crust is approximately 4 parts in 10^6. All uranium isotopes are radioactive. The three natural uranium isotopes are: U^{234} (natural abundance 0.006%), U^{235} (0.719%) and U^{238} (99.28%). While theoretically all three are fissionable by fast neutrons, U^{235} is readily fissioned by slow neutrons as well; this and the fact that it is capable of sustaining an efficient fission chain reaction make it a very practical reactor "fuel" and atomic bomb material. Symbol: U.

uranium I (UI)

A radioactive isotope of uranium ($_{92}U^{238}$).

uranium II (UII, U_2)

A radioactive isotope of uranium ($_{92}U^{234}$).

uranium X_1 (UX_1)

A radioactive isotope of thorium ($_{90}Th^{234}$).

uranium X_2 (UX_2)

A radioactive isotope of protactinium ($_{91}Pa^{234}$).

uranium-actinium series

A natural radioactive series which starts with the uranium isotope $_{92}U^{235}$ (actinium-uranium) and ends with the stable lead isotope $_{82}Pb^{207}$ (actinum D). Designated also as ($4n + 3$) *series,* due to the fact that the mass number of every one of its members equals 3 + an integral multiple of 4. Also called *actinium family, uranium-actinium family* and *($4n + 3$) series* or *family.*

uranium-radium series

A natural radioactive series, which starts with the uranium isotope $_{92}U^{238}$ (uranium I) and ends with the stable lead isotope $_{82}Pb^{206}$ (radium G). The designation ($4n + 2$) *series* is due to the fact that the mass number of every one of its members equals 2 + an integral multiple of 4. Also called *uranium family, uranium-radium family* and *($4n + 2$) series* or *family.*

uranium reactor

A *pile* in which uranium is used as the feeder.

V

V

The symbol for the element *vanadium.*

vacuum

A space where the pressure is much lower than the normal atmospheric pressure of 760 mm. of mercury. While a "perfect" or "absolute" vacuum (the absence of all air, gas and other forms of matter) cannot be produced, a "partial vacuum" (reduction of the gas pressure to between 10^{-5} and 10^{-6} mm. of mercury) can be obtained with ease. With the aid of mercury-vapor pumps, the pressure can be reduced to even as little as 10^{-10} mm. of mercury.

vacuum tube

Any device consisting of an evacuated tube or enclosure containing a number of electrodes, between two or more of which conduction of electricity may take place through the partial vacuum.

valence

The chemical combining power of the atom, i.e. its capacity to combine with other atoms to form a molecule, expressed in terms of the number of hydrogen atoms ($_1H^1$), or their equivalent, with which it will combine or replace. An atom unable to combine with other atoms is called *zerovalent* (having the valence 0), one which can combine with one hydrogen atom or its equivalent is said to be *monovalent* or having the valence 1; one which can combine with two hydrogen atoms or their equivalent is referred to as *divalent* (or *bivalent*) or having the valence 2; an atom having the valence 3 is *trivalent,* one with the valence 4 is *tetravalent,* atoms with the valence 5 are called *pentavalent,* and those having the valence 6 are referred to as *hexavalent.* According to the theory proposed by G. N. Lewis in 1916 and subsequently developed and extended by I. Langmuir, there are three types of valence, viz.: *positive valence* (representing the number of electrons which an atom can yield), *negative valence* (the number of electrons which an atom can take up), and *covalence* (the number of electron pairs which an atom can share with other atoms). According to Langmuir, positive and negative valence differ from each other fundamentally in algebraic sign only, so that both may be included under the collective term *electrovalence,* as positive and negative electrovalence. Also called *valency.*

valence bond

The force holding two or more atoms together in a molecule. Also called *chemical bond.* When two monovalent atoms are linked, the linking force is a *single valency bond;* two divalent atoms are joined by a *double valency bond.*

valence electrons

The orbital electrons in the outermost electron shell of an atom, which determine its chemical properties.

valence shell

The outermost electron shell of an atom, which contains the *valence electrons.*

valve

(1) A device for controlling the flow of a fluid. (2) The British term for a *vacuum tube.*

van de Graaff generator

An electrostatic generator, designed for producing potentials of millions of volts. It consists fundamentally of an endless belt made of a material of good insulating properties, placed within a large hollow metal sphere, and set in motion, on two rollers; as it moves, it collects electric charges by induction and discharges them within the metal sphere.

van der Waals' equation

An expression of the thermodynamic behavior of real gases and of their deviation from the ideal-gas law. It is stated in the form of the equation: $P + \left(\frac{a}{V^2}\right)(V - b) = RT$, where P is the pressure in atmosphere, V is the volume (expressed in units of the volume of the gas at normal temperature and pressure), R is the universal gas constant, T is the absolute temperature, and a and b are constants.

van der Waals forces

Weak forces of interatomic and intermolecular attraction.

vanadium

A metallic element in Group V of the periodic table, discovered by Del Rio in 1801, although Sefström, who studied its properties extensively, is frequently credited with its discovery, too. Atomic number 23, atomic weight 50.95, melting-point 1710-1720°C., specific gravity 5.685, valence 5. Symbol: V.

vanish

Said of a function when its value is zero.

vanishing fraction

In mathematics, a fraction which approaches the value zero.

vanishing line

A line converging to a *vanishing point.*

vanishing point

The point to which all parallel lines appear to the eye to recede. In perspective drawing, the point at which a horizontal line on an object, when extended, disappears at the horizon.

Van't Hoff law

When the temperature of a system in equilibrium is raised, the equilibrium point is displaced in the direction which absorbs heat.

V-particles

Observed in 1948 in a cloud chamber. There is a positive V, a negative V, and two variations of neutral V particles.

vapor

The gaseous phase of any substance below its critical temperature, where increasing pressure can still produce condensation in it. (Above the critical temperature, the substance becomes a *gas,* and no amount of pressure will produce any condensation in it.)

vapor density

The density of a gas referred to the density of hydrogen or air as unity. If the density of hydrogen is taken as 2, the vapor density is approximately the same as the molecular weight.

vapor pressure

The pressure exerted by a *vapor* either by itself or in a mixture of gases. The term is often taken to mean saturated vapor pressure, which is the vapor pressure of a vapor in contact with its liquid form. The saturated vapor pressure increases with rise of temperature.

vapor pressure of a liquid

The pressure exerted by a vapor when in equilibrium with the liquid.

vapor tension
The vapor-forming tendency of a liquid. (The term was used formerly to designate the *vapor pressure.*)

vaporimeter
An apparatus in which the volatility of oils is estimated by heating them in a current of air.

vaporization
Change of a substance from the liquid to the gaseous state.

variability
The scattering of the values of a frequency distribution, measured frequently by the *standard deviation.*

variable
A quantity which, throughout a mathematical calculation or investigation, is capable of varying, or does vary in value. A quantity which may assume a succession of values. (See *dependent variable, independent variable.*)

variable angle
An angle described by a line which is continually revolving in the positive (counterclockwise) direction.

variance
A measure of the amount of the dispersion or scattering of a set of values around their mean, determined by computing the mean value of the squares of the deviations from the mean, and thus equal to the square of the *standard deviation.*

variation of signs in a polynomial
The condition that exists between two successive terms of a *polynomial* which have different signs.

vary directly
To be *directly proportional.*

vary inversely
To be *inversely proportional.*

vary jointly
A quantity varies jointly as the product of a number of others which is multiplied by a constant. Thus $A = mBC$ in which A varies as the product of B and C.

vector
(1) A line which has both magnitude and direction. Its direction is defined relative to some fixed directed line when it is in a given plane, and relative to three fixed directed lines when it is in space. (2) A synonym for *vector quantity.*

vector analysis
The study of vectors, relations between vectors, and their applications.

vector quantity

A quantity which has both magnitude and direction. Examples of physical quantities that are vectors: displacement, velocity, force, magnetic intensity. (Also called *vector*.)

vector ratio

The ratio between two alternating quantities, e.g. currents, in which both relative amplitudes and phases are expressed in the form of *vectors*.

vectorial angle

One of the polar coordinates of a point on a plane. It is the angle enclosed by the initial line and the radius vector.

velocity

The rate of motion or speed of a body at any instant; usually measured in miles per hour or feet per second or minute. Velocity is a vector quantity, denoting both the time rate and the direction of the motion.

velocity constant

The speed of a chemical reaction, in gram-molecules of change per liter per second, when the active masses of all the reactants are unity.

velocity of light

A universal constant for the velocity of all electromagnetic waves, equal to $(2.99796 \pm 0.00004) \times 10^{10}$ centimeters per second in a vacuum (186,326 miles per second).

velocity of propagation

The velocity at which a disturbance is radiated as a wave through a medium.

velocity of sound

Measurements of the velocity of sound in air obtained by various physicists are in close agreement, giving a value of 331.7 meters per second (1120 feet per second, or 760 miles per hour).

velocity ratio

In mechanics, a synonym for the *ideal mechanical advantage* of a machine.

velocity sensitivity

In metallurgy, a measure of the capacity of a metal to deform suddenly and rapidly before failure, as revealed by impact testing.

Venturi tube

A short tube of varying cross section, designed to measure the rate of flow of fluids. The flow through the tube produces a drop of pressure in the smallest section, the magnitude of the drop being a function of the velocity of the flow. (Frequently called *venturi* for short.)

vergence

A generic term for the convergence and divergence of radiation.

verify

In mathematics, to prove the correctness of a solution by substituting in the sum the value obtained for the unknown quantity.

vernier

A short auxiliary scale made to slide along the divisions of a graduated instrument, to indicate fractions of the divisions of the larger scale.

versed cosine

See *coversed sine.*

versed sine

A trigonometric function of an acute angle, vers $\theta = 1 - \cos\theta$. It is that part of the radius of a unit circle (radius = 1) between the end of the cosine and the arc. Vers $A = 1 - \cos A = 2\sin^2\frac{A}{2}$. (Also called *versine.*)

versine

Sec *versed sine.*

vertex (*plural:* vertices)

The point of intersection of any number of concurrent lines. In topology, the end point of an *arc,* or the point of intersection of two or more arcs.

vertex of a hyperbola

Either extremity of the transverse axis.

vertex of a parabola

The point of the intersection of the parabola and its main axis.

vertex of a polygon

The point of intersection of two of the sides.

vertex of a polyhedral angle

The point at which the *faces* meet.

vertex of a polyhedron

The point of intersection of *edges.*

vertex of a reflecting or refracting surface

The point where the axis of the optical system intersects the surface.

vertex of a triangle

The apex opposite the line chosen as the *base.*

vertex of an angle
The point of intersection of the two straight lines or arms which form the angle.

vertex of an ellipse
Either of the two points in which the major axis intersects the ellipse.

vertical
At right angles to the plane of the horizon or earth's surface. In a line with gravity.

vertical angles
Angles having the same vertex; the arms of one being the prolongation of the arms of the other angle. Also called *vertically opposite angles* or *opposite angles.*

vertically opposite angles
See *vertical angles.*

vibrate
To execute a swinging to-and-fro motion or oscillation.

vibration
A rapid to-and-fro or back-and-forth motion or oscillation.

vinculum
A mathematical symbol, consisting of a straight line drawn over an expression instead of enclosing it in brackets and indicates that the expression has to be simplified before proceeding with the solution. Thus $\overline{a + 2a} \times b$ means $3a \times b = 3ab$; or $a - \overline{b + c}$ means $a - b - c$.

vinegar
See *acetic acid.*

virtual image
The image formed by a *divergent lens.* The virtual image of an object-point is the point from which light, originating in the object-point and having traversed an optical system, appears to be diverging. A virtual image can be seen by the eye or by the camera, but cannot be received on a screen.

viscosimeter
An instrument for measuring *viscosity.*

viscosity
The internal friction of a fluid, due to the molecular cohesion within it; it is the manifestation of the internal properties of the fluid which offer resistance to shear or flow. It equals the ratio of the shearing stress to the time rate of shear, and varies with the temperature. Measured in multiples or submultiples of the unit of viscosity called *poise,* equal to one dyne-second

per square centimeter. (See *kinematic viscosity* and *relative viscosity*.)

viscous

Having an appreciable viscosity.

visibility

The ratio of the *luminous flux* (measured in lumens) to the total radiant energy (measured in ergs per second or in watts).

visual angle

The angle subtended by an object to the eye of the observer.

vitamin

A generic term for a group of organic compounds found in foods and necessary for the normal metabolism, growth and health of human and animal bodies.

vitreous

Glassy, glass-like; having undergone *vitrification.*

vitrification

The conversion, at high temperatures, of a refractory material into glass or a *vitreous* form; it is usually associated with a reduction in porosity.

vitrification point

(1) The temperature at which *vitrification* of a refractory material occurs. (2) The temperature at which clays reach the condition of maximum density and shrinkage.

vitriol

Concentrated sulphuric acid, H_2SO_4; *blue vitriol* is copper sulphate $CuSO_4.5H_2O$; *green vitriol* is ferrous sulphate, $FeSO_47H_2O$; *white vitriol* is zinc sulphate, $ZnSO_4.7H_2O$.

volatile

Readily evaporating at relatively low temperatures.

volatile oils

See *essential oils.*

volatility

The tendency of a substance to vaporize at the temperature under consideration.

volatilize

(1) To convert into a gas or vapor. (2) To become a gas or vapor.

volt (v)

The practical unit of electric potential, potential difference or electromotive force or electrical pressure: 1 volt = the difference of potential which when steadily applied to a conductor of a resistance of 1 ohm, will send through the latter a current of 1 ampere; it is the potential difference between points such that 1 joule of work is needed to transfer 1 coulomb of charge between them. (1 volt = 10^8 abvolts.)

voltage

(1) The presence or action of potential, potential difference or electromotive force or electric pressure between two points. (2) The value of potential, potential difference, electromotive force or electric pressure, expressed in volts.

voltage drop

The drop in electric potential between two given points in an electric circuit.

voltaic cell

A source of electrical energy depending on chemical action and complete in itself, e.g. a primary cell, or an accumulator. It has two electrodes, each of which is immersed in an electrolyte and reacts therewith to produce an *e.m.f.*

voltaic current

An electric current resulting from chemical action.

voltaic pile

A source of direct current supply. It comprises a battery of primary cells in series, arranged in the form of a pile of disks, sucessive disks being of dissimilar metals separated by a pad soaked in the chemical agent.

voltameter

An instrument for measuring a current by means of the amount of metal deposited, or the gas liberated, from an electrolyte in a given time owing to the passage of the current.

voltammeter

A combination voltmeter and ammeter.

volt-ampere

A unit of electrical measurement, equal the product of one volt and one ampere.

voltmeter

An instrument for measuring potential difference in an electrical circuit, graduated in volts and multiples and fractions thereof.

volume

The portion of space enclosed by the closed surface forming a solid, expressed as a number of units, each unit representing a cube each edge of which measures one linear unit.

volume expansion
The expansion of a body, solid or liquid, in all three dimensions.

volume resistivity
See *specific resistance.*

volume strain
In compression, the fractional change produced in the volume of the object acted upon.

volume stress
In compression, the force per unit area.

volumenometer
An apparatus for the determination of the density of a substance by measuring the volume a given quantity of exactly known weight.

volumetric
Pertaining to the measurement by volume as opposed to weight *(gravimetric).*

volumetric analysis
A method of chemical analysis in which the substance to be measured, which must be in the form of a solution, is determined by the addition of a known quantity of a reagent of standardized strength. This standardized solution is usually added by means of a graduated *burette* and in certain cases, e.g. the addition of a standard alkali to an acid, an indicator such as *phenol phthalein* is added which changes color when the *end point,* i.e. neutrality, is reached. The process is known as *titration.* The amount of the element sought is calculated from the volume of standard solution required to reach neutrality.

volumetric modulus of elasticity
When a body is subjected to external forces all around it (such as are due to total immersion in a liquid) there is a slight reduction in the *volume.* The volumetric modulus of elasticity is the number by which the stress upon the exterior of the substance must be divided to give the unit diminution in volume or cubical strain. (Also referred to as *bulk modulus* or *modulus of cubic compressibility.*)

volumetry
The measurement of substances by comparison of volumes by *volumetric analysis.*

vulgar fraction
A synonym for *common fraction* (q.v.).

W

W

The symbol for the element *tungsten.*

wake

The region behind an obstacle in a fluid stream which is disturbed by its passage.

water

A chemical compound, each molecule of which consists of two hydrogen atoms and one oxygen atom; found in nature in the liquid (water), solid (ice, snow, etc.) and gaseous (water vapor) states. Symbol: H_2O. Its melting point is 0°C. (32°F.), its boiling point 100°C. (212°F.), its density at standard temperature and pressure is 0.99987 gram per cubic centimeter (it is at maximum density at 4°C., i.e., 39.2°F., when it is 1.00 gram per milliliter, by definition of the liter).

water bath

A vessel containing boiling water, used as a source of uniformly distributed heat, as for heating another vessel above it, especially for evaporating a substance placed in the upper vessel.

water equivalent

That mass of water to which a given body is thermally equivalent, represented numerically by the mass of the body multiplied by its *specific heat.*

water gas

The general name for a mixture of gases obtained by the decomposition of steam by incandescent carbon. It usually contains from 43% to 44% carbon monoxide, 48% to 49% of hydrogen, 3% to 4% of carbon dioxide, and 3% to 4% of nitrogen. Used for heating and lighting.

water glass

Sodium silicate, a viscous liquid which when mixed with powdered fireclay forms a refractory cement.

water of crystallization

The water present in crystals or in hydrated salts. These substances, when crystallized from solution in water, retain a definite amount of water, e.g. ferrous sulphate, $FeSO_4.7H_2O$. This water, the molecules of which retain their identity as H_2O, can be removed by heating. (Also called *crystal water.*)

water of hydration

See *water of crystallization.*

water of plasticity

The water required to bring a clay to a good working consistency.

water vapor

The gaseous form of water; one of the most important constituents of the atmosphere.

watt (w)

A unit of measure of electrical power; the power required to keep a current of one ampere flowing under a potential drop of one volt; 1/746 of one horsepower; ten million (10^7) ergs per second; one joule per second.

watt loss

See *core loss.*

wattage

Electrical power measured in *watts.*

wattful current

The active component of an *alternating current.*

watt-hour

A unit of measure of electrical energy; the work done by one watt acting for one hour. It equals 36 billion ergs, or 3,600 joules.

watt-hour efficiency

The ratio of the amount of energy available during the discharge of an accumulator to the amount of energy charged.

watt-hour meter

A device for measuring and registering electric energy in watt-hours or kilowatt-hours.

wattmeter

An instrument for measuring electrical power. Wattmeters are provided with a scale, usually graduated in watts or kilowatts.

watts horsepower

The unit of power used in the United States and Great Britain, equal to 550 foot-pounds per second at sea level and 50° latitude. It is 1.0139 times the *metric horsepower.*

wave

A disturbance which is propagated in a medium in such a

manner that at any point in the medium the displacement is a function of the time, while at any instant the displacement at a point is a function of the position of the point. Any physical quantity which has the same relationship to some independent variable (usually time) that a propagated disturbance has, at a particular instant, with respect to space, may be called a wave.

wave character of matter

See *wave mechanics.*

wave front

A surface at right angles to a bundle of rays as they proceed from a source; it passes through those parts of the waves which are in the same phase and travel in the same direction. For parallel rays, the wave front is a plane; for rays which diverge from or converge toward a point, the wave front is spherical.

wave mechanics

A physical theory, introduced by E. Schrödinger in 1926, which forms an integral part of quantum phenomena and physical processes in terms of wave forms.

wave motion

The propagation of a periodic succession of disturbances or deformations through a medium which itself is capable of deformation. (See *wave.*)

wave normal

The direction normal to the *wave* front and toward the direction of propagation.

wave number

The reciprocal of the *wavelength.*

wave pulse

A single disturbance or deformation of a substance or medium.

wave theory of light

(1) The currently generally accepted theory that light is a wave phenomenon, and not any stream of material particles, as held by the formerly accepted *corpuscular theory,* although the concept of the photon may be regarded as a return, to some extent at least, to that theory. (2) Sunlight is propagated as wave motion in a medium called the ether. The distance between two successive crests or valleys of a wave is known as the wavelength. Short waves produce on the retina of the eye the sensation of blue, intermediate waves the sensation of red. The simultaneous perception of the entire range of waves that affect the eye produces white.

wave train
A continuous series of periodic disturbances or deformations propagated in a medium capable of deformation.

waveform
The shape of the curve obtained when instantaneous values of an alternating-current quantity are plotted against time in a rectangular coordinate system.

wavelength
The distance between two successive peaks of a propagated disturbance, i.e. the distance travelled by the propagated disturbance during one cycle.
The following four units are used for measuring the wavelength of radiant energy other than radio waves:

Name	*Abbreviation*	*Value*
Micron	μ	10^{-3} mm.
Millimicron	$m\mu$	10^{-6} mm.
Angstrom (unit)	A	10^{-7} mm.
X unit	XU	10^{-10} mm.

weak acid
An acid which undergoes merely a slight ionization in water solutions.

weathering
The result of the destructive effect of the atmosphere and of climatic factors on natural geologic formations as well as artificial structures.

weber
The practical unit of *magnetic flux;* one weber = 10^8 (one hundred million) maxwells.

wedge photometer
A photometer in which a graduated wedge of an absorbent material is inserted into the light beam issuing from the brighter source to the point where the intensity of the beam is reduced to equality with that of the beam from the other source. The ratio of the reduction by the wedge scale, is a measure of the ratio of the luminous intensity of the two sources.

wedge spectrograph
(*Slitless Spectrograph.*) An x-ray spectrograph in which the beams of incident and diffracted x-rays are limited by a wedge brought into contact with a face of the crystal.

weight

(1) The force which a mass experiences owing to its being in a gravity field, in particular that of the earth, i.e., the force by which the mass is attracted to the earth. Since the acceleration of gravity decreases with the distance from the center of the earth, the same body weighs less on a mountain than at sea level. (2) A body of known mass, used in measuring the masses of other bodies.

weight-density

Weight per unit volume.

welding

In general, a process of uniting two pieces of metal or alloy by raising the temperature of the surfaces of the metal or alloy to be joined to a plastic or molten condition with or without the application of pressure and with or without the use of added metal. This definition excludes the more recently developed method of *cold welding* in which pressure alone is used. Cold welding, however, has a limited application and is used principally for aluminum and its alloys and not for steel. There are numerous methods of welding among which the following are some of the more important: Argon-Arc, Argonaut, atomic hydrogen arc, carbon arc, flash-butt, forge, fusion, gas, metal-arc, oxy-acetylene, oxy-hydrogen, oxy-propane, pressure, projection, resistance, seam, spot, stitch, stud, submerged-arc.

wet analysis

A term frequently applied to qualitative chemical analysis employing solutions containing reagents, as opposed to *spectrographic* (called *dry*) analysis.

wet assay

The determination of the metal content of an ore by chemical analysis which involves taking the ore into solution.

wet cell

A cell the electrolyte of which is in liquid form and free to flow and move.

wet metallurgy

See *hydrometallurgy*.

wettability

The extent to which a solid is wetted by a liquid, measured in terms of the tension of adhesion between the solid and liquid phases.

Wheatstone bridge
A divided electrical circuit used for the measurement of electrical resistance.

white light
Light, such as sunlight and daylight, which is composed of all the different wavelengths of the solar spectrum.

whole number
An *integer* (q.v.)

whole number rule
The atomic weight of every element is found to be very nearly a whole number when determined by the mass spectrometer.

width
See *breadth.*

Wien's displacement constant
The constant, 0.2892 centimeter-degree, appearing in the mathematical expression of *Wien's displacement law.*

Wien's displacement law
A radiation law stating that the wavelength of the maximum intensity of radiation from a *black body* is inversely proportional to the absolute temperature of that body, viz.: $\lambda_m = \frac{a}{T}$ where λ_m is the wavelength measured in microns, T is the temperature in degrees absolute, and a is a constant (called Wien's displacement constant, equal to 0.2892 centimeter-degree).

Wilson cloud chamber
See *cloud chamber.*

Wimshurst machine
A device for generating static charges.

witch
A plane curve the equation of which, in rectangular coordinates is $x^2y = 4a^2(2a - y)$. Also referred to as *witch of Agnesi.*

wolfram, wolframium
Alternative names of the element *tungsten.*

Wollaston prism
A prism, constructed of quartz, for producing *plane polarization* of light.

work
The result of a force acting against resistance to produce motion in a body; a scalar product of the force by the linear distance through which the point of application moves. In a

physical sense, work is the exertion of a force over a given distance and is defined as the product of the force times the distance. It is expressed in foot-pounds. (It should be noted that the element of time does not enter into consideration concerning work. If a weight of 50 pounds is raised 20 feet, 1000 foot-pounds of work have been accomplished. Whether it required a few minutes or several weeks to raise the weight makes no difference. When all of the 50 pounds has been raised to the height of 20 feet, the work has been done irrespective of the time required.)

X

X

A chemical symbol indicating any *halogen* or acid radical in general.

x axis

In a *rectangular coordinate system,* the horizontal axis.

x coordinate

A synonym for *abscissa.*

x-radiation

X-rays.

x-ray

An electromagnetic radiation, similar to light but of a much shorter wavelength (10^{-10} to 10^{-7} cm), discovered by W. K. Roentgen in 1895. X-rays are highly penetrating; they penetrate all solids, to varying degrees, and produce secondary radiation when stopped by material bodies. They ionize gases through which they pass, affect photographic plates and fluorescent screens. X-rays are usually produced by allowing a stream of high-energy electrons to impinge on a metal target; due to the electronic bombardment, the atoms of the target lose energy which is radiated as x-ray quanta. Over-exposure to x-rays may cause serious damage in living human and animal tissue. In medical literature the term *Roentgen ray* is used in preference to *x-ray.*

x-ray analysis

See *x-ray crystallography.*

x-ray crystallography

As Max von Laue first showed in 1921, the planes of atoms in crystals act as a diffraction grating to X-rays, which are scattered by them and provide an accurate means of determining the details of the internal atomic structure. These X-ray photographs of metals provide information which in many cases cannot be obtained by ordinary microscopic methods. The lines produced by each element, or phase, are characteristic and their general pattern enables the crystalline structure to be identified. The scale of the pattern can be used to determine accurately the size of the unit cell and therefore the distance apart of the individual atoms; and from the relative intensity of the lines can be deduced the distribution throughout the unit cell of the various types of atoms in an alloy, or the degree of preferred orientation in the material. In addition, the sharpness of the lines provides information on both the state of strain and the grain size of the material. (Also referred to as *x-ray analysis*.)

x-ray diffraction

See *Debye-Scherrer-Hull method.*

x-ray goniometer

The instrument used to determine accurately the position of the electrical axes of a quartz crystal, making use of the reflection of the x-rays from the atomic planes of the crystal.

x-ray micrography

Micrography employing *x-ray photography.*

x-ray photography

The process of making photographs with x-rays. Such photographs are used extensively in medicine and dentistry, and also in metallurgy for material testing and analysis.

x-ray spectrograph

An apparatus for recording photographically the x-ray diffraction patterns produced by crystals.

x-ray spectrometer

(1) An instrument for producing an x-ray spectrum and measuring the wavelengths of its components.– (2) A type of spectrometer used to measure the angles of diffraction of x-rays produced by reflection from the surface of a crystal.

x-ray spectrum

An array of wavelengths or frequencies, showing lines, the position of which identify the characteristic x-rays emitted by a body exposed to electronic bombardment.

x-ray tube

A vacuum tube, in which the gas pressure is of the order of 10^{-3} mm. of mercury, used to generate x-rays by directing the streams of electrons emitted from the cathode on to a metal target, the atoms of which react to the electronic bombardment by releasing a stream of x-ray photons.

x-ray vacuum

Gas pressure below 0.01 mm.

X unit (XU)

A wavelength unit; 1 XU = 0.001 A = 10^{-10} mm.

Xe

The symbol for the element *xenon.*

xenon

A zerovalent gaseous element in Group 0 of the periodic table, one of the rare gases, discovered by Ramsay and Travers in 1898. Atomic number 54, atomic weight 131.3, melting-point −112°C., boiling-point 106.9°C., density 5.85 grams per liter (or 4.53 when air is 1.00), abundance in the atmosphere about 1 part in 17×10^7 by volume. Symbol: Xe.

xeroradiography

A rapid, all electric method of obtaining permanent *X-ray* images, which employs an electrically conductive metal plate, often of aluminum, in place of the usual film.

XU

The abbreviation for *X unit.*

Y

Y
The symbol for the element *yttrium.*

y axis
In a *rectangular coordinate system,* the vertical axis.

y coordinate
A synonym for *ordinate.*

Y-particles
See *hyperons.*

Yagi aerial
A type of directional aerial used in radio astronomy, in which only one dipole is connected to the feeder that leads to the receiver, and the other dipoles are "parasitic" (without direct connection to the receiver terminals).

Yb
The symbol for the element *ytterbium.*

yield point
The *stress* in a material at which there occurs a marked increase in *strain* without an increase in stress.

Young's modulus
The ratio within the limits of elasticity of the stress to the corresponding strain, in tension and compression, e.g. the load in tons per sq. in. divided by the deformation in fractions of an inch for each inch of the gauge length of the test piece. (Usually called simply *Young's modulus,* and referred to also as *modulus of elasticity.*)

Yt
An alternative symbol for the element *yttrium.*

ytterbium
A trivalent metallic element, one of the rare group of the periodic table, discovered by Marignac in 1878. Atomic number 70, atomic weight 173.04, melting-point 1800°C. Symbol: Yb.

yttrium
A trivalent metallic element in Group III of the periodic table, discovered by Gadolin in 1794. Atomic number 39, atomic weight 88.92, melting-point 1490°C., specific gravity 3.8. Symbol: Y.

Z

Z (atomic number)
The number of *protons* in the *nucleus.*

z plane
See *Argand diagram.*

Zeeman effect
The effect of a strong magnetic field on spectral lines: when a source of a line spectrum, such as a gas flame, is placed within the magnetic field its spectral lines are split up into three or more lines each. When the single line is broken up into three lines, the effect is called the *normal Zeeman effect;* when more than three lines appear in the place of one, the phenomenon is referred to as the anomalous *Zeeman effect.*

zenith
A point on the celestial sphere vertically above any place or observer and opposite to the *nadir.*

zero
A figure, character or symbol (written 0) that has the same meaning as *cipher, naught, nought* or *nothing.* It means the absence of all magnitude or quantity, however small, except when in combination with other digits or numbers.

zero gravity
(1) The condition wherein a body is away from the gravitational field of the earth. (2) Unobstructed and unrestricted motion.

zero of a function
That value of the independent variable of a given function for which the value of the function becomes zero.

zero-order reaction
A chemical reaction the rate of which is independent from and unaffected by the concentration of any reactant, and is either constant or depends on some other factor or effect.

zerovalent
Refers to inert elements which do not combine and therefore have no valence.

Zeta
Short for Zero *E*nergy *T*hermonuclear *A*ssembly, a machine

used in atomic *fusion* experiments to produce short-lived temperatures as high as 5,000,000°C.

zinc

A divalent metallic element in Group II of the periodic table; atomic number 30, atomic weight 65.38, melting-point 419.4°C., boiling-point about 910°C., specific gravity 7.14. Symbol: Zn.

zirconium

A metallic element in Group IV of the periodic table, discovered by Klaproth in 1879, and isolated for the first time by Berzelius in 1824. Atomic number 40, atomic weight 91.22, melting-point above 1750°C., specific gravity 6.5. Symbol: Zr.

Zn

The symbol for the element *zinc.*

zonal aberration

See *coma.*

zone

A part of a sphere, situated between and bounded by two parallel planes which pass through the sphere; the circular areas of the intersections of the sphere with the planes are called the *bases* of the zone. (A zone situated by a plane that intersects the sphere and another plane which is tangent to it, the zone is said to be a *zone of one base.*)

zone of one base

A part of a sphere situated between two parallel planes, one of which passes through the sphere, while the other one is tangent to it. The circular area representing the intersection of the sphere with the first plane is called the *base* of the zone.

Zr

The symbol for the element *zirconium.*

zwitterion

An ion that is charged both positively and negatively. Also called *amphoteric ion* and *dipole ion.*

1.0080 **H** 1 1

PERIODIC TABLE

LIGHT METALS **HEAVY METALS**

I A	II A	III B	IV B	V B	VI B	VII B	VIII B	
6.940 **Li** 3 2 1	9.013 **Be** 4 2 2							
22.991 **Na** 11 2 8 1	24.32 **Mg** 12 2 8 2							
39.100 **K** 19 2 8 8 1	40.08 **Ca** 20 2 8 8 2	44.96 **Sc** 21 2 8 9 2	47.90 **Ti** 22 2 8 10 2	50.95 **V** 23 2 8 11 2	52.01 **Cr** 24 2 8 13 1	54.94 **Mn** 25 2 8 13 2	55.85 **Fe** 26 2 8 14 2	58.94 **Co** 27 2 8 15 2
85.48 **Rb** 37 2 8 18 8 1	87.63 **Sr** 38 2 8 18 8 2	88.92 **Y** 39 2 8 18 9 2	91.22 **Zr** 40 2 8 18 10 2	92.91 **Nb** 41 2 8 18 12 1	95.95 **Mo** 42 2 8 18 13 1	[99] **Tc** 43 2 8 18 14 1	101.1 **Ru** 44 2 8 18 15 1	102.91 **Rh** 45 2 8 18 16 1
132.91 **Cs** 55 2 8 18 18 8 1	137.36 **Ba** 56 2 8 18 18 8 2	57 to 71 Lanthanide Series Rare Earth Elements	178.58 **Hf** 72 2 8 18 32 10 2	180.95 **Ta** 73 2 8 18 32 11 2	183.86 **W** 74 2 8 18 32 12 2	186.22 **Re** 75 2 8 18 32 13 2	190.2 **Os** 76 2 8 18 32 14 2	192.2 **Ir** 77 2 8 18 32 17 0
[223] **Fr** 87 2 8 18 32 18 8 1	226.05 **Ra** 88 2 8 18 32 18 8 2	89 to 102 Actinide Series Rare Earth Elements						

Lanthanide Series	138.92 **La** 57 2 8 18 18 9 2	140.13 **Ce** 58 2 8 18 20 8 2	140.92 **Pr** 59 2 8 18 21 8 2	144.27 **Nd** 60 2 8 18 22 8 2	[145] **Pm** 61 2 8 18 23 8 2	150.35 **Sm** 62 2 8 18 24 8 2	152.0 **Eu** 63 2 8 18 25 8 2	157.26 **Gd** 64 2 8 18 25 9 2
Actinide Series	227 **Ac** 89 2 8 18 32 18 18 9 2	232.05 **Th** 90 2 8 18 32 18 18 10 2	231 **Pa** 91 2 8 18 32 20 9 2	238.07 **U** 92 2 8 18 32 21 9 2	[237] **Np** 93 2 8 18 32 23 8 2	[242] **Pu** 94 2 8 18 32 24 8 2	[243] **Am** 95 2 8 18 32 25 8 2	[245] **Cm** 96 2 8 18 32 25 9 2

A value given in brackets denotes the mass number of the isotope of longest known half-life.

OF THE ELEMENTS

NONMETALS

	I B	II B	III A	IV A	V A	VI A	VII A	INERT GASES VIII A
								4.003 **He** 2 (2)
			10.82 **B** 5 (2-3)	12.011 **C** 6 (2-4)	14.008 **N** 7 (2-5)	16.0000 **O** 8 (2-6)	19.00 **F** 9 (2-7)	20.183 **Ne** 10 (2-8)
			26.98 **Al** 13 (2-8-3)	28.09 **Si** 14 (2-8-4)	30.975 **P** 15 (2-8-5)	32.066 **S** 16 (2-8-6)	35.457 **Cl** 17 (2-8-7)	39.944 **A** 18 (2-8-8)
58.71 **Ni** 28 (2-8-16-2)	63.54 **Cu** 29 (2-8-18-1)	65.38 **Zn** 30 (2-8-18-2)	69.72 **Ga** 31 (2-8-18-3)	72.60 **Ge** 32 (2-8-18-4)	74.91 **As** 33 (2-8-18-5)	78.96 **Se** 34 (2-8-18-6)	79.916 **Br** 35 (2-8-18-7)	83.8 **Kr** 36 (2-8-18-8)
106.7 **Pd** 46 (2-8-18-18-0)	107.880 **Ag** 47 (2-8-18-18-1)	112.41 **Cd** 48 (2-8-18-18-2)	114.82 **In** 49 (2-8-18-18-3)	8.70 **Sn** 50 (2-8-18-18-4)	121.76 **Sb** 51 (2-8-18-18-5)	127.61 **Te** 52 (2-8-18-18-6)	126.91 **I** 53 (2-8-18-18-7)	131.30 **Xe** 54 (2-8-18-18-8)
195.09 **Pt** 78 (2-8-18-32-17-1)	197.0 **Au** 79 (2-8-18-32-18-1)	200.61 **Hg** 80 (2-8-18-32-18-2)	204.39 **Tl** 81 (2-8-18-32-18-3)	207.21 **Pb** 82 (2-8-18-32-18-4)	209.00 **Bi** 83 (2-8-18-32-18-5)	210 **Po** 84 (2-8-18-32-18-6)	[211] **At** 85 (2-8-18-32-18-7)	222 **Rn** 86 (2-8-18-32-18-8)

RARE EARTH ELEMENTS

158.93 **Tb** 65 (2-8-18-27-8-2)	162.51 **Dy** 66 (2-8-18-28-8-2)	164.94 **Ho** 67 (2-8-18-29-8-2)	167.27 **Er** 68 (2-8-18-30-8-2)	168.94 **Tm** 69 (2-8-18-31-8-2)	173.04 **Yb** 70 (2-8-18-32-8-2)	174.99 **Lu** 71 (2-8-18-32-9-2)

[245] **Bk** 97 (2-8-18-32-27-8-2)	[248] **Cf** 98 (2-8-18-32-28-8-2)	[255] **E** 99 (2-8-18-32-29-8-2)	[252] **Fm** 100 (2-8-18-32-30-8-2)	[256] **Mv** 101 (2-8-18-32-31-8-2)	**No** 102 (2-8-18-32-32-8-2)	

Reproduced by special permission of the publishers, Henry Holt and Company, Inc. from Dull, Metcalfe, Williams: MODERN CHEMISTRY, 1958 edition.

ACKNOWLEDGMENTS

The editor and the publisher of the present volume express herewith their grateful acknowledgment and sincere thanks to the following publishers for their kind permission to use copyrighted material from their respective publications listed below in the preparation of this Dictionary:

American Geological Institute, Washington, D. C., publisher of *Glossary of Geology and Related Sciences,* by J. V. Howell, co-ordinating chairman (1957).

Barnes & Noble, Inc., New York, N. Y., publisher of the *College Outline Series,* and specifically of the following Volumes:
- No. 5—*First Year College Chemistry,* by John R. Lewis, 7th edition (1951);
- No. 6—*Organic Chemistry,* by Ed. F. Degering, 6th edition (rev. 1957);
- No. 13—*Geology,* by Richard M. Field, 4th edition (1951);
- No. 21—*Physics,* by Clarence E. Bennett, 5th edition (1952);
- No. 38—*Algebra,* by Gerald E. Moore (1951);
- No. 45—*Plane and Spherical Geometry,* by Kaj. L. Nielsen and John H. Vanlonkhuyzen (1954);
- No. 48—*The Calculus,* by C. O. Oakley (1957);
- No. 68—*Analytic Geometry,* by C. O. Oakley, (1957);
- No. 97—*Physical Chemistry,* by Scott L. Kittsley (1955);
- No. 101—*Descriptive Geometry,* by Steve M. Slaby (1956).

Chemical Rubber Publishing Co., Cleveland, Ohio, publisher of *Handbook of Chemistry and Physics,* by Charles D. Hodgman (ed.), 39th edition (1958).

Harper & Brothers, New York, N. Y., publishers of *Astronomy,* by John C. Duncan, 5th edition (1955).

Longmans Green and Co., New York, N.Y., publishers of *Dictionary of Physics,* by H. J. Gray (ed.) (1958).

McGraw-Hill Book Company, Inc., New York, N. Y., publisher of:
- *Chemistry for our Times,* by Elbert C. Weaver and Laurence S. Foster (1954);
- *College Physics,* by Robert L. Weber, Marsh W. White and Kenneth V. Manning, 2nd edition (1952);

General College Mathematics, by W. L. Ayres, Cleota G. Fry and H. F. S. Jonah (1952);

Hackh's Chemical Dictionary, by Julius Grant, 3rd edition (1944);

Introductory Organic Chemistry, by E. Wertheim and Harold Jeskey, 3rd edition (1956);

Plane Geometry, by Leroy H. Schnell and Mildred G. Crawford, 3rd edition (1953);

Plane Trigonometry, by C. R. Wylie, Jr. (1955);

Solid Geometry, by Leroy H. Schnell and Mildred G. Crawford (1935).

Penguin Books Ltd., Harmondsworth, Middlesex, England, publisher of *A Dictionary of Science,* by E. B. Uvarov and D. R. Chapman, revised edition (1951).

Philosophical Library, Inc., New York, N. Y., publisher of:

A Short Dictionary of Mathematics, by C. H. McDowell (1957);

An Encyclopaedia of the Iron & Steel Industry, by A. K. Osborne (1956);

Concise Encyclopedia of Atomic Energy, by Frank Gaynor (1950);

General Principles of Geology, by J. F. Kirkaldy (1956);

Illustrated Technical Dictionary, by Maxim Newmark (1944);

Mathematical Tables and Formulae, by F. J. Camm (1958);

Nuclear Physics, by Werner Heisenberg (1951);

Phenomena, Atoms and Molecules, by I. Langmuir (1950);

Physics of the 20th Century, by P. Jordan (1944);

Science since 1500, by H. T. Pledge (1947);

The Atom and the Energy Revolution, by Norman Lansdell (1958).

G. P. Putnam's Sons, New York, N. Y., publishers of *Field Book of the Skies,* by W. T. Olcott, revised by R. N. Mayall and M. W. Mayall (Copyright for revised edition 1955 by R. Newton and Margaret W. Mayall).

Research Studies Institute, Air University, Maxwell Air Force Base, Ala., publisher of *Interim Glossary of Aero-Space Terms* (1958). (The credit for the definitions in this Glossary is due to Dr. Woodford A. Heflin.)

Technical Information Division, U. S. Army Signal Research & Development Laboratory, Fort Monmouth, N. J., for the permission to quote from the *Dictionary of Technical and Scien-*

tific Terms, Interim Edition, published by *Signal Corps Engineering Laboratories* (1947).
U. S. Government Printing Office, Washington, D. C., publisher of *Weather Glossary,* compiled by Alfred H. Thiessen, W. B. No. 1445, reprinted May 1949.
U. S. Naval Institute, Annapolis, Md., publisher of *Navigation and Nautical Astronomy,* 7th edition, original edition by Commander Benjamin Dutton, U. S. Navy, revisions by the Department of Seamanship and Navigation, U. S. Naval Academy (1942).